SUSTAINABLE AQUACULTURE

Recent Advances and Techniques

SUSTAINABLE AQUACULTURE

Recent Advances and Techniques

Dhanaji W. Patil, PhD
Sandip S. Markad, PhD
Editors

APPLE ACADEMIC PRESS

First edition published 2027

Apple Academic Press Inc.
1265 Goldenrod Circle, NE,
Palm Bay, FL 32905 USA

760 Laurentian Drive, Unit 19,
Burlington, ON L7N 0A4, Canada

CRC Press
2385 NW Executive Center Drive,
Suite 320, Boca Raton FL 33431

4 Park Square, Milton Park,
Abingdon, Oxon, OX14 4RN, UK

Apple Academic Press exclusively co-publishes with CRC Press, an imprint of Taylor & Francis Group, LLC

Library and Archives Canada Cataloguing in Publication

CIP data on file with Canada Library and Archives

Library of Congress Cataloging-in-Publication Data

CIP data on file with US Library of Congress

ISBN: 978-1-77964-110-6 (hbk)
ISBN: 978-1-77964-111-3 (ebk)
DOI: 10.1201/9781779641113

About the Editors

Dhanaji W. Patil, PhD
Head and Associate Professor, Department of Fishery Science, Toshniwal Arts, Commerce and Science College, Sengaon, Hingoli, Maharashtra, India

Dhanaji W. Patil, PhD, has been working as Head and Associate Professor in the Department of Fishery Science at Toshniwal Arts, Commerce, and Science College, Sengaon, District Hingoli, Maharashtra, India, since October 2012. He graduated with a Bachelor of Fishery Science (B.F.Sc) in 2002 from Dr. Balasaheb Sawant Konkan Krishi Vidyapeeth, Dapoli, Maharashtra, India, and also completed a Master of Fishery Science (M.F.Sc.) from the same institute in 2004 with a specialization in Aquaculture. He qualified for the NET (Fishery Science) conducted by the National Agriculture Recruitment Board (ICAR), New Delhi, in 2012. He completed his PhD in Fishery Science from Swami Ramanand Teerth Marathwada University, Nanded, Maharashtra, in 2020. He successfully ran and completed two research projects funded by the Rajiv Gandhi Science and Technology Commission, Mumbai. He has published 21 research articles and one edited book, along with two book chapters and five extension publications. He made participated in the organization of a one-day National Conference on "Advances in Fisheries, Biological and Allied Research (AFBAR-2019)" in March 2019, serving as a convener. He has organized and conducted four training programs on aquaculture and fish seed production technologies for farmers and fishermen and two faculty development programs on educational video creation and e-content development for the teaching community across India. He contributed as a resource person in 14 faculty development programs conducted by different institutions across the country. He is also engaged in extension activities for fishermen and farmers of the region through the Department of Fisheries, Government of Maharashtra. He has also participated in several seminars and conferences and has presented his research work. He has developed e-content through a YouTube channel (FisheriesOnly). He is a member of the Asian Fisheries Society.

Sandip S. Markad, PhD
Department of Fishery Science, Toshniwal Arts, Commerce and Science College, Sengaon, Hingoli, Maharashtra, India

Sandip Markad, PhD, has been serving as an Assistant Professor in the Department of Fishery Science at Toshniwal Arts, Commerce and Science College, Sengaon, District Hingoli, Maharashtra, since December 2013. He graduated with a Bachelor of Fishery Science (BFSc) in 2007 from Dr. Balasaheb Sawant Konkan Krishi Vidyapeeth, Dapoli, Maharashtra, India, and acquired a Master of Fishery Science (MFSc) in 2009 with a specialization in Fisheries Resource Management and Extension Education. He completed his PhD in Fishery Science from Swami Ramanand Teerth Marathwada University, Nanded, Maharashtra, India, in 2022. He qualified for the NET (Fishery Science) conducted by the National Agriculture Recruitment Board (ICAR), New Delhi, in 2012. He worked as a research fellow for research projects from 2009 to 2013. He has published his research work through 22 research articles and two books, along with two book chapters and eight popular extension articles. He has successfully completed two research projects sanctioned by the Rajiv Gandhi Science and Technology Commission, Mumbai. He played a key role as the organizing secretary for a one-day National Conference on "Advances in Fisheries, Biological and Allied Research (AFBAR-2019)" in March 2019. He has organized four training programs on fish seed production and aquaculture techniques for fishermen and farmers, and two faculty development programs (FDPs) on educational video creation and e-content development for teaching faculties across the country. He contributed as a resource person in 10 faculty development programs conducted by different institutions in India. He is also engaged in extension activities for fishermen and farmers of the region through the Department of Fisheries, Government of Maharashtra. He has participated in various conferences and workshops and has presented his research work, earning one best oral presentation award. He has developed e-content through a YouTube channel (Servavid) as well as the institutional website. He is a permanent active member (PAM) of the Asian Fisheries Society.

Contents

Contributors

Sajid Abdullah
Department of Zoology, Wildlife and Fisheries, University of Agriculture, Faisalabad, Punjab, Pakistan

Gyamfua Afriyie
Department of Water Resources and Sustainable Development, School of Sustainable Development, University of Environment and Sustainable Development, Somanya, Eastern Region, Ghana

Muhammad Sarfraz Ali
Department of Biosciences, COMSATS University Islamabad (CUI), Islamabad, Pakistan

Prachi Asgolkar
Fish Genetics and Biotechnology Division, ICAR–Central Institute of Fisheries Education, Mumbai, Maharashtra, India

Christian Larbi Ayisi
Department of Water Resources and Aquaculture Management, School of Sustainable Development, University of Environment and Sustainable Development, Somanya, Eastern Region, Ghana

Sana Aziz
Department of Zoology, University of Jhang, Jhang, Punjab, Pakistan

V. C. Bajaniya
Department of Aquaculture, College of Fisheries Science, Kamdhenu University, Veraval, Gujarat, India

Mohit Bamaniya
Department of Fish Processing Technology, College of Fisheries Science, Kamdhenu University, Veraval, Gujarat, India

Pinak Bamaniya
Department of Aquatic Environment Management, College of Fisheries Science, Kamdhenu University, Veraval, Gujarat, India

Ishita J. Bambhaniya
Department of Aquaculture, College of Fisheries Science, Kamdhenu University, Veraval, Gujarat, India

Mirza Masum Beg
Department of Fisheries Science, Ramnagar College, Vidyasagar University, West Bengal, India

Mayur Bhadarka
Department of Aquatic Environment Management, College of Fisheries Science, Kamdhenu University, Veraval, Gujarat, India

Nisar Ahmad Bhatti
Department of Zoology, University of Jhang, Jhang, Punjab, Pakistan

Sayan Biswas
Department of Aquaculture, College of Fisheries Science, Kamdhenu University, Veraval, Gujarat, India

Sudipta Sankar Bora
DBT-North East Centre for Agricultural Biotechnology (DBT-NECAB), Assam Agricultural University, Jorhat, Assam, India

Madhurjya Protim Borah
Department of Biosciences and Bioengineering, Indian Institute of Technology Jammu, Jammu and Kashmir, India

Jugabrata Das
College of Horticulture and Farming System Research, Assam Agricultural University, Nalbari, Assam, India

T. H. Dave
Department of Fisheries Resource Management, College of Fisheries Science, Kamdhenu University, Veraval, Gujarat, India

Priyadarshini Deka
Hereditary Biosciences, Guwahati, Assam, India

Sahabhop Dokkaew
Department of Aquaculture, Faculty of Fisheries, Kasetsart University, Bangkok, Thailand

Samim Dullah
Nanda Nath Saikia College, Titabar, Assam, India

R. R. Gamit
Department of Aquaculture, College of Fisheries Science, Kamdhenu University, Veraval, Gujarat, India

Ankita Godavarikar
Fish Genetics and Biotechnology Division, ICAR–Central Institute of Fisheries Education, Mumbai, Maharashtra, India

Rahul Gogoi
Department of Agricultural Biotechnology, Assam Agricultural University, Jorhat, Assam, India

Chih-Yang Huang
Department of Aquaculture, National Taiwan Ocean University, Keelung, Taiwan

Pathan Imran
Department of Aquaculture, College of Fisheries Science, Kamdhenu University, Veraval, Gujarat, India

Gowhar Iqbal
ICAR–Central Institute of Fisheries Education, Panch Marg, Mumbai, Maharashtra, India

Saima Jan
Faculty of Fisheries, Sher-e-Kashmir University of Agricultural Sciences and Technology of Kashmir (SKUAST-K), Rangil, Ganderbal, Jammu and Kashmir, India

Bamaniya Pinak Kamleshbhai
Department of Aquatic Environment Management, College of Fisheries Science, Kamdhenu University, Veraval, Gujarat, India

Phunsin Kantha
Department of Aquaculture, Faculty of Fisheries, Kasetsart University, Bangkok, Thailand

Sutanu Karmakar
Department of Aquatic Environment Management, Faculty of Fishery Science, West Bengal University of Animal and Fishery Sciences, Kolkata, West Bengal, India

Neha Kharadi
Department of Fish Processing Technology, College of Fisheries Science, Kamdhenu University, Veraval, Gujarat, India

Taeho Kim
Smart Aquaculture Research Centre, Chonnam National University, Yeosu, South Jeolla, Republic of Korea

Jora Kirankumar
Department of Aquaculture, College of Fisheries Science, Kamdhenu University, Veraval, Gujarat, India

Suwaree Kitikiew
Department of Animal Production Technology and Fisheries, Faculty of Agricultural Technology, King Mongkut's Institute of Technology Ladkrabang, Bangkok, Thailand

Shakeela Manzoor
Department of Zoology, Wildlife and Fisheries, University of Agriculture, Faisalabad, Punjab, Pakistan

Sandip Markad
Department of Fishery Science, Toshniwal Arts, Commerce and Science College, Sengaon, Hingoli, Maharashtra, India

Gertrude Dzifa Mensah
CSIR College of Science and Technology, Cantonments, Accra, Ghana

B. Mukilarasi
Department of Applied Mechanics and Biomedical Engineering, Indian Institute of Technology Madras, Chennai, Tamil Nadu, India

Shafiya Mushtaq
Faculty of Fisheries, Sher-e-Kashmir University of Agricultural Sciences and Technology of Kashmir (SKUAST-K), Rangil, Ganderbal, Jammu and Kashmir, India

A. Natarajan
Department of Applied Mechanics and Biomedical Engineering, Indian Institute of Technology Madras, Chennai, Tamil Nadu, India

Hua Thai Nhan
College of Aquaculture, Can Tho University, Can Tho, Vietnam

P. V. Parmar
Department of Aquaculture, College of Fisheries Science, Kamdhenu University, Veraval, Gujarat, India

Philominal Pathinathan
College of Fishery Science, Andhra Pradesh Fisheries University, Narasapuram, West Godavari, Andhra Pradesh, India

Dhanaji Patil
Department of Fishery Science, Toshniwal Arts, Commerce and Science College, Sengaon, Hingoli, Maharashtra, India

Modi Kiran Piyushbhai
Institute of Fisheries Post Graduate Studies, Tamil Nadu Dr. J. Jayalalithaa Fisheries University, OMR Campus, Vaniyanchavadi, Tamil Nadu, India

Dabhi Rajkumar
Department of Aquaculture, College of Fisheries Science, Kamdhenu University, Veraval, Gujarat, India

S. Ramakrishnan
Department of Applied Mechanics and Biomedical Engineering, Indian Institute of Technology Madras, Chennai, Tamil Nadu, India

Kiran D. Rasal
Fish Genetics and Biotechnology Division, ICAR–Central Institute of Fisheries Education, Mumbai, Maharashtra, India

Aparna Rathore
Department of Botany, Kanoria PG Mahila Mahavidyalaya, Jaipur, Rajasthan, India

K. T. Hafeef Roshan
College of Fishery Science, Andhra Pradesh Fisheries University, Narasapuram, West Godavari, Andhra Pradesh, India

Subha M. Roy
Smart Aquaculture Research Centre, Chonnam National University, Yeosu, Jeonnam, Republic of Korea

Kishan Sakariya
Department of Aquaculture, College of Fisheries Science, Kamdhenu University, Veraval, Gujarat, India

S. Saravanan
Department of Prosthodontics, Saveetha Dental College and Hospital, Saveetha Institute of Medical and Technical Sciences (SIMATS), Saveetha University, Chennai, Tamil Nadu, India

Madhurjya Ranjan Sharma
Department of Agricultural Biotechnology, Assam Agricultural University, Jorhat, Assam, India

Hardik Sikotariya
Department of Fisheries Resource Management, College of Fisheries Science, Kamdhenu University, Veraval, Gujarat, India

A. K. Singh
Department of Fisheries Engineering and Technology, College of Fisheries Science, Birsa Agricultural University, Ranchi, Jharkhand, India

Arya Singh
ICAR–Central Institute of Fisheries Education, Panch Marg, Mumbai, Maharashtra, India

Soranuth Sirisuay
Department of Aquaculture, Faculty of Fisheries, Kasetsart University, Bangkok, Thailand

Jebarson Solomon
ICAR–Central Institute of Fisheries Education, Mumbai, Maharashtra, India

S. B. Lakshmi Swarubhaa
Department of Applied Mechanics and Biomedical Engineering, Indian Institute of Technology Madras, Chennai, Tamil Nadu, India

S. Swathi
Department of Applied Mechanics and Biomedical Engineering, Indian Institute of Technology Madras, Chennai, Tamil Nadu, India

Janshi Tandel
Department of Fisheries Resource Management, College of Fisheries Science, Kamdhenu University, Veraval, Gujarat, India

Ritika A. Tandel
Department of Fish Processing Technology, College of Fisheries Science, Kamdhenu University, Veraval, Gujarat, India

Vivek R. Tandel
Department of Aquatic Environment Management, College of Fisheries Science, Kamdhenu University, Veraval, Gujarat, India

Riesti Triyanti
Research Center for Behavioral and Circular Economics, National Research and Innovation Agency (BRIN), Jakarta, Indonesia

Yuh Tzean
Department of Plant Pathology & Microbiology, National Taiwan University

Kailash Vadher
Department of Aquaculture, College of Fisheries Science, Kamdhenu University, Veraval, Gujarat, India

D. T. Vaghela
Department of Aquatic Environment Management, College of Fisheries Science, Kamdhenu University, Veraval, Gujarat, India

Tincy Varghese
ICAR – Central Institute of Fisheries Technology, Kochi, Kerala, India

S. Vimalraj
Department of Biotechnology, Faculty of Biomedical Sciences and Technology, Sri Ramachandra Institute of Higher Education and Research (SRIHER) (Deemed to be University), Porur, Chennai, Tamil Nadu, India

M. S. Vishnupriya
Department of Fisheries Engineering and Technology, College of Fisheries Science, Birsa Agricultural University, Ranchi, Jharkhand, India

Rizki Aprilian Wijaya
Research Center for Behavioral and Circular Economics, National Research and Innovation Agency (BRIN), Jakarta, Indonesia

Tsung-Meng Wu
Department of Aquaculture, National Pingtung University, Pingtung, Taiwan

Abbreviations

A2M	alpha-2-macroglobulin
ACNNV	atlantic cod NNV
ADP	abiotic depletion potential
AFGP	antifreeze glycoproteins
AFP	antifreeze proteins
AHLs	acylated homoserine lactones
ALA	α-linolenic acid
AMR	antimicrobial resistance
AP	acidification potential
AP1	activating protein-1
AqGR	aquatic genetic resources
ARA	arachidonic acid
ARDEC	Aquaculture Research and Development Centre
ASG-10	atlantic salmon gill epithelial cell line
BALOs	bdellovibrio and like organisms
BF-2	bluegill fry
BFOS	biofloc organisms
BFT	biofloc technology
BHPP	butyl-hydroxyl-phenyl propanoic acid
BIV	bohle virus
BKD	bacterial kidney disease
BOD	biochemical oxygen demand
BS	bacillus subtilis
CAC	Competent Approval Committee
CAT	catalase
CCRF	code of conduct for responsible fisheries
CCV	channel catfish virus
CEAA	conditionally essential amino acids
CEFs	chicken embryo fibroblasts
CF	condition factor

CH_4	methane
CHSE-214	chinook salmon embryo
CITES	convention on international trade in endangered species of wild fauna and flora
CMFRI	Central Marine Fisheries Research Institute
CMS	convention on migratory species
CND	consumer-driven nutrient dynamics
CO	carbon monoxide
CO	corn oil
CO_2	carbon dioxide
COD	chemical oxygen demand
CRISPR	clustered regularly interspaced palindromic repeats
CSCA	circular stepped cascade aerator
DAAs	dispensable amino acids
DCF	diclofenac
DHA	docosahexaenoic acid
DI	direct immersion
DIY	do it yourself
DO	dissolved oxygen
DWC	deep water culture
EAA	ecosystem approach to aquaculture
EAF	ecosystem approach to fisheries
EBA	ecosystem-based aquaculture
EBFM	ecosystem-based fisheries management
EC	electrical conductivity
EFA	essential fatty acids
ELISA	enzyme-linked immunosorbent assay
EPA	eicosapentaenoic acid
EPC	*Epithelioma papulosum cyprini*
EST	expressed sequence tags
FAO	food and agriculture organization
FAO-UN	Food and Agriculture Organization of the United Nations
FCL	fish cell lines
FCR	feed conversion ratio
FHM	fathead minnow
FI	feed intake

FO	fish oil
FSH	follicle-stimulating hormone
FSMs	farm-scale models
FuFAs	functional feed additives
FW	final weight
GAP	gnrh-associated peptide
GCRV	grass carp reovirus
GEBV	genomic estimated breeding values
GF	grunt fin
GH	growth hormone
GHG	greenhouse gases
GI	galvanized iron
GIT	gastrointestinal tract
GnRH	gonadotropin-releasing hormone
GnRHa	gonadotropin-releasing hormone analogs
GPx	glutathione peroxidases
GS	genomic selection
GTHs	gonadotropins
GWAS	genome-wide association studies
GWP	global warming potential
H_2	hydrogen
HDL	high-density lipoprotein
HDPE	high-density polyethylene
HEAV	hemagglutinin-esterase virus
HI	hypertonic immersion
HIRRV	hirame rhabdovirus
HSI	hepatosomatic index
HTC	hydrothermal carbonization
HUFAs	highly unsaturated fatty acids
IAAs	indispensable amino acids
ICAR-CMFRI	Indian Council of Agricultural Research-Central Marine Fisheries Research Institute
IFN	interferon
IG	immunoglobulins
IgM	immunoglobulin M
IHNV	infectious hematopoietic necrosis virus

IL-1β	interleukin 1β
ILs	interleukins
IMTA	integrated multitrophic aquaculture
IPN	infectious pancreatic necrosis
IPNV	infectious pancreatic necrosis virus
IRAK4	il-1-receptor-associated kinase 4
IRF	interferon-related factor
IROMP	iron-regulated outer membrane protein
ISAV	infectious salmon anemia virus
ISKNV	infectious spleen and kidney necrosis virus
ISO	International Organization for Standardization
ISW	inland saline water
IUU	illegal, unreported, and unregulated
JLGs	joint liability groups
KHV	koi herpes virus
KIMST	Korea Institute of Marine Science and Technology Promotion
LA	linoleic acid
LCA	life cycle assessment
LCDV	lymphocystis disease virus
LC-PUFAs	long-chain polyunsaturated fatty acids
LEDs	light emitting diodes
LFBH	lemon fin barb hybrid
LGP2	laboratory of genetics and physiology 2
LH	luteinizing hormone
LO	linseed oil
LPS	large polyp stony
LPS	lipopolysaccharide
LS	living shorelines
MAS	marker-assisted selection
MBGB	media-based growing bed
MDCK	madin darby canine kidney
MMT	million metric tons
MO	mixed oil
MP	microbial proteins
MPB/MPF/MPK	mylopharyngodon piceus bladder/fin/kidney

MS	mild steel
MX	myxovirus resistance
NFT	nutrient film technique
NF-κB	nuclear factor kappa b
NH_3	ammonia
NH_3-N	ammonia nitrogen
NNV	nervous necrosis virus
NO_2	nitrite
NO_2-N	nitrite nitrogen
NO_3	nitrate
NO_3-N	nitrate nitrogen
NOS	nitric oxide synthase
NRC	National Research Council
OA	oleic acid
OO	olive oil
PA	propeller aerator
PAMPs	pathogen-associated molecular patterns
PASFs	polychaete-assisted sand filters
PCP	pentachlorophenol
PCSCA	pooled circular stepped cascade aerator
PD	pancreatic disease
PER	protein efficiency ratio
PHB	poly-β-hydroxybutyrate
PMMSY	Pradhan Mantri Matsya Sampada Yojana
PO_4–	active phosphate
PO_4	phosphorus
POM	particulate organic matter
PRRs	pattern recognition receptors
PRXS	peroxiredoxins
PSB	photosynthetic bacteria
PTA	perforated tray aerator
PVC	polyvinyl chloride
PWA	paddle wheel aerator
QQ	quorum quenching
QS	quorum sensing
QTL	quantitative trait loci

RAS	recirculation aquaculture systems
rDNA	recombinant dna technology
rFsh	recombinant follicle stimulating hormone
rGH	recombinant growth hormone
RH	radiation hybrid
RIA	radio immune assays
RIG-I	retinoic acid-inducible gene-I
rLh	recombinant luteinizing hormone
RO	rice oil
ROS	reactive oxygen species
RSIV	red sea bream iridovirus
RTG-2	rainbow trout gonad
SA	spiral aerator
SALT	skin-associated lymphoid tissue
SBO	soybean oil
SC	*Saccharomyces cerevisiae*
SCFA	short-chain fatty acids
SCP	self-contained proposal
SCP	single-cell protein
SDGs	sustainable development goals
SEM	standard error of the mean
SEO	search engine optimization
SFPAs	sustainable fisheries partnership agreements
SGR	specific growth rate
SHGs	self-help groups
SMS	spent mushroom substrate
SNPs	single nucleotide polymorphisms
SO	sunflower oil
SOD	superoxide dismutase
SPS	small polyp stony
SS	suspended solids
SSO	safflower seed oil
STAT	signal transducer and activator of transcription
SVC	spring viremia of carp
SVCV	spring viremia of carp virus
TDS	total dissolved solids

TiLV	tilapia lake virus
TLR	toll-like receptor
TNF-α	tumor necrosis factor-α
TO	tea oil
TRAF6	TNF receptor-associated factor-6
TSA	tryptic soy agar
TSS	total suspended solids
TT	tecloftalam
UNCED	United Nations Conference on Environment and Development
UNCLOS	United Nations Convention on the Law of the Sea
UNEP	United Nations Environment Program
UNICPOLOS	Unofficial Consultative Process on Oceans and the Law of the Sea
UTs	Union Territories
VHSV	viral hemorrhagic septicemia virus
VLPs	virus-like particles
VNN	viral nervous necrosis
VO	vegetable oils
WG	weight gain
WHO	World Health Organization
WSSV	white spot syndrome virus

Preface

This book is designed to provide comprehensive insight into diverse aspects of aquaculture as well as recent technological advances in the field.

Aquaculture is a major food-producing sector of the world, delivering nutritious food to people around the globe and generating substantial entrepreneurship and job opportunities. Continuous efforts are being made by researchers, entrepreneurs, and fishermen toward the modernization and diversification of fisheries activities to improve production and profitability with minimal environmental impact. The editors sought contributions from diverse researchers for the book and incorporated the most relevant work according to predefined themes. Efforts were made to include work that focused on development and sustainability in aquaculture to yield equitable outcomes. The content of the book will educate beginning aquaculturists on the status, scope, and roles of aquaculture, while professional aquaculturists will be able to improve and enhance their knowledge and technical know-how.

Introductory information and research work encompassing the status and scope of aquaculture, which provide an overview and synoptic understanding of the sector, are covered in the initial two chapters of the book. The book proceeds with chapters elaborating on the use of probiotics and microbial approaches in aquaculture. The current potential and forthcoming prospects of mariculture, with respect to cage farming, are discussed in the fifth chapter. Biofloc technology is addressed for its adaptability, lower space requirements, and higher yields, while aquaponics, the integration of aquaculture with a hydroponic system, is elaborated for its resource recycling, environmentally friendly, and economical traits. Some of the important aspects of aquaculture management, such as water quality management and aeration management, are also covered by researchers in the eighth and ninth chapters.

Research work on genetics as well as biotechnological approaches and advancements in the aquaculture sector is discussed in the biotechnology section of the book, which includes the tenth to fifteenth chapters. This section presents a detailed discussion on the application of genetic and biotechnological approaches for fish health management and the use of recombinant proteins in the aquaculture sector. The use of biotechnological approaches such as cell lines, cell culture, vaccination, nutrient recycling, and environmental considerations for the sustainability of the sector is detailed in the next four chapters of the section. The literature contributed by researchers on aquaculture sustainability approaches, such as ecosystem-based aquaculture and *biochar* production techniques, is included in the sixteenth and seventeenth chapters of the book. Detailed discussions in the form of case studies on certain species and technological advancements are included in the eighteenth to twenty-second chapters of the book.

Ultimately, the book will not only enlighten enthusiasts and hobbyists regarding different aspects of aquafarming but also satisfy the thirst of farmers, researchers, and industry experts for comprehensive information and technical know-how about advanced technologies in the aquafarming sector.

The editors have sought to cover all aspects of aquaculture in the book, ranging from fishery resources and scope to recent innovations and management practices, in the hope that it will serve as a reference guide. We are deeply indebted to all the contributors who made efforts in the compilation of this book. We would also like to extend our sincere gratitude to Apple Academic Press, in exclusive worldwide co-publication with CRC Press, a Taylor & Francis Group.

CHAPTER 1

Scope and Status of Aquaculture

HARDIK SIKOTARIYA,[1,*] T. H. DAVE,[1] MAYUR BHADARKA,[2] SAYAN BISWAS,[3] and MOHIT BAMANIYA[4]

[1]*Department of Fisheries Resource Management, College of Fisheries Science, Kamdhenu University, Veraval, Gujarat, India*

[2]*Department of Aquatic Environment Management, College of Fisheries Science, Kamdhenu University, Veraval, Gujarat, India*

[3]*Department of Aquaculture, College of Fisheries Science, Kamdhenu University, Veraval, Gujarat, India*

[4]*Department of Fish Processing Technology, College of Fisheries Science, Kamdhenu University, Veraval, Gujarat, India*

*Corresponding author

ABSTRACT

This chapter explores the dynamic world of aquaculture by examining its historical roots, current status, and future trajectories. It begins by defining aquaculture and highlighting its pivotal role in global food security, setting the stage for a comprehensive examination. The text traces its evolution through time, from ancient civilizations to modern practices, and underscores the enduring significance of aquaculture. The scope of aquaculture is meticulously dissected, covering aspects such as species diversity, production systems, geographic distribution, and operational scales, thereby providing a holistic understanding of its expansive reach. Moreover, the chapter emphasizes aquaculture's economic importance, quantifying its contribution to the global food supply, employment generation, and international trade. Technological advancements in

Sustainable Aquaculture: Recent Advances and Techniques. Dhanaji W. Patil, Sandip S. Markad (Eds.)

DOI: 10.1201/9781779641113-1

aquafeed formulation, water management, and genetic enhancements are spotlighted, showcasing the industry's relentless pursuit of innovation. Concluding with a reflection on challenges and prospects—including disease management, regulatory frameworks, and emerging trends—the chapter offers a nuanced perspective on the trajectory of aquaculture. Through rigorous research and insightful analysis, this chapter serves as indispensable literature for policymakers, researchers, and practitioners keen to delve into the complex landscape of aquaculture.

1.1 INTRODUCTION

Diving into the world of aquaculture unveils a realm where protein production thrives, ensuring global food security. Across the map, from bustling China to distant shores, aquaculture's growth knows no bounds, transforming into a major industrial force (Houfang et al., 2023). Yet, the quest for suitable aquaculture sites becomes a gripping challenge, with every inch of potential water space under scrutiny (Chen and Gao, 2023). As appetites soar, aquaculture steps up to the plate, supplying nearly 21 kg of fish per person annually.

But in Ethiopia, the journey is filled with hurdles—from age-old farming methods to the harsh realities of climate change and overfishing (Sebastian, 2022). However, hope swims strong, buoyed by visions of technological marvels, renewable energy solutions, and ingenious management strategies (Delilo, 2020). Despite the environmental challenges stemming from increased production in the sector, aquaculture holds significant promise in addressing poverty and nutrition concerns. This is evident from the fact that developing countries contribute almost 80% of total global aquaculture production, while more than 80% of the world's aquaculture output originates from small-scale aquaculture fish farms typically managed by family members (FAO, 2014; Phillips et al., 2016).

1.2 HISTORICAL OVERVIEW

Aquaculture's journey of growth and evolution mirrors the expanding demand for protein and food security, especially amidst our burgeoning global population (Houfang et al., 2023). With a surge in inland aquaculture across Asia leading the charge, the global aquaculture landscape has flourished. However, amidst this growth lie hurdles that demand attention for sustainable progress. These hurdles encompass high capital and operational expenses, cutthroat market competitiveness, disparities in production area distribution, and the reverberations of the COVID-19 pandemic (Chen and Gao, 2023).

Furthermore, sustainability woes loom large, from grappling with pathogens, parasites, and pests to navigating the relentless impacts of climate change (Daniel, 2017). Yet, strides have been made in fortifying the governance, technology, siting, and management of aquaculture, fostering encouraging trends in environmental stewardship (Naylor et al., 2021). Despite these advances, the call for continued exploration and innovation resounds, particularly in realms such as feed nutrition, additives, and the integration of renewable energy sources. In this ceaseless pursuit of progress, aquaculture charts a course toward a more resilient, responsible, and nourished future.

1.3 SCOPE OF AQUACULTURE

Aquaculture emerges as a beacon of hope, holding immense promise for bolstering protein production and fortifying food security amidst the dual challenges of population growth and poverty. At its core lies artificial propagation, a cornerstone of fish culture, where advancements in our understanding of fish reproductive physiology pave the way for refined breeding techniques (Chaube, 2023). Operating akin to agriculture, aquaculture involves the meticulous cultivation of aquatic organisms under controlled conditions (Sahitya, 2015).

With arable land at a premium, aquaculture emerges as a viable alternative, leveraging the remarkable efficiency of fish as converters of feed into high-quality protein (Khan et al., 2011). Moreover, beyond nourishing our plates, aquaculture catalyzes job creation and foreign exchange generation (Bjom et al., 2019). Innovative aquaculture methods, like cylindrical cages, have emerged to enhance efficiency and streamline maintenance. Aquaculture presents promising avenues for sustainable protein generation and food security, yet effective management and consideration of environmental and social dynamics are essential for its viability.

1.3.1 SPECIES DIVERSITY

Aquaculture boasts remarkable species diversity compared to terrestrial farming, with fish, mollusks, and crustaceans leading the industry. *Carp* stands out as the most prevalent freshwater fish, while the western white leg shrimp dominates marine crustacean production (Ton, 2022). Biodiversity plays a pivotal role in optimizing aquaculture systems, as evidenced by the positive correlation between macroalgal species richness and bioremediation efficacy in wastewater treatment (Knoop et al., 2022). Polyculture and integrated aquaculture methods are vital for sustainable progress in the industry, fostering synergies among cultivated species, bolstering system resilience, and delivering social advantages (Kim et al., 2022). In aquaculture populations like the *yellowtail kingfish*, controlling inbreeding rates through marker panels for parentage determination ensures long-term sustainability (Martinez et al., 2023). Overall, aquaculture offers a rich array of species, farming techniques, and environments, rendering it a versatile and promising sector for food production (Metian et al., 2020).

1.3.2 PRODUCTION SYSTEMS

Aquaculture has emerged as a major segment of the global food supply, characterized by diverse production systems and a drive toward intensification and enhanced management (Verdegem et al., 2023). Regional disparities are evident in aquaculture systems, with inland aquaculture prevailing in Asia and Africa, while marine aquaculture dominates in the Americas, Europe, and Oceania (Burcu, 2023). The substantial growth in aquaculture production evident from 2000 can be ascribed to factors such as improved feeds, heightened aquaculture management practices, and increased focus on biosecurity measures (Chiquito-Contreras et al., 2022). Nonetheless, aquaculture presents environmental hurdles, notably concerning water usage and nutrient discharge (Dong and Dong, 2023; Jim and Qi, 2020). Addressing these challenges necessitates greater emphasis on scaling, meticulous site selection, and holistic consideration of the broader production ecosystem's health. Moreover, endeavors to render aquaculture climate-neutral and mitigate its environmental footprint are imperative. Integrating aquaculture into local food systems and fostering the development of sustainable, circular food systems should be pivotal areas of focus for advancing aquaculture systems in the future.

1.3.3 GEOGRAPHIC DISTRIBUTION

Aquaculture exhibits a widespread global presence, with Asia, particularly China, serving as a focal point. Asia holds the largest share of aquaculture distribution, with China alone contributing the majority. In China, aquaculture ponds are notably concentrated within a 30-kilometer buffer region along the coastline. Other countries such as Vietnam and India also boast substantial aquaculture areas, although their extents differ by less than 10% compared to China. Over the years, there has been a noticeable expansion of aquaculture ponds globally, with Asia and South America witnessing a doubling in size since 1984. However, the overall growth rate of aquaculture pond areas worldwide has decelerated. Remote sensing techniques have been instrumental in monitoring the global distribution of aquaculture ponds, providing valuable insights into their spatial-temporal trajectory and potential impacts (Kang et al., 2019; Lim and Neo, 2014; Naylor et al., 2023; Wang et al., 2022).

1.3.4 SCALE OF OPERATIONS

Aquaculture operations can span various scales, from individual farms to regional and global levels. At the farm scale, integrated mathematical models known as farm-scale models (FSMs) play a crucial role in simulating farm operations and tackling sustainability issues in aquaculture. These models adeptly quantify input flows and predict outputs like fish biomass and waste. The development of FSMs is tailored to the specific farming system under study and the objectives of the research. Various modeling techniques are available to simulate fish farm processes, each with its unique set of advantages and limitations. FSMs contribute to addressing sustainability challenges in aquaculture by facilitating precision fish farming, modeling Integrated multi-trophic aquaculture (IMTA), managing spatial considerations, and conducting life cycle assessments. However, there are existing limitations and research gaps in FSMs that require attention (Chary et al., 2022).

1.4 ECONOMIC IMPORTANCE

Aquaculture serves as a cornerstone of economic development for many nations, offering avenues for livelihoods, poverty reduction, and overall economic expansion. Recognized as a potent generator of income and employment, it catalyzes the growth of ancillary industries while providing affordable and nutritious food options. Particularly crucial for impoverished populations, aquaculture enriches diets with vital vitamins, proteins, micronutrients, and minerals (Jolly and Clonts, 2020). Its economic significance extends to global food security efforts, as it augments protein production without imposing undue strains on ecosystems (Ursula da Silva et al., 2022). Sustainable aquaculture hinges on prudent economic planning and management, coupled with meticulous attention to long-term viability for both individual producers and community initiatives (Pradeepkiran, 2019). Moreover, the application of energy analysis in aquaculture systems aids in gauging sustainability levels and underscores the importance of water in production and resource stewardship. Aquaculture is often lauded as a pro-poor economic activity, serving as a means of income generation to uplift livelihoods in rural areas of low- and middle-income countries (Dey and Ahmed, 2005; Mohamed and Dodson, 1998; Olaganathan and Kar Mun, 2017). However, the limited empirical evidence on this matter paints a more nuanced picture, where the impact varies depending on local production and consumption dynamics within the sector. Recent studies conducted in Ghana have indicated that aquaculture can indeed contribute to economic growth and poverty reduction at a national level (Kassam and Dorward, 2017; Rashid et al., 2019). Nevertheless, evidence has also surfaced suggesting that the promotion of aquaculture may primarily benefit larger and more affluent farms, potentially exacerbating inequality (Ahmed et al., 1995; Kassam and Dorward, 2017).

1.4.1 EMPLOYMENT OPPORTUNITIES

Aquaculture serves as a key driver for employment opportunities and socioeconomic progress in countries. Recognized for its potential to bolster protein production and enhance food security, it is identified as a pivotal sector (Yadala et al., 2017). Employment generated by aquaculture spans various levels, with farm-level engagement surpassing other value chain segments (Phillips et al., 2016). Small-scale aquaculture producers notably contribute more to employment compared to medium- and large-scale counterparts. Moreover, women play a crucial role in fisheries, contributing significantly to postharvest activities, preservation, processing, and marketing of seafood (Vipinkumar, 2020). The expansion of aquaculture, particularly in community water bodies, has empowered rural farmers, including women, by providing livelihoods and income opportunities (Nune, 2008). Overall, aquaculture emerges as a vital contributor to employment and economic growth, underlining its significance in fostering sustainable development.

1.4.2 EXPORT EARNINGS

Export earnings in aquaculture play a pivotal role in the economic advancement of nations such as India. The marine fisheries sector in India has witnessed substantial growth, evolving into a capital-intensive industry that provides employment and income to millions engaged in fisheries, aquaculture, processing, and related activities (Rajyasri, 2010). Recognized as a potent income and employment generator, the fisheries sector stimulates the growth of ancillary industries and bolsters the country's foreign exchange reserves (Goswami and Zade, 2015). In Belize, aquaculture, particularly shrimp farming, has significantly propelled development by boosting export production, revenue, employment, and foreign exchange earnings (Yadala et al., 2017). The economic feasibility of freshwater aquaculture in tropical countries, including India, has been established, positioning it as a crucial farm diversification strategy and commercial pursuit (Myvett and Quintana, 2002). Collectively, export earnings in aquaculture exert a substantial impact on the economic landscape of the fisheries sector, fostering income generation, employment opportunities, and bolstering foreign exchange reserves.

1.5 TECHNOLOGICAL ADVANCEMENTS

1.5.1 INNOVATIONS IN AQUAFEED

Innovations in aquafeed within aquaculture have become increasingly vital to foster sustainable and efficient production practices. The traditional reliance on feed ingredients such as fishmeal and fish oil is giving way to alternative sources, including plant-based raw materials, microorganisms, and non-conventional feed ingredients like insects

(George et al., 2023; Iulian et al., 2023; Midhun and Arun, 2023; Sarker, 2023). These innovations aim to mitigate the ecological, food security, and economic challenges associated with conventional feed ingredients. Nanotechnology is also being leveraged to enhance feed quality and increase the availability of micronutrients (Hameed et al., 2022). The utilization of non-conventional feed ingredients like insects holds promise for replacing fishmeal and reducing the strain on natural resources. These advancements not only contribute to sustainable food production but also facilitate the efficient utilization of organic waste and nutrient recovery. Further research is imperative to assess the nutritional value and physiological response of fish to these novel feed ingredients.

1.5.2 WATER MANAGEMENT TECHNIQUES

Water management techniques in aquaculture encompass diverse approaches aimed at maintaining optimal water quality to support the growth and survival of fish. One strategy involves leveraging machine learning techniques, such as decision tree algorithms, to minimize the mortality rate of fish fingerlings in outdoor ponds (Matondang et al., 2022). Another effective method entails deploying a water quality management apparatus capable of predicting future water quality based on measured values, thereby determining the necessity for water exchange (Kasai et al., 2023). Fuzzy logic techniques can also be employed to assess the status of freshwater pumps and aerators by analyzing water parameters (Nagothu, 2021). Moreover, a comprehensive management system for aquaculture can automate the control of oxygenators, heat pumps, and refrigerators to effectively regulate the water environment (Yang et al., 2020). Additionally, the monitoring and prediction of water quality parameters, such as temperature and dissolved oxygen levels, can be achieved using sensors and neural networks, thereby mitigating losses in aquaculture production (Quintero et al., 2022).

1.5.3 GENETIC IMPROVEMENTS

Genetic advancements in aquaculture play a pivotal role in fostering sustainable development and aligning with the UN's Sustainable

Development Goals. Efforts are underway to monitor and characterize farmed aquatic genetic resources (AqGR) while developing information systems for their effective management (Li and Jiang, 2023). Genomic technologies, including genotyping and genome-wide association studies (GWAS), are instrumental in identifying genetic markers linked to economically significant traits, subsequently integrated into breeding programs via marker-assisted selection (MAS) (Sonesson et al., 2023). Genomic selection, leveraging genomic estimated breeding values (GEBV), demonstrates promising outcomes in accelerating genetic progress for polygenic traits in aquaculture species (Yanez et al., 2023). However, the adoption of selective breeding and genetic technologies in aquaculture varies across species and regions, with certain extensively produced species lacking well-established breeding programs (Houston et al., 2022). Emerging gene editing and gene silencing techniques, such as *CRISPR/Cas9*, hold immense potential to revolutionize aquaculture by enhancing disease resistance, growth, and reproduction in fish and crustaceans (Gutasi et al., 2023). Transgenic technology has been increasingly utilized in various fish species in recent years, albeit primarily for research purposes. Recent studies conducted on salmonids have revealed that while incorporating growth hormone gene constructs into slow-growing wild strains led to remarkable growth improvements, the same constructs did not yield comparable results when introduced into fast-growing domesticated stock (Devlin et al., 2023). Continuous research, development, and adoption of these genetic technologies are imperative for ensuring the long-term success and sustainability of the aquaculture industry.

1.6 CHALLENGES AND FUTURE OUTLOOK

Aquaculture confronts a myriad of challenges alongside promising prospects. As the global population burgeons, the demand for fish and seafood escalates, necessitating aquaculture's response to environmental concerns such as pollution, resource inefficiency, and chemical usage (Mair et al., 2023). Disease outbreaks pose a significant threat to aquaculture operations, leading to economic losses. While vaccines and functional feeds offer avenues for disease prevention and control, their efficacy varies depending on species and conditions (Picado et al., 2023). To meet the escalating demand sustainably, aquaculture must prioritize reducing its environmental impact and embracing a circular economy approach (Cain, 2022). Biotechnological tools play a pivotal role in augmenting fish production through diverse approaches, including environmental manipulations, breeding techniques, diagnostics, and feed fermentation (Sahoo and Paul, 2023). Furthermore, the development of self-sufficient multi-trophic systems and optimization of cleaner fish species farming can contribute to mitigating environmental impact and enhancing aquaculture practices (Bacenetti et al., 2019). In summary, addressing these challenges and harnessing technological advancements can empower aquaculture to satisfy the mounting demand for fish while minimizing its environmental footprint.

1.6.1 DISEASE MANAGEMENT

Disease management and biosecurity in aquaculture encounter significant challenges that profoundly impact industry sustainability and productivity. Disease outbreaks present a formidable risk, resulting in substantial economic losses and necessitating effective control strategies (Cain, 2022). However, the emergence of treatment-resistant parasites and pathogens poses a worrisome concern that demands attention (Coates, 2023). Globally, antimicrobial resistance (AMR) presents a pressing health issue, prompting the exploration of alternative strategies to traditional antimicrobial products (Wright et al., 2023). In seaweed cultivation, robust biosecurity measures are essential to prevent economic losses and uphold ecological balance (Sugumaran et al., 2022). In western Kenya, inadequate knowledge and preparedness among fish farmers regarding fish health management and biosecurity measures pose a formidable threat to aquaculture production (Kyule-Muendo et al., 2022). To surmount these challenges, it is imperative to delve into parasite evolutionary biology, develop sustainable disease management strategies, and orchestrate coordinated awareness campaigns. The future trajectory of disease management and biosecurity in aquaculture hinges on continual research, innovation, and the widespread adoption of sustainable practices.

1.6.2 EMERGING MARKET TRENDS AND CONSUMER PREFERENCES

Emerging market trends and consumer preferences in aquaculture are propelling the demand for sustainable and eco-friendly practices. Despite a rising need for fish products, conventional fishing methods pose sustainability challenges and ecological harm (Kaimakoudi, 2024). Consequently, aquaculture is being advocated as a remedy, yet encountering hurdles in consumer perception and awareness (Evangelou and Lyridis, 2021). Consumers exhibit a bias towards wild-caught fish over farmed varieties, often due to a lack of understanding regarding aquaculture processes (Valente, 2022). To mitigate this, initiatives are underway to enhance consumer education and communication regarding aquaculture practices (Bacenetti et al., 2019). Furthermore, there is a pressing need to diminish the reliance on marine ingredients in aquafeeds and explore alternative sources (Baldi and Mancuso, 2015). However, it is crucial to assess the nutritional and functional attributes of these alternatives to uphold fish health and safety. Ultimately, aquaculture must position itself as a sustainable and nutritious option to meet the escalating demand for fish while safeguarding the environment.

1.6.3 POTENTIAL FOR INNOVATION AND SUSTAINABLE GROWTH

The global aquaculture industry is witnessing rapid expansion, fueled by shifting consumption patterns and advancements in technology. However, environmental concerns underscore the need for a transition towards sustainable practices (Perez-Viveros et al., 2023). Innovation across the aquaculture value chain holds promise for poverty alleviation, promoting the consumption of nutritious aquatic foods, and fostering sustainable production (Bunting et al., 2023). While aquaculture fulfills a crucial role in meeting the dietary needs of a burgeoning global population, its growth has brought forth environmental and social challenges (Mair et al., 2023). In response, innovative aquaculture technologies such as recirculatory aquaculture systems, biofloc technology, aquaponics, and cage culture have been developed to promote sustainable development (Chauhan and Mishra, 2022). The future trajectory of aquaculture hinges on reducing reliance on marine ingredients in aquafeeds, evaluating alternative feed ingredients' nutritional and functional value, and adopting a comprehensive approach to sustainable farming (Valente, 2022). Embracing technological innovation, implementing sustainable practices, and enacting supportive legislation emerge as imperative steps for ensuring the long-term success of the aquaculture sector worldwide.

1.7 CONCLUSIONS

The exploration of aquaculture presented in this chapter unveils a realm teeming with historical significance, contemporary challenges, and promising prospects for the future. Delving into the roots of aquaculture, from ancient practices to modern innovations, underscores its enduring importance in sustaining global food security. As aquaculture continues to evolve, its scope encompasses a diverse array of species, production systems, geographic distributions, and operational scales, offering a comprehensive understanding of its multifaceted landscape.

The economic importance of aquaculture cannot be overstated, as it contributes significantly to global food supply, employment generation, and international trade. However, this growth is not without its challenges, including environmental sustainability, disease management, and regulatory frameworks. Yet, technological advancements in aquafeed, water management, and genetic improvements hold promise for addressing these challenges and fostering sustainable growth in the industry.

Looking ahead, the future trajectory of aquaculture is shaped by emerging market trends, consumer preferences, and innovation potential. As consumer awareness grows and demands for sustainable practices increase, aquaculture must adapt to meet these expectations while continuing to provide nutritious food sources for a growing global population. By embracing technological innovation, implementing sustainable practices, and enacting supportive legislation, aquaculture can chart a course toward a resilient, responsible, and nourished future for generations to come.

KEYWORDS

- **aquaculture**
- **aquaculture value chain**
- **aquafarming**
- **aquaponics**
- **biofloc technology**
- **global aquaculture industry**
- **poverty alleviation**
- **recirculatory aquaculture systems**
- **sustainable production**

REFERENCES

Ahmed, M., Rab, M. A., & Bsimbao, M. A. P. (1995). *Aquaculture Technology Adoption in Kapasia Thana, Bangladesh: Some Preliminary Results from Farm Record-Keeping Data* (43p No. 44). International Center for Living Aquatic Resources Management, Manila, Philippines.

Bacenetti, J., Haaland-Gaeta, F., Facchinetti, D., & Parolini, M. (2019). Approaches to tackle emerging challenges in European aquaculture. In *16th International Conference on Environmental Science and Technology*, Rhodes, Greece, 4 to 7 September 2019.

Baldi, L., & Mancuso, T. (2015). Aquaculture production and use of insects for feed: A consumer perspective. *Proceedings in Food System Dynamics*, *2015*, 215–225.

Bjøm, A., Brimsholm, E., & Haugland, E. (2019). *Aquaculture system for cultivating aquatic organisms* (U.S. Patent No. US20190345678A1). United States Patent and Trademark Office.

Bunting, S. W., Bostock, J., Leschen, W., & Little, D. C. (2023). Evaluating the potential of innovations across aquaculture product value chains for poverty alleviation in Bangladesh and India. *Frontiers in Aquaculture*, *2*, 1111266.

Burcu, K. (2023). *Aquaculture Ecosystem*. https://doi.org/10.1007/978-981-19-5486-3_2 (accessed on 10th July 2025).

Cain, K. (2022). The many challenges of disease management in aquaculture. *Journal of the World Aquaculture Society*, *53*(6), 1080–1083.

Chary, K., Brigolin, D., & Callier, M. D. (2022). Farm-scale models in fish aquaculture–An overview of methods and applications. *Reviews in Aquaculture*, *14*(4), 2122–2157.

Chaube, R. (2023). An update on induced breeding methods in fish aquaculture and scope for new potential techniques. *Frontiers in Aquaculture Biotechnology*, 55–68.

Chauhan, R. S., & Mishra, A. (2022). New innovative technologies for sustainable aqua production. In *Biodiversity* (pp. 97–111). CRC Press.

Chen, W., & Gao, S. (2023). Current status of industrialized aquaculture in China: A review. *Environmental Science and Pollution Research*, *30*(12), 32278–32287.

Chiquito-Contreras, R. G., Hernandez-Adame, L., Alvarado-Castillo, G., Martínez-Hernández, M. D. J., Sánchez-Viveros, G., Chiquito-Contreras, C. J., & Hernandez-Montiel, L. G. (2022). Aquaculture—Production system and waste management for agriculture fertilization—A review. *Sustainability*, *14*(12), 7257.

Coates, A. (2023). Research pre-empting parasite adaptation is key to sustainable disease management in aquaculture. *Aquaculture Environment Interactions*, *15*, 35–43.

Daniel, N. (2017). Status of aquaculture with respect to nutrition and feed. *International Journal of Fisheries and Aquatic Studies*, *5*(1), 333–345.

Delilo, G. (2020). Fisheries production and aquaculture development: Current status and future directions in Ethiopia. *Global Journal of Fisheries Science*, *2*(2), 14–29.

Devlin, R. H., Sakhrani, D., Tymchuk, W. E., Rise, M. L., & Goh, B. (2009). Domestication and growth hormone transgenesis cause similar changes in gene expression in coho salmon (*Oncorhynchus kisutch*). *Proceedings of the National Academy of Sciences*, *106*(9), 3047–3052.

Dey, M. M., & Ahmed, M. (2005). Aquaculture—Food and livelihoods for the poor in Asia: A brief overview of the issues. *Aquaculture Economics & Management*, *9*(1, 2), 1–10.

Dong, S. L., & Dong, Y. W. (2023). Sustainability of aquaculture production systems. In *Aquaculture Ecology* (pp. 491–530). Springer Nature Singapore.

Evangelou, P., & Lyridis, D. (2021). Challenges and market trends in the offshore aquaculture industry. In *Int. Conf. Environ. Sci. Technol.*

FAO. (2014). *Sustainable Fisheries and Aquaculture for Food Security and Nutrition.* A report by the High-Level Panel of Experts on Food Security and Nutrition of the Committee on World Food Security. Food and Agriculture Organization of the United Nations, Rome.

George, D., Lakshmi, S., Sharma, A., Prakash, S., Siddiqui, M., Malavika, B. R., & Elumalai, P. (2023). Nanotechnology: A novel tool for aquaculture feed development. In *Nanotechnological Approaches to the Advancement of Innovations in Aquaculture* (pp. 67–88). Springer International Publishing.

Goswami, C., & Zade, V. S. (2015). Analysis of international trade and economic commercial scope of ornamental fishes. *International Journal of Engineering and Applied Sciences*, *2*(5), 257929.

Gutasi, A., Hammer, S. E., El-Matbouli, M., & Saleh, M. (2023). Recent applications of gene editing in fish species and aquatic medicine. *Animals*, *13*(7), 1250.

Hameed, A., Majeed, W., Naveed, M., Ramzan, U., Bordiga, M., Hameed, M., & Rana, N. (2022). Success of aquaculture industry with new insights of using insects as feed: A review. *Fishes*, *7*(6), 395.

Houfang, S., & Sang-Moo, K. (2023). Current status and prospects of the seafood industry. *Food Science and Industry*. https://doi.org/10.23093/fsi.2023.56.2.122 (accessed on 10th July 2025).

Houston, R. D., Kriaridou, C., & Robledo, D. (2022). Animal board invited review: Widespread adoption of genetic technologies is key to sustainable expansion of global aquaculture. *Animal*, *16*(10), 100642.

Iulian, V., Florin, N., Mihai, M., Remus, O., & Vlad-Nicolae, A. (2023). Experimental research on production and use of floating pellets in fish feed in aquaculture systems. In *22nd International Scientific Conference "Engineering for Rural Development": Proceedings: [Jelgava, Latvia,] May 24–26, 2023* (Vol. 22, pp. 216–223).

Jim, D., & Liu, Q. (2020). *Aquaculture Production Systems*. https://doi.org/10.1007/978-981-15-8836-5_17 (accessed on 10th July 2025).

Jolly, C. M., & Clonts, H. A. (2020). *Economics of Aquaculture*. CRC Press.

Kaimakoudi, E. (2024). Policy initiatives towards enhancing consumer knowledge and tackling consumer confusion in aquaculture sector. *Aquaculture International*, *32*(1), 1–9.

Kang, J., Sui, L., Yang, X., Liu, Y., Wang, Z., Wang, J., & Ma, Y. (2019). Sea surface-visible aquaculture spatial-temporal distribution remote sensing: A case study in Liaoning Province, China from 2000 to 2018. *Sustainability*, *11*(24), 7186.

Kasai, S., Oshiro, M., Fukuda, T., Endo, T., Fujii, Y., & Imakubo, K. (2023). *U.S. Patent No. 11,753,312*. Washington, DC: U.S. Patent and Trademark Office.

Kassam, L., & Dorward, A. (2017). A comparative assessment of the poverty impacts of pond and cage aquaculture in Ghana. *Aquaculture*, *470*, 110–122.

Khan, M. A., Khan, S., & Miyan, K. (2011). Aquaculture as a food production system: a review. *Biology and medicine*, *3*(2), 291–302.

Kim, D. Y., Shinde, S. K., Kadam, A. A., Saratale, R. G., Saratale, G. D., Kumar, M., & Ghodake, G. S. (2022). Advantage of species diversification to facilitate sustainable development of aquaculture sector. *Biology*, *11*(3), 368.

Knoop, J., Barrento, S., Lewis, R., Walter, B., Griffin, J. N., Knoop, J., & Griffin, J. N. (2022). Incorporating concepts of biodiversity into modern aquaculture: Macroalgal species richness enhances bioremediation efficiency in a lumpfish hatchery. *Algae*, *37*(3), 213–226.

Kyule-Muendo, D., Otachi, E., Awour, F., Ogello, E., Obiero, K., Abwao, J., & Munguti, J. (2022). Status of fish health management and biosecurity measures in fish farms, cages and hatcheries in Western Kenya. *CABI Agriculture and Bioscience*, *3*(1), 18.

Li, C., & Jiang, J. (2023). High-throughput genotyping in estimating genetic resources and detecting pathogens in aquaculture. *Journal of the World Aquaculture Society*, *54*(3), 558–565.

Lim, G., & Neo, H. (2014). The economic geographies of aquaculture. *Geography Compass*, *8*(9), 665–676.

Mair, G. C., Halwart, M., Derun, Y., & Costa-Pierce, B. A. (2023). A decadal outlook for global aquaculture. *Journal of the World Aquaculture Society*, *54*(2), 196–205.

Martinez, V., Galarce, N., & Setiawan, A. (2023). Developing methods for maintaining genetic diversity in novel aquaculture species: The case of *Seriola lalandi*. *Animals*, *13*(5), 913.

Matondang, P. A. S., Taparhudee, W., Yoonpundh, R., & Jongjaraunsuk, R. (2022). Water quality management guidelines to reduce mortality rate of red tilapia (*Oreochromis niloticus* × *Oreochromis mossambicus*) fingerlings raised in outdoor earthen ponds with a recirculating aquaculture system using machine learning techniques. *ASEAN Journal of Scientific and Technological Reports*, *25*(4), 30–41.

Metian, M., Troell, M., Christensen, V., Steenbeek, J., & Pouil, S. (2020). Mapping diversity of species in global aquaculture. *Reviews in Aquaculture*, *12*(2), 1090–1100.

Midhun, S. J., & Arun, D. (2023). Alternative feed technology in aquaculture. In *Recent Advances in Aquaculture Microbial Technology* (pp. 291–306). Academic Press.

Mohamed, N., & Dodson, B. (1998). Sustainable rural livelihoods? Evaluating the potential of small-scale aquaculture in the Western Cape. *Development Southern Africa*, *15*(1), 103–121.

Myvett, G., & Quintana, R. (2002). "The Status of Aquaculture in Belize 2001," *Belize Audubon Society Newsletter, 34*, 1–5.

Nagothu, S. K. (2021). Intelligent control of aerator and water pump in aquaculture using fuzzy logic. In *Microelectronic Devices, Circuits and Systems: Second International Conference, ICMDCS 2021, Vellore, India, February 11–13, 2021, Revised Selected Papers 2* (pp. 160–171). Springer Singapore.

Naylor, R. L., Hardy, R. W., Buschmann, A. H., Bush, S. R., Cao, L., Klinger, D. H., & Troell, M. (2021). A 20-year retrospective review of global aquaculture. *Nature*, *591*(7851), 551–563.

Naylor, R., Fang, S., & Fanzo, J. (2023). A global view of aquaculture policy. *Food Policy*, *116*, 102422.

Nune, S. R. (2008). Role of aquaculture in poverty reduction and empowerment of women in India through the medium of self-help groups. In A. L. Shriver (Ed.), *Proceedings of the Fourteenth Biennial Conference of the International Institute of Fisheries Economics & Trade: Achieving a sustainable future—Managing aquaculture, fishing, trade and development* (pp. 1–12). International Institute of Fisheries Economics & Trade.

Olaganathan, R., & Kar Mun, A. T. (2017). Impact of aquaculture on the livelihoods and food security of rural communities. *International Journal of Fisheries and Aquatic Studies*, *5*(2), 278.

Pérez-Viveros, K. J., Cadena-Ramírez, A., Castro-Rosas, J., Villagómez-Ibarra, J. R., Chávez-Urbiola, E. A., & Gómez-Aldapa, C. A. (2023). Development of Mexican and global aquaculture as a sustainable activity [Preprint]. *Preprints*. https://doi.org/10.20944/preprints202306.1631.v1.

Phillips, M., Subasinghe, R., Tran, N., Kassam, L., & Chan, C. Y. (2016). *Aquaculture Big Numbers* (No. 601). FAO Fisheries and Aquaculture Technical Paper. FAO, FAO Rome.

Picado, A., Pereira, H., Sousa, M. C., & Dias, J. M. (2023). Ria de Alvor suitability for aquaculture: Future challenges. *Journal of Marine Science and Engineering*, *11*(5), 1009.

Pradeepkiran, J. A. (2019). Aquaculture role in global food security with nutritional value: A review. *Translational Animal Science*, *3*(2), 903–910.

Quintero, R., Parra, J., & Félix, F. (2022). Water quality assurance in aquaculture ponds using machine learning and IoT techniques. In *2022 IEEE Mexican International Conference on Computer Science (ENC), August 2022* (pp. 1–6). IEEE.

Rajyasri, N. (2010). Export performance of marine fisheries sector in India—An appraisal. *Advances in Management*.

Rashid, S., Minot, N., & Lemma, S. (2019). Does a "blue revolution" help the poor? Evidence from Bangladesh. *Agricultural Economics*, *50*(2), 139–150.

Sahitya, K. (2015). Analysis on aquaculture. *Research and Reviews: Journal of Engineering and Technology*, *3*(1), July–November. e-ISSN: 2321–6190 / p-ISSN: 2347–2294.

Sahoo, P. K., & Paul, A. (2023). Opportunities and challenges in aquaculture biotechnology. *Frontiers in Aquaculture Biotechnology*, 15–23.

Sarker, P. K. (2023). Microorganisms in fish feeds, technological innovations, and key strategies for sustainable aquaculture. *Microorganisms*, *11*(2), 439.

Sebastian, S. (2022). *An Introduction to Global Aquaculture*. https://doi.org/10.1016/b978-0-12-812211-2.00002-0 (accessed on 10th July 2025).

Snelder, T. (2022). *Overview of Aquaculture Species Diversity*. https://doi.org/10.1093/oso/9780198850229.003.0007 (accessed on 10th July 2025).

Sonesson, A. K., Hallerman, E., Humphries, F., Hilsdorf, A. W. S., Leskien, D., Rosendal, K., & Mair, G. C. (2023). Sustainable management and improvement of genetic resources for aquaculture. *Journal of the World Aquaculture Society*, *54*(2), 364–396.

Sugumaran, R., Padam, B. S., Yong, W. T. L., Saallah, S., Ahmed, K., & Yusof, N. A. (2022). A retrospective review of global commercial seaweed production—Current challenges, biosecurity and mitigation measures and prospects. *International Journal of Environmental Research and Public Health*, *19*(12), 7087.

Ursula da Silva Morales, Ú., Rotta, M. A., Fornari, D. C., & Streit Jr, D. P. (2022). Aquaculture sustainability assessed by emergy synthesis: The importance of water accounting. *Agriculture*, *12*(11), 1947.

Valente, L. M. (2022). Global challenges to feed the blue revolution in a sustainable way. In *Biology and Life Sciences Forum*, *13*(1), 92. MDPI.

Verdegem, M., Buschmann, A. H., Latt, U. W., Dalsgaard, A. J., & Lovatelli, A. (2023). The contribution of aquaculture systems to global aquaculture production. *Journal of the World Aquaculture Society*, *54*(2), 206–250.

Vipinkumar, V. P. (2020). Fishery-based entrepreneurial capacity building through self-help groups: A short glimpse. In *The blue bonanza: A manual for on-the-job training programme for VHSE students on advances in fisheries & aquaculture techniques* (pp. 150–176). ICAR-Central Marine Fisheries Research Institute.

Wang, Z., Zhang, J., Yang, X., Huang, C., Su, F., Liu, X., & Zhang, Y. (2022). Global mapping of the landside clustering of aquaculture ponds from dense time-series 10 m Sentinel-2 images on Google Earth Engine. *International Journal of Applied Earth Observation and Geoinformation*, *115*, 103100.

Wright, A., Li, X., Yang, X., Soto, E., & Gross, J. (2023). Disease prevention and mitigation in US finfish aquaculture: A review of current approaches and new strategies. *Reviews in Aquaculture*.

Yadala, S., Sowjanya, I., & Annapurna, I. (2017). Economic status: Women workers in marginalized small-scale aquaculture sector. *Imperial Journal of Interdisciplinary Research*.

Yanez, J. M., Barria, A., Lopez, M. E., Moen, T., Garcia, B. F., Yoshida, G. M., & Xu, P. (2023). Genome-wide association and genomic selection in aquaculture. *Reviews in Aquaculture*, *15*(2), 645–675.

Yang, P., Han, Q., Wang, W., Luo, Y., & Zou, W. (2020). *Management system for aquaculture and using method thereof* [Chinese patent application]. China National Intellectual Property Administration.

CHAPTER 2

Status of Aquarium Fish and Plants in Asia

PHUNSIN KANTHA,[1,*] SAHABHOP DOKKAEW,[1] SORANUTH SIRISUAY,[1] SUWAREE KITIKIEW,[2] HUA THAI NHAN,[3] TSUNG-MENG WU,[4] YUH TZEAN,[5] and CHIH-YANG HUANG[6]

[1]*Department of Aquaculture, Faculty of Fisheries, Kasetsart University, Bangkok, Thailand*

[2]*Department of Animal Production Technology and Fisheries, Faculty of Agricultural Technology, King Mongkut's Institute of Technology Ladkrabang, Bangkok, Thailand*

[3]*College of Aquaculture, Can Tho University, Can Tho, Vietnam*

[4]*Department of Aquaculture, National Pingtung University, Pingtung, Taiwan*

[5]*Department of Plant Pathology & Microbiology, National Taiwan University*

[6]*Department of Aquaculture, National Taiwan Ocean University, Keelung, Taiwan*

*Corresponding author

ABSTRACT

The ornamental aquarium market thrives on live organisms from wild captures and breeding programs, with over 90% of freshwater species and half of traded species sourced from breeding initiatives. Breeding marine species such as clownfish, marine angelfish, tang/surgeonfish, and corals

Sustainable Aquaculture: Recent Advances and Techniques. Dhanaji W. Patil, Sandip S. Markad (Eds.)

DOI: 10.1201/9781779641113-2

has recently expanded, reflecting a shift towards sustainability and ecological conservation. In Asia, aquarium hobbyists blend ornamental fish and aquatic plants into miniature aquatic ecosystems, marking aquascaping's evolution from traditional fishkeeping to intricate underwater landscapes. Rich biodiversity, complex layouts, and integration of natural processes like photosynthesis and the nitrogen cycle characterize Asian aquascaping. International recognition of Asian aquascapers at competitions like the World Aquascape Championships (IAPLC) underscores the global community's pivotal role in the art's continuous growth and innovation. This chapter further examines breeding techniques, ecological roles, and aesthetic contributions of aquatic life in Asia, highlighting sustainable practices and responsible stewardship essential for industry viability. Emphasizing ecological conservation, the chapter advocates for a balanced approach respecting both artistic expression and environmental sustainability, urging continued research, collaboration, and active community involvement to ensure an ecologically responsible future for this unique intersection of art and nature.

2.1 INTRODUCTION

Ornamental aquatic organisms have become popular among the general public, a trend that has increased over time (Palmtag, 2017). Nowadays, the technology of aquarium tools has enhanced the capability of the Internet of Things. For example, aquarium lighting has become easier to use, with the ability to control it via apps or adjust the intensity to suit aquatic plants and animals. Through the assessment of the internet community, people can learn how to take care of their aquarium tanks by themselves, making aquascaping more accessible. The trend of aquascaping and aquariums can be defined in three ways: complete set tanks, designs based on customer needs, and Do It Yourself (DIY). Some beginners will purchase a complete set tank when they start with aquascaping. However, once they become familiar with the system, hobbyists prefer to choose DIY since it allows them to set their budget and layout.

The ornamental aquarium industry serves multifunctional roles, including leisure and entertainment, pet keeping, dissemination of scientific knowledge, and life education. It showcases the biodiversity of various aquatic animals and plants through both commercial and public aquariums.

In recent years, there has been a trend toward creating habitats that mimic natural behaviors and ecological contexts of species, allowing individuals to glimpse the rich diversity of underwater ecosystems through the keeping or appreciation of aquariums (Mignani, 2022). It is conservatively estimated that the global industry employs over 1.5 million people, with at least 3.5 million aquarium enthusiasts worldwide (Dey, 2010). The COVID-19 pandemic, which impacted global industries and human health since the end of 2019, has highlighted the stress-relieving and comforting functions of ornamental aquaria. Personalized leisure keeping can meet the basic human needs for biophilia and nature affinity, especially under conditions of home isolation or remote work, fulfilling both psychological and leisure entertainment needs.

In recent years, aquascaping has gained global recognition among aquarium enthusiasts. Aquascaping involves the arrangement of aquatic plants, rocks, stones, or driftwood to create a unique layout. The result is a splendid arrangement of visually appealing underwater landscapes in aquariums, essentially mimicking gardening beneath the water's surface. It lies at the intersection of aquaculture, environmental sustainability, and the cultural significance of aquascaping in Asia. It encapsulates the historical background, biodiversity, ecological considerations, market dynamics, and future prospects of aquascaping, emphasizing the need for a harmonious balance between aesthetic appeal and ecological responsibility. This summary aims to engage a wide audience, from aquascaping enthusiasts and environmentalists to researchers and practitioners in the field, highlighting the chapter's relevance and contribution to the understanding of this unique and evolving practice. This discussion extends to the environmental impacts and the delicate balance between aquascaping practices and natural aquatic ecosystems. The commercial aspect of aquascaping in Asia is also explored, analyzing market trends and the influence of community and online platforms in shaping the hobby. This includes a look at how enthusiasts share knowledge, the role of technological advancements, and the growing global interest in Asian aquascaping styles and practices.

2.2 AQUARIUM TOOLS AND ACCESSORIES

With the help of aquascaping, you can create a unique and captivating environment within your aquarium. You have the ability to craft your own interpretation of an underwater environment, rather than simply placing rocks or artificial plants randomly.

2.2.1 AQUARIUM TANK

The most critical component of setting up an aquarium is the tank, which holds the water essential for the survival of aquatic animals and plants. In today's market, there is a wide variety of aquarium tanks available, ranging in size from as small as a teacup to as large as a room. These tanks cater to different needs, preferences, and spaces, allowing enthusiasts to create a suitable environment for their aquatic life.

Aquarium tanks come in various shapes, materials, and configurations, each offering unique benefits and challenges. Glass and acrylic are the most common materials, with glass being more scratch-resistant and acrylic being lighter and more impact-resistant. The shape of the tank can also significantly affect the health and happiness of its inhabitants. Traditional rectangular tanks provide a stable environment and maximize the surface area for gas exchange, while bow-front and cylindrical tanks offer aesthetic appeal and a unique viewing experience.

When selecting a tank, it is important to consider the specific requirements of the aquatic life you plan to keep. Freshwater and saltwater species have different needs in terms of water chemistry, temperature, and filtration. The tank's size influences the biological load it can support and how stable the water parameters will be over time. Larger tanks generally provide a more forgiving environment for beginners, as they are less susceptible to rapid changes in water quality.

In addition to the tank itself, a successful aquarium setup requires appropriate filtration, heating, and lighting systems. Filtration systems keep the water clean and healthy by removing waste products and promoting beneficial bacterial growth. Heaters are necessary to maintain a consistent temperature suited to the species being kept. Lighting not only enhances the visual appeal of the aquarium but is also vital for the growth of plants and corals.

Creating a thriving aquarium is both an art and a science. It involves understanding the complex interplay between the tank's physical environment and the biological needs of its inhabitants. By carefully selecting and maintaining the right aquarium tank and equipment, enthusiasts can ensure a healthy and vibrant aquatic ecosystem that provides enjoyment and education for years to come.

2.2.2 OXYGENATION AND FILTRATION

In an aquarium setting, maintaining a continuous flow of water is crucial for the oxygen exchange process. This is because aquatic animals consume oxygen, making water circulation a key aspect of their survival. The market offers a diverse range of air pumps and water filtration systems designed for use in aquarium tanks, catering to the specific needs of various aquatic species. While some species, such as *Bettas* or *Guppies*, may not require constant aeration and can sometimes be kept outdoors without such systems, ensuring proper oxygen levels remains essential for most aquarium inhabitants.

Air pumps for aquariums are available in several types, with at least three main categories based on their size and functionality. One common type is the mini air pump, characterized by its modest power consumption of approximately 1.5 watts and its ability to deliver around 0.42 liters of air per minute. This type of air pump typically features a single magnetic motor and a valve rubber designed to facilitate air movement. A significant portion of the air pumps available in the Asian market are manufactured in mainland China, offering a wide range of brands and models to meet the diverse requirements of aquarium enthusiasts.

It is important to select the appropriate air pump and filtration system based on the specific needs of the aquatic species being kept, as well as the size and setup of the aquarium. Proper water circulation and oxygenation are essential for creating a healthy and sustainable environment for aquatic life.

2.2.3 BASED SOIL AND SUBSTRATE

Aquarium soil and substrates can be categorized into two grades. The first is the local substrate, which is generally sold at a lower price, such as gravel, beach sand, or colored gravels. The price could be 1–2 USD per kilogram. This kind of base substrate is used for regular aquariums, such as *Guppy* or *Goldfish* tanks. In contrast, aquascape tanks may require special plant soil, which is usually brown or black in color. This soil typically costs around 3–5 USD per kilogram.

Aquarium soil alters the hydrochemical content of water by causing acidification and reducing water hardness, which promotes the proliferation of plants and enhances the overall ecology. The process of producing aquarium soil involves several stages to ensure its suitability for plant cultivation and its safety for aquatic organisms. First, a foundational substance such as organic potting soil that is free of chemicals, pesticides, and synthetic fertilizers should be selected. The potting soil should then be sterilized by heating it in an oven at 200?F for about 30 minutes to eliminate any harmful germs.

Commercially produced aquarium soil is composed of specialized clay materials, natural compressed earth, or volcanic materials that are rich in trace elements and nutrients. These materials undergo processing to form granular particles that are lighter than gravel, making it easier to plant and establish aquatic plants. The soil color usually varies between shades of brown and black and may include substances such as fertilizers or activated carbon.

2.2.4 LIGHT AND ILLUMINATION

The lighting regimen in aquariums is an essential consideration for the welfare of the aquarium environment, with various kinds and brands available. Fluorescent lights and Light Emitting Diodes (LEDs) are the most commonly used light sources in aquarium setups. LEDs have gained popularity due to their energy efficiency, low heat generation, and adaptability in installation.

Lighting in aquariums serves three primary functions. First, lighting provides illumination to the tank and aquascape, enhancing their visual appeal. In this function, the emphasis is not necessarily on the hue and spectrum of the light; rather, it caters to the enjoyment and observation of the aquarium by the owner and spectators. Second, sufficient illumination is crucial for replicating the natural cycle of daylight and darkness, thereby creating a more authentic habitat for the inhabitants. This simulation improves the biological rhythms of aquatic organisms, encouraging healthier and more natural behavioral patterns. Finally, the growth of photosynthetic organisms, such as aquatic plants, is highly dependent on appropriate light supply. These organisms require specific light spectra for photosynthesis, which is essential for their development and the overall balance of the aquarium ecosystem.

The selection of a suitable light source for an aquarium is not only a matter of aesthetic preference but also a biological necessity, as different lighting options produce varying light spectra.

Essentially, the selection of lighting in an aquarium is crucial since it not only affects the aesthetic appeal of the tank but also the well-being and development of its occupants. LEDs have become a popular option among aquarium hobbyists due to their energy efficiency and low heat emission. It is crucial to understand the precise requirements of your aquarium's ecology, such as the necessary duration of light exposure and the exact range of light wavelengths required for photosynthetic species. This information is vital when choosing the most appropriate lighting solution.

2.3 AQUARIUM BUSINESS

The aquarium business comprises many sectors, such as retail pet shops that specialize in selling aquariums and related materials, firms that specialize in designing and installing customized aquariums, breeders and distributors of fish and aquatic plants, and maintenance services for both residential and commercial aquariums. This enterprise is intricately connected with marketing tactics to effectively reach and captivate prospective clients, as well as to differentiate itself in a competitive industry.

Product uniqueness is of utmost importance in the aquarium industry. This includes offering an extensive range of fish, plants, aquarium sizes, and opportunities for customization. Promoting these distinctive features may appeal to clients seeking specific products or services. Branding and positioning are crucial for firms to build a strong market presence. This may include targeting specific demographics, such as enthusiasts or business customers, and establishing the brand as a leader in terms of excellence, originality, or customer support.

Several marketing approaches are commonly used in the aquarium business to enhance visibility, attract clients, and promote long-term growth:

1. **Digital Marketing:** Online channels are essential for targeting prospective clients. This involves establishing a robust presence on

social media platforms, using search engine optimization (SEO) techniques for the company's website, and potentially engaging in content marketing by creating blogs or videos focused on aquarium care and setup. These techniques help educate prospective clients and establish the organization as a knowledgeable authority in the industry. Client engagement and loyalty initiatives, such as newsletters, reward programs, and community activities, may be effective in cultivating a devoted client base. Offering specialized guidance, interactive sessions, and post-purchase assistance may further bolster client loyalty and amplify positive word-of-mouth promotion.

2. **Partnerships and Collaborations:** Engaging in collaborations with influential individuals in the aquatic hobby, forming partnerships with other businesses, or participating in trade fairs may effectively broaden market reach and enhance brand exposure.
3. **Visual Marketing:** The visually appealing nature of aquariums enhances the effectiveness of marketing tactics that rely on imagery, such as Instagram posts, YouTube videos, and Pinterest boards. Displaying impressive aquarium installations may inspire prospective clients and boost sales.

Market research and consumer feedback are essential for understanding market trends and client preferences. This data can provide valuable insights for product development, improving services, and implementing targeted marketing strategies.

Asia is known for its popularity in the tropical and subtropical zones, which provide the ideal temperature for aquatic life. Southeast Asia has the most biodiversity among its constituent areas, including Thailand, Vietnam, Malaysia, the Philippines, Indonesia, Myanmar, Cambodia, and Laos. Nevertheless, several species continue to be captured from the wild. Marine organisms for aquariums are often obtained from coastal regions using unlawful methods such as cyanide or fish bombs. However, enforcing the law to prevent these practices remains very challenging. The acoustic emissions generated by fish bombs are well confined underwater, making them quite difficult to detect from the surface. When fishermen are likely to be caught, they may promptly dispose of any evidence of fish bombing by throwing it overboard, resulting in very few prosecutions. It is crucial for fishermen to engage in conscious thinking as part of government campaign practices. Engaging in unlawful fishing practices not only devastates the ecosystem but also poses a threat to the internal organs of aquatic animals. Exposure to toxic substances or high levels of noise may result in persistent effects on fish behavior and organ function, potentially leading to mortality over time.

2.3.1 THE REGIONAL DEVELOPMENT PROCESS AND ROLE

The global trade value of the ornamental aquarium industry is estimated to be between $15 to $30 billion annually (Biondo and Burki, 2019; Evers et al., 2019; Penning et al., 2009), with over 120 (128) countries participating in the production and trade of ornamental aquatics. The supply of live organisms is primarily driven by Asia and developing countries (Loh et al., 2020; Nuryanto et al., 2020; Raja et al., 2019), with Southeast Asian countries producing about 60% of both captive-bred and wild-caught ornamental fish (Evers et al., 2019).

Taiwan's aquarium industry has developed for over 50 years, undergoing phases of inception, growth, stability, fluctuation, and restructuring, each lasting approximately a decade. With market trends, technological innovations in equipment, and the rotation of bred species and strains, the industry has shifted from a focus on fishkeeping to landscape and ecological habitat creation. The subjects of aquaculture have gradually integrated landscape planting that complements habitats and ecology, and the use of shrimp, snails, or specific fish species for functions such as disease sentinels, biological control, and water quality monitoring.

Before the 1970s, Taiwan's industry was characterized by exporting wild-caught marine fish and breeding viviparous fish, supplying European, American, and domestic markets. In the following decade, it earned foreign exchange by breeding a variety of African and South American cichlids for export. In the past 20 years, Taiwan has actively developed trade markets, primarily in Europe and America, while also becoming a major driving force in the rapid growth of China's aquarium market, with information, technology, and the supply of live animals and plants playing a crucial role.

Additionally, Taiwan serves as an industrial hub linking Southeast Asia with China and Europe and America. Possessing the most professional Chinese-language aquarium information, Taiwan significantly influences aquarium market development in Asian Chinese-speaking regions. Although Taiwan does not host large-scale international exhibitions like Singapore in the past or China more recently, the active engagement between industry players and the consumer market, along with the diversity of industry and market developments, positions Taiwan as a key driver in popularity and breeding trends, as well as in cultivation techniques within the Chinese-speaking regions.

2.3.2 INDUSTRY STRUCTURE

The vertical structure of the ornamental aquarium industry encompasses the production, wholesale, and retail of live animals and plants, peripheral equipment for breeding, and related information, while also providing specialized services at different stages. However, when viewed in terms of its three-dimensional scale, the industry structure more closely resembles a globe. At its core globally are the live production, storage, transport, and sales of organisms, branching out to the peripheral equipment and consumables required for breeding. The industry is supported centrally by professionalism and driven by information, propelling continuous expansion.

In Taiwan, a variety of freshwater and marine ornamental species are imported from origins including Southeast Asia, Africa, and South America, including wild-caught individuals and captive-bred specimens or artificial strains primarily cultivated in Southeast Asia or Europe and America, to meet domestic market demands. Furthermore, Taiwan, leveraging its superior breeding technology, exports various bred species such as small cyprinids, cichlids (e.g., dwarf cichlids, African cichlids, and freshwater angelfish), pleco, freshwater stingrays, and aquatic plants. Marine species exports are led by clownfish and hard corals (SPS/LPS, small and/or large polyp stony).

The production of "moss cakes," combining various mosses with driftwood or mesh, along with the growing export of amphibian and reptile breeding strains in recent years, has shifted Taiwan's industry from domestic sales to exports. The production and supply of species have expanded from narrowly defined to broadly encompassing the scope of the ornamental aquarium, covering both aquatic and terrestrial habitats.

A notable success story is the Amano shrimp (*Caridina multidentata*), supplied globally to aquarium landscaping enthusiasts for algae control. Through stable breeding techniques, cultivation has replaced wild capture, simultaneously achieving resource conservation, ecological maintenance, and industry development. This example epitomizes the successful integration of conservation and commerce within the ornamental aquarium industry.

2.3.2.1 FRESHWATER ORNAMENTAL SPECIES

Approximately 60% of freshwater ornamental fish species and 90% of the quantities supplied are bred and cultivated (Tlusty, 2002), whereas up to at least 90% of marine ornamental species originate from wild capture (Andrews, 1990). Taiwan is notable for its unique ornamental species, which, over the past 50 years of development, have included:

i. **Inception Phase:** Livebearers (viviparous fish), goldfish, and koi.

ii. **Growth Phase:** African cichlids.

iii. **Stability Phase:** Aquatic plants and marine ecological landscaping.

iv. **Fluctuation Phase:** Freshwater angelfish, guppies, freshwater stingrays, and Asian arowanas.

v. **Restructuring Phase:** Various freshwater and saltwater breeding species, along with amphibians and reptiles as special pets.

In the last decade, diverse ecological landscapes, freshwater ornamental shrimp (*Caridina* spp. and *Neocaridina* spp.), and pleco (Loricariidae) various strains have become the main categories for export markets. Production methods have shifted from small-scale to professional greenhouses, with breeding and cultivation replacing wild capture. Selective breeding of aquarium strains has become a key to enhancing product added value.

Currently, Taiwan's major export freshwater aquatic species, both in terms of volume and value, are dominated by freshwater ornamental shrimp, followed by pleco, South American cichlids, small cyprinids, African cichlids, and aquatic plants (including plants, mosses, and algae). Some species bred or wild-caught in Southeast Asia, Africa, and South America are exported to primarily European and American markets through cargo consolidation and trade transshipment, with some being re-exported via Singapore.

2.3.2.2 MARINE ORNAMENTAL SPECIES

Prior to the year 2000, less than 5% of marine species were supplied through breeding and cultivation. Post-2020, a variety of species—predominantly clownfish, along with damselfish, marine angelfish, gobies, tang/surgeonfish, and dottybacks, as well as conservation species like hard corals and giant clams (*Tridacna* spp. and *Hippopus* spp.) and seahorses (*Hippocampus* spp.)—are not only produced through breeding and cultivation but also selected for specific morphological traits to establish fixed strains, or even interspecies hybridization, to enhance their aesthetic value.

Additionally, for species with limited populations, unclear status, long-term fishing pressure, restricted distribution areas, or those that are endangered, efforts are made to breed or propagate them in captivity to reduce the overexploitation of wild animal resources and to avoid the frequent disturbances, microhabitat destruction, and habitat loss caused by collection.

Therefore, the breeding and cultivation of marine ornamental species not only meet consumer market demands, stimulate industry activity and normalized development, and ensure the reasonable and legal use of resources but also safeguard habitat and promote ecological stability and sustainability (Hoseinifar et al., 2023). Surrounded by the sea on all sides, Taiwan is home to nearly 10% of the global marine fish species diversity. Coupled with relatively advanced aquaculture technology and a stable, meticulously organized industry structure, merely shifting the focus from edible species to ornamental aquarium species—with slight adjustments and optimizations—positions Taiwan to gain a relative advantage in development speed and production capacity. This establishes Taiwan as a major supplier to Asian and even European and American markets, effectively meeting their demands.

2.3.2.3 EXOTIC PETS

In the broader ornamental aquarium market, amphibians, reptiles, and terrestrial arthropods such as insects and arachnids—referred to as exotic or special pets—share similar supply channels, equipment and technology use, display venues, and consumer markets with the narrowly defined aquarium market. Thus, they have experienced long-term stable development and continuous expansion in Europe, America, and Asia.

Similar to the development of the freshwater and marine ornamental species industries and consumer markets, the exotic pet market has evolved significantly over the past two decades due to rapid resource depletion, increasingly stringent international conservation regulations (e.g., CITES), and a more mature consumer market with a stronger focus on species conservation and preservation. As a result, wild-caught individuals originally sourced from their habitats are gradually being replaced by artificially bred and selectively bred strains.

Taiwan, with its advanced breeding technology relative to neighboring countries for amphibians such as horned frogs, poison dart frogs, and salamanders, and for reptiles such as snakes, turtles, and chameleons, has become a major source of supply for artificially bred horned frogs, poison dart frogs, and chameleons, second only to the United States and Germany. The breeding and production of species like leopard geckos (*Eublepharis macularius var.*), knob-tailed geckos (*Nephrurus* spp.), and various types of day geckos (*Phelsuma* spp.) from Madagascar have also shown remarkable performance in terms of quantity and strain production, making Taiwan a major supply source for the regional market in Asia, outside of China.

2.3.3 DOMESTIC MARKETING

The ornamental fish market in Asia is characterized by its diverse structure, ranging from basic to advanced levels, and is integral to both local economies and cultural practices. The retail landscape varies significantly, with shops ranging in size and often located in every district, sometimes operating as standalone establishments or integrated within pet stores. In Thailand, Bangkok is a pivotal city for the tropical fish farming industry, reflecting a deep-rooted appreciation for aquarium fish. This city is not only a hub for local enthusiasts but also plays a crucial role in the exportation of freshwater species.

Thailand's thriving tropical fish scene is exemplified by the presence of various species in public and private spaces, from large *Osphronemus gourami* in city fountains to luxurious aquariums in restaurant entrances, showcasing species like *Arowana* and hybrid cichlids. A notable highlight for enthusiasts and tourists alike is the Chatuchak Weekend Market, the largest of its kind globally. Spanning over 35 acres, this market is a bustling center of commerce, especially crowded on weekends, offering an array of goods, including a dedicated section for ornamental fish and aquatic plants. Here, one can find stalls with unique fish varieties, some exhibiting genetic variations that are highly sought after, despite being viewed differently in other parts of the world, such as the US and Europe. This reflects a cultural divergence in the perception and valuation of line-bred or genetically modified fish.

Asian cultures have played a significant role in the cultivation and diversification of species like the comet goldfish, showcasing a long history of breeding for unique forms and appearances. The market also features exotic species, such as "batman" freshwater rays, distinguished by their unique development, and Flowerhorns, which, despite waning interest, remain in demand. Another fascinating aspect is the high regard for *bettas*, especially the *pla-kat* (fighter *betta*) and *koi bettas*, which can fetch premium prices even for average-quality specimens.

The Chatuchak market not only offers a glimpse into the rich diversity and cultural significance of ornamental fish in Bangkok but also highlights the intricate balance between commerce, breeding ethics, and aesthetic appreciation within this industry. As such, it stands as a testament to the city's and, by extension, Asia's enduring fascination and engagement with the ornamental fish trade, offering insights into the nuanced perspectives that shape this vibrant market.

2.3.4 INTERNATIONAL MARKET CROSS-BORDER TRADE

International trade plays a vital role in this business, involving over 133 million ornamental fish traded globally each year. Thailand is a well-known producer of decorative fish and aquatic plants, supplying markets across Southeast Asia, including Laos, Cambodia, and Vietnam, as well as East Asian countries like Japan, South Korea, and Taiwan. Because of its advantageous geographical position and highly advanced maritime infrastructure, Singapore plays a crucial role as a central transfer point for the shipping of these decorative fish exports bound for Western nations.

Thailand cultivates a wide range of species, but the *Betta* fish, often known as the Siamese fighting fish, is particularly popular. The rise of the ornamental fish industry in Thailand may be attributed to the active engagement of breeders and substantial assistance from the government sector. This support has played a critical role in fostering industry growth. The government provides financial assistance for research and development, training programs for breeders, and actively promotes the ornamental fish trade at international fairs. These efforts help to enhance the country's status as a prominent supplier of top-quality ornamental fish.

2.4 AQUARIUM LIFES

The optimal aquarium setup generally comes in two primary aquatic animal layouts: the single fish tank and the multiple fish tank configurations. A single fish tank typically consists of a large space or compact aquarium with full water capacity. On the other hand, multiple fish tanks are designed to accommodate larger volumes of water and various fish species that coexist harmoniously. The key to a successful multi-tank setup is fish compatibility, ensuring that all species can thrive together without causing disturbances or engaging in aggressive behavior.

Numerous aquascape layouts in aquariums are available for enthusiasts, each tailored to different preferences and objectives. Among these, the reef-type layout stands out, allowing aquarists to recreate a vibrant underwater ecosystem with living corals, aquatic plants, and various marine species coexisting in harmony. This setup mimics natural habitats, promoting a balanced environment for marine life.

Alternatively, the fish-only layout appeals to those primarily interested in fish. This arrangement focuses on creating a naturalistic setting without the inclusion of plants or invertebrates, potentially requiring more time to reach full maturity due to its simpler setup. Additionally, display aquariums exist purely for aesthetic purposes, featuring artistic designs without housing any aquatic life, serving as unique decorative elements.

2.4.1 AQUARIUM FISH

Aquarium fish are often categorized according to the salinity of the water in which they live, resulting in two primary classifications: marine and freshwater aquariums. Distinguishing between aquaculture-sourced and wild-caught species in marine aquariums is difficult owing to the intricate nature of marine aquaculture. Recent advancements and research have demonstrated substantial progress in the breeding of various marine species, suggesting an increasing inclination toward sustainable practices in marine aquaculture (Bruckner, 2005).

In contrast, freshwater species are often more manageable to reproduce in commercial environments. However, certain distinctive or rare species still pose challenges for large-scale cultivation owing to their lack of economic viability. Due to this constraint, these species are often only accessible during certain seasons or in accordance with the prevailing trends and needs in the aquarium trade (Tlusty, 2002).

Notable for their ability to adapt to farm breeding settings, species from families such as *Cyprinidae, Osteoglossinae, Cichlidae, Characidae*, and *Poeciliidae* are often grown for aquariums (Figure 2.1). The effective production of these species in farms highlights the potential for implementing sustainable methods in the aquarium fish trade, thereby decreasing dependence on wild-caught specimens and promoting conservation initiatives (Evers et al., 2019). The cichlid fishes are a remarkable example of nature's extensive evolutionary diversification, as described in the book (Kornfield, 2002).

Algae growth in an aquascape aquarium is considered undesirable due to its negative impact on visual appeal. To resolve this issue, aquarists frequently introduce algae-consuming aquatic organisms from both crustacean and mollusk groups. These organisms not only aid in algae removal but also play a vital role in sustaining the nutrient cycle of the aquarium. Figures 2.2 and 2.3 depict some species that are appropriate for this purpose.

2.4.2 AQUARIUM PLANT

Ornamental plants are often utilized in both marine and freshwater aquariums for aquascaping purposes. They enhance the natural aesthetics of the aquarium, oxygenate the water, and provide shelter for fish—particularly fry—as well as invertebrates. Certain species of fish and invertebrates also consume live plants as part of their diet.

The early phase of Taiwan's ornamental freshwater plant market was predominantly characterized by the sale of coontail (*Ceratophyllum*

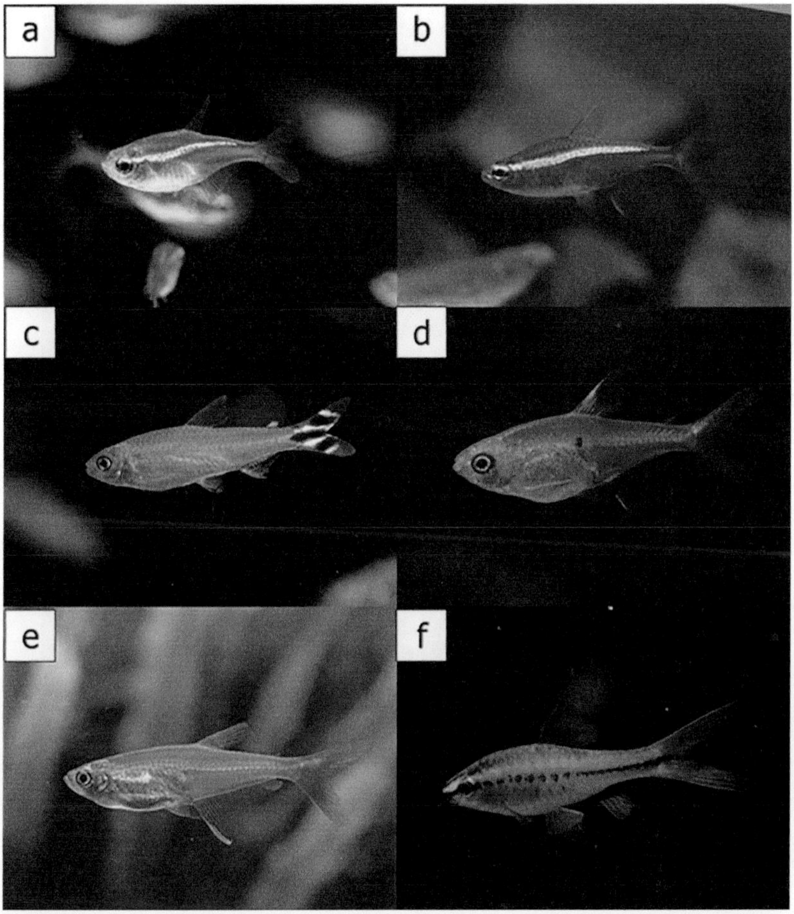

FIGURE 2.1 Ornamental schooling fish commonly cultivated in aquascape tanks. (a) *Paracheirodon innesi* (Neon tetra); (b) *Paracheirodon axelrodi* (Cardinal tetra); (c) *Petitella rhodostoma* (Rummy-nose tetra); (d) *Hyphessobrycon amandae* (Ember tetra); (e) *Aphyocharax anisitsi* (Bloodfin tetra); and (f) *Puntius titteya* (Cherry barb).

demersum) and Brazilian waterweed (*Egeria densa*). However, in the early 1980s, Taiwanese entrepreneurs began importing a diverse range of aquatic plant species from Europe, thereby enriching the domestic selection. The subsequent introduction of German aquatic plant cultivation techniques in the 1990s positioned Taiwan as a major exporter of over 200 plant varieties to key markets such as Europe, the United States, China, and Japan. As a result, Taiwan's aquaculture industry gained international

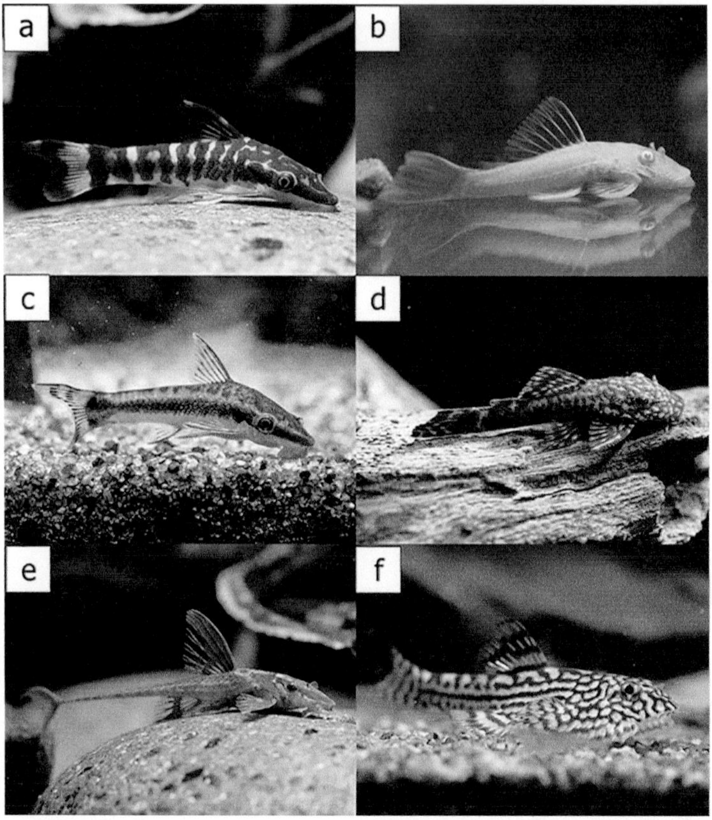

FIGURE 2.2 Algae-feeding fish commonly cultivated in aquascape tanks. (a) *Otocinclus vittatus* (Dwarf sucker, zebra variant); (b) *Ancistrus* sp. (Black-eyed yellow bristlenose); (c) *Otocinclus vittatus* (Dwarf sucker, normal variant); (d) *Ancistrus* sp. (Black-eyed bristlenose); (e) *Sturisoma panamense* (Farlowella whiptail catfish); and (f) *Sewellia lineolata* (Tiger Hillstream Loach).

Source: Photo by Cube Aquarium, Surachet Kriengnarongdech (2024).

recognition for its expertise in cultivating plants like the dwarf *Anubias* (Chang, 2017). Most aquatic plants in Taiwan are produced using natural topography cultivation and protected horticulture methods to produce either emersed-leaf or submerged-leaf forms. These are then transported and sold in pots, sheets, or boxes according to the species.

Over the past 10 years, the Taiwanese aquatic plant sector has undergone a substantial shift due to growing global competition and regulatory challenges. Although the prevalent reliance on asexual propagation provides

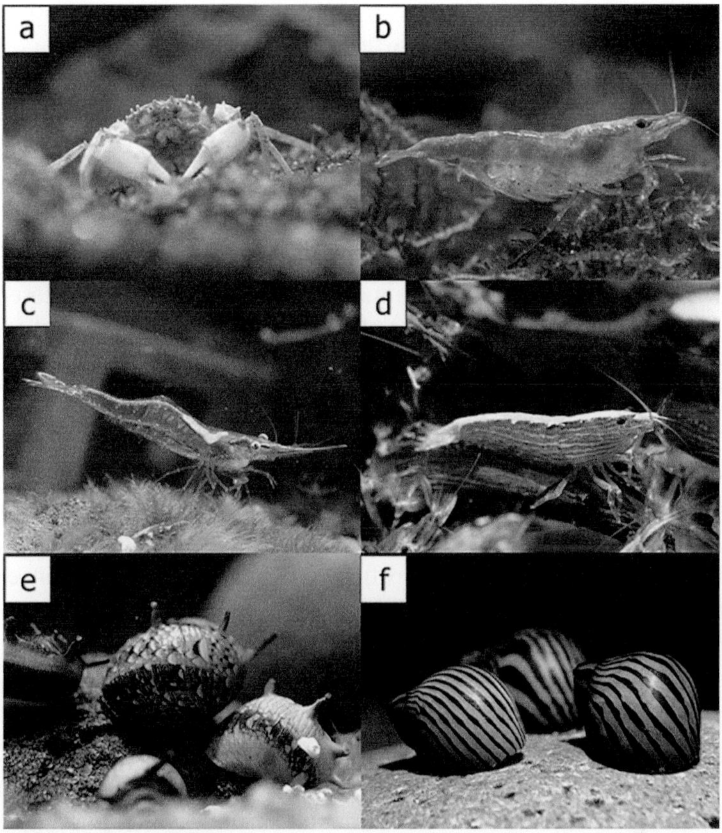

FIGURE 2.3 Algae-feeding invertebrates commonly cultivated in aquascape tanks. (a) *Limnopilos naiyanetri* (Thai micro crab); (b) *Neocaridina* sp. (Orange-sushi shrimp); (c) *Caridina gracilirostris* (long-nose, red-nose shrimp); (d) *Atyopsis moluccensis* (Bamboo shrimp); (e) *Clithon corona* (Bumblebee Horn Snail); and (f) *Neritina natalensis* (Zebra Nerite Snail).

Source: Photo by Cube Aquarium, Surachet Kriengnarongdech (2024).

a consistent output of plantlets, this method has led to seasonal supply fluctuations, phenotypic inconsistencies, and infestations of epiphytes and nematodes. These issues have caused consumer dissatisfaction and further complicated adherence to international quarantine standards. Reflecting the urgency for strategic change, the industry's revenue declined dramatically from 1.46 million USD to just 350,000 USD between 2011 and 2022 (Fisheries Agency, 2023).

More recently, the cultivation and global distribution of aquatic plants in Taiwan have evolved significantly, marked by the advancement of plant tissue culture technology. Tissue culture offers numerous advantages, including higher spatial and land utilization efficiency, the capability for vertical and intensive production systems, and reduced susceptibility to seasonal and climatic variation. Most importantly, it enables the production of pathogen-free seedlings, which are critical for facilitating international trade and transportation. As Taiwan adopts this technology, the sale of tissue-cultured aquatic plants is steadily increasing, bolstered by a growing number of international orders. This shift toward refined product offerings represents a strategic move by Taiwan's aquaculture industry to overcome present challenges and embrace new opportunities for growth and sustainability.

The development of the ornamental aquaculture industry in Taiwan began in the early 1960s, initially thriving on the cultivation and trade of species such as goldfish, koi, and cichlids (e.g., Blood Parrot cichlids), thereby generating substantial profits and establishing a golden era for the industry. However, as aquaculture technologies in Southeast Asian countries advanced, Taiwan's industry—previously dominant through mass production—lost its competitive edge and its status as a pivotal international transshipment hub.

Since 2000, the Taiwanese ornamental aquaculture industry has shifted toward a more refined development model. This transition was significantly influenced by the rise of ornamental shrimp, which not only diversified the market beyond traditional fishkeeping but also accelerated the industry's move toward miniaturization, sophistication, ecological integration, professionalization, and informatization. Integrating the concept of horticultural landscaping into aquariums has shifted the focus from fish being the main attraction to a complementary relationship between aquatic animals and plants, thereby increasing the demand for aquatic plants.

In recent years, the live ornamental aquaculture industry in Taiwan has generated revenues between 31.78 and 38.13 million USD, with the overall industry valuation exceeding 133 million USD (Li, 2020).

The early phase of Taiwan's ornamental plant market was predominantly characterized by the sale of coontail (*Ceratophyllum demersum*) and Brazilian waterweed (*Egeria densa*). However, in the early 1980s, Taiwanese entrepreneurs began importing diverse aquatic plant species from Europe, thereby enhancing the variety of offerings within the domestic sphere.

The subsequent introduction of German aquatic plant cultivation techniques in the 1990s positioned Taiwan as a major exporter of over 200 plant varieties to key markets such as Europe, the United States, China, and Japan. As a result, Taiwan's aquaculture industry gained international recognition for its expertise in cultivating plants like the dwarf *Anubias* (Chang, 2017). Most aquatic plants in Taiwan are produced using natural topography cultivation and protected horticulture methods to produce either emersed-leaf or submerged-leaf forms. They are then transported and sold in pots, sheets, or boxes according to the species.

Amidst intensifying global competition and escalating regulatory challenges, the Taiwanese aquatic plant industry has undergone a significant transformation over the past decade. The prevalent reliance on asexual propagation of aquatic plants provides a consistent output of plantlets; however, this method has also led to seasonal supply fluctuations, phenotypic inconsistencies, and infestations of epiphytes and nematodes. These issues have resulted in consumer dissatisfaction and compounded the difficulties in adhering to international quarantine standards. Reflecting the need for an urgent strategic shift, the industry's revenue declined dramatically from 1.46 million USD to a mere 350,000 USD between 2011 and 2022 (Fisheries Agency, 2023).

More recently, the cultivation and global distribution of aquatic plants in Taiwan have undergone significant evolution, marked by advancements in plant tissue culture technology. Plant tissue culture offers numerous advantages, including higher spatial and land utilization efficiency, the capability for vertical and intensive production systems, and reduced susceptibility to seasonal and climatic variations. Importantly, it enables the production of pathogen-free seedlings, which are crucial for facilitating international trade and transportation. As Taiwan embraces this technology, the sale of tissue-cultured aquatic plants is witnessing a steady increase, bolstered by a growing number of international orders. This pivot toward refined product offerings represents a strategic move by Taiwan's aquaculture industry to navigate present challenges and harness new opportunities for growth and sustainability.

2.4.3 EXOTIC REPTILE AND OTHERS

The cultivation of exotic reptiles in terrariums has become increasingly popular in Asia, resulting in a wide variety of species being carefully preserved in controlled environments. These species include a diverse range of reptiles and amphibians, each with distinct habitat requirements for optimal survival. Examples include the visually impressive horned frogs, known for their prominent cranial protrusions; various species of turtles, admired for their longevity and diversity; the water dog, also known as the axolotl—a type of salamander highly regarded for its exceptional regenerative capabilities; and poison dart frogs, recognizable by their brightly colored skin, which indicates toxicity in the wild.

This approach to the care of reptiles and amphibians not only highlights the delicate balance required for ecosystem management but also showcases the extensive biodiversity found within Asia's ecosystems. It serves both as an educational tool and a conservation initiative, offering insight into the natural behaviors and habitats of these animals while emphasizing the importance of environmental stewardship.

Since 2000, the Taiwanese ornamental aquaculture industry has shifted toward a more refined development model. This transition was significantly influenced by the rise of ornamental shrimp, which not only diversified the market beyond traditional fishkeeping but also accelerated the industry's move toward miniaturization, sophistication, ecological integration, professionalization, and informatization.

Integrating the concept of horticultural landscaping into aquariums has shifted the focus from fish being the main attraction to a complementary relationship between aquatic animals and plants, thereby increasing the demand for aquatic plants. In recent years, the live ornamental aquaculture industry in Taiwan has generated revenues between 31.78 and 38.13 million USD, with the overall industry valuation exceeding 133 million USD (Li, 2020).

The importing and sale of exotic reptiles in Asian nations are subject to complex procedures dictated by both international and local regulations aimed at safeguarding these species and their native environments. The trade in exotic reptiles is influenced by key elements such as the Convention on International Trade in Endangered Species of Wild Fauna and Flora (CITES), which is followed by several Asian nations. This convention ensures that the trade is sustainable and does not pose a threat to the survival of these reptiles in their natural habitats.

In order for a reptile species to be brought into an Asian country, it is often required to undergo health examinations to prevent the transmission of diseases and to verify that it was obtained in an ethical and sustainable manner. Overcollection can significantly impact the populations of species such as horned frogs, axolotls, and poison dart frogs in their natural habitats. The import process usually includes quarantine periods and veterinary inspections to ensure biosecurity and safeguard native species from invasive risks.

The marketing of exotic reptiles in Asia is primarily driven by the pet industry, with a strong focus on educating and raising awareness among potential pet owners. Retailers, breeders, and conservation organizations often emphasize the importance of responsible pet ownership, which includes understanding the specialized care requirements for each species, such as their dietary needs, habitat design, and healthcare requirements. This educational approach is intended to prioritize animal welfare and ensure the satisfaction of pet owners, promoting a market that values the ethical care and preservation of these distinctive species.

2.5 CONCLUSION

The ornamental fish live trade market was valued at $5.4 billion in 2021 and is expected to expand at a compound annual growth rate of 8.5% from 2022 to 2030 (Hoseinifar et al., 2023). It is conservatively estimated that the global industry employs over 1.5 million people, with at least 3.5 million aquarium enthusiasts (Dey, 2010). Just as floriculture represents the epitome of sophisticated agriculture, ornamental aquaculture generates the highest unit price and value within aquaculture, resonating with humans' inherent biophilia and affinity for nature.

Previously considered merely a leisure or hobby activity, ornamental aquarium keeping now also encompasses stress relief, science education, life lessons, and environmental conservation advocacy. Moreover, in some countries, aquarium keeping ranks among the top five leisure activities, with the industry growing at a minimum rate of 4% annually (Kumar et al., 2015), supported by global market demand.

Therefore, integrating information, leveraging regional advantages and niche markets with professional leadership, and promoting the responsible use of animal resources—while adding value through information and technology—should be prioritized. At the same time, the industry should focus on ecological and habitat protection, while gradually implementing animal welfare, as core directions for collaborative and sustainable progress.

Social media and internet platforms have become essential tools for marketing and raising awareness of exotic reptiles. They provide a forum for enthusiasts to exchange care tips, share experiences, and offer advice on sustainable practices. The presence of such online community support is crucial in promoting ethical pet ownership and advancing conservation efforts.

Moreover, there is a growing trend of incorporating conservation-focused initiatives and collaborations among zoos, aquariums, and conservation groups in Asia into the exotic reptile trade. These programs aim to educate the general public about the importance of biodiversity and the roles these species play in their natural habitats. This is often achieved through captive breeding projects, which help reduce pressure on wild populations.

In Asian nations, the import and marketing of exotic reptiles are complex processes that require balancing the demand from the pet sector with the need to protect and ethically manage these species.

KEYWORDS

- **aquaculture industry**
- **aquascape**
- **aquatic plant**
- **biosecurity**
- **fish keeping**
- **global industry employment**
- **ornamental fish**
- **pathogen-free seedlings**

REFERENCES

Andrews, C. (1990). The ornamental fish trade and fish conservation. *Journal of Fish Biology*, *37*, 53–59.

Biondo, M. V., & Burki, R. P. (2019). Monitoring the trade in marine ornamental fishes through the European Trade Control and Expert System TRACES: Challenges and possibilities. *Marine Policy*, *108*, 103620.

Bruckner, A. (2005). The importance of the marine ornamental reef fish trade in the wider Caribbean. *Revita de Biology Tropical*, *53*, 127–137.

Chang, C. W. (2017). *Comprehensive Analyses of Ornamental Aquarium Industry in Taiwan* (Master's thesis). https://hdl.handle.net/11296/9b7fw7 (accessed on 10th July 2025).

Chen, J. Y., Zeng, C., Jerry, D. R., & Cobcroft, J. M. (2020). Recent advances of marine ornamental fish larviculture: Broodstock reproduction, live prey and feeding regimes, and comparison between demersal and pelagic spawners. *Reviews in Aquaculture*, *12*, 1518–1541.

Corfield, J., Diggles, B., Jubb, C., McDowall, R. M., Moore, A., Richards, A., & Rowe, D. K. (2008). *Review of the Impacts of Introduced Ornamental Fish Species That Have Established Wild Populations in Australia* (277 pp.). Australian Government, Department of the Environment, Water, Heritage and the Arts.

Dey, V. (2010). Ornamental fish trade–Recent trends in Asia. In *Souvenir, Ornamental* (pp. 39–45). Department of Fisheries, Government of Kerala: Kerala, India.

Dudgeon, D., Arthington, A. H., Gessner, M. O., Kawabata, Z. I., Knowler, D. J., Lévêque, C., Naiman, R. J., Prieur-Richard, A. H., Soto, D., Stiassny, M. L. J., & Sullivan, C. A. (2006). Freshwater biodiversity: Importance, threats, status and conservation challenges. *Biological Reviews*, *81*(2), 163–182.

Evers, H. G., Pinnegar, J. K., & Taylor, M. I. (2019). Where are they all from? –Sources and sustainability in the ornamental freshwater fish trade. *Journal of Fish Biology*, *94*(6), 909–916.

Fahmi, M., Ginanjar, R., & Kusumah, R. V. (2015). Keragaman ikan hias di lahan gambut Cagar Biosfer Bukit-Batu, Propinsi Riau. *Pros Sem Nas Masy Biodiv Indon*, *1*(1), 51–58. https://doi.org/10.13057/psnmbi/m010108 (accessed on 10th July 2025).

Fisheries Agency. (2023). *The Ornamental Fish Industry's Value and Production Volume*. Ministry of Agriculture, Taiwan.

Haenen, O. L. M., Way, K., Bergmann, S. M., & Ariel, E. (2004). The emergence of koi herpesvirus and its significance to European aquaculture. *Bulletin of the European Association of Fish Pathologists*, *24*(6), 293–307.

Hensen, R. R., Ploeg, A., & Fossa, S. A. (2010). *Standard Names for Freshwater Fishes in The Ornamental Aquatic Industry* (OFI Educational Publication 5, 146 p.). Ornamental Fish International, The Netherlands.

Hill, J., Tuckett, Q. M., Hardin, S., Lawson, L. L., Lawson, K. M., Ritch, J. L., & Partridge, L. (2017). Risk screen of freshwater tropical ornamental fishes for the conterminous United States. *Transactions of the American Fisheries Society*, *146*, 927–938.

Hoseinifar, S. H., Maradonna, F., Faheem, M., Harikrishnan, R., Devi, G., Ringø, E., Van Doan, H., Ashouri, G., Gioacchini, G., & Carnevali, O. (2023). Sustainable ornamental fish aquaculture: The implication of microbial feed additives. *Animals*, *13*, 1583. https://doi.org/10.3390/ani13101583 (accessed on 10th July 2025).

Khoironi, F. E., & Saskara, I. A. N. (2017). Analisis pengaruh kurs dollar, inflasi, dan produksi terhadap ekspor ikan hias di Provinsi Bali. *E-Journal EP Universitas Udayana*, *6*(3), 337–361.

Kolar, C. S., & Lodge, D. M. (2001). Progress in invasion biology: Predicting invaders. *Trends in Ecology & Evolution*, *16*(4), 199–204.

Kornfield, I. (2002). The cichlid fishes: Nature's grand experiment in evolution. In *JSTOR*.

Kumar, T. A., Gunasundari, V., & Prakash, S. (2015). Breeding and rearing of marine ornamentals. In *Advances in Marine and Brackishwater Aquaculture* (pp. 101–107). Springer: Berlin/Heidelberg, Germany.

Li, J. H. (2020). Taiwan swims into the blue sea of ornamental fish. In Z. C. Miu (Ed.), *Fisheries Extension* (Vol. 406, pp. 99–114). Fisheries Agency, Ministry of Agriculture: Taipei, Taiwan.

Liang, S. H., Chuang, L. C., & Chang, M. H. (2006). The pet trade as a source of invasive fish in Taiwan. *Taiwania*, *51*(2), 93–98.

Loh, R., Vukcevic, J., & Bastos, G. (2020). Current status of aquatic veterinary services for ornamental fish in Australasia. *New Zealand Veterinary Journal*, *68*(3), 145–149. https://doi.org/10.1080/00480169.2020.1718564 (accessed on 10th July 2025).

Lyons, T., Tuckett, Q. M., Durland Donahou, A., & Hill, J. (2020). Risk screen of lionfishes, *Pterois*, *Dendrochirus*, and *Parapterois*, for southeastern United States coastal waters of the Gulf of Mexico and Atlantic Ocean. *Biological Invasions*, *22*, 1573–1583.

Mignani, H. (2022). Significance of ornamental fishes in aquaculture. *Journal of Aquaculture Research & Development*, *13*(9), 1, 2. https://doi.org/10.35248/2155-9546.22.13.700 (accessed on 10th July 2025).

Nuryanto, A., & Bhagawati, D. (2020). Evaluation of conservation and trade status of marine ornamental fish harvested from Pangandaran Coastal Waters, West Java, Indonesia. *Biodiverse*, *21*(2), 512–520. https://doi.org/10.13057/biodiv/d210212 (accessed on 10th July 2025).

Palmtag, M. R. (2017). The marine ornamental species trade. In *Marine Ornamental Species Aquaculture* (pp. 3–14).

Papavlasopoulou, I., Vardakas, L., Perdikaris, C., Kommatas, D., & Paschos, I. (2013). Ornamental fish in pet stores in Greece: A threat to biodiversity? *Mediterranean Marine Science*. https://doi.org/10.12681/mms.484 (accessed on 10th July 2025).

Penning, M., Reid, G. McG., Koldewey, H., Dick, G., Andrews, B., Arai, K., Garratt, P., Gendron, S., Lange, J., Tanner, K., Tonge, S., Van den Sande, P., Warmolts, D., & Gibson, C. (2009). *Turning the Tide: A Global Aquarium Strategy for Conservation and Sustainability*. World Association of Zoos and Aquariums, Bern, Switzerland.

Prakash, S., Kumar, T. T. A., Raghavan, R., Rhyne, A., Tlusty, M. F., & Subramoniam, T. (2017). Marine aquarium trade in India: Challenges and opportunities for conservation and policy. *Marine Policy*, *77*, 120–129.

Raghavan, R., Dahanukar, N., Tlusty, M. F., Rhyne, A. L., Kumar, K. K., Molur, S., & Rosser, A. M. (2013). Uncovering an obscure trade: Threatened freshwater fishes and the aquarium pet markets. *Biological Conservation*, *164*, 158–169.

Raja, K., Aanand, P., Padmavathy, S., & Sampathkumar, J. S. (2019). Present and future market trends of Indian ornamental fish sector. *International Journal of Fisheries and Aquatic Studies*, *7*(2), 6–15.

Ramachandran, A. (2002). Freshwater indigenous ornamental fish resources in Kerala and their prospects for international marketing. In B. Meenakumari, J. Joseph, T. V. Sankar, P. Pravin, & L. Edwin (Eds.), *Riverine and Reservoir Fisheries of India* (pp. [pages not provided]). Central Institute of Fisheries Technology and Society of Fisheries Technologists, India.

Rhyne, A. L., Tlusty, M. F., Schofield, P. J., Kaufman, L., & Morris Jr., J. A. (2012). Revealing the appetite of the marine aquarium fish trade: The volume and biodiversity of fish imported into the United States. *PLoS ONE*, *7*(5), e35808.

Satam, S. B., Sawant, N. H., Ghughuskar, M. M., Sahastrabuddhe, V. D., Naik, V. V., Pagarkar, A. U., Chogale, N. D., Metar, S. Y., Shinde, K. M., Sadawarte, V. R., Sawant, A. N., Singh, H., Sawant, P. A., Patil, V. K., Rane, D., Haldankar, P. M., & Bhattacharyya, T. (2018). Ornamental fisheries: A new avenue to supplement farm income. *Advances in Agricultural Research & Technology Journal*, *2*(2), 193–197.

Sharma, M. (2020). Ornamental fish rearing and breeding—A new dimension to aquaculture entrepreneurship in Himachal Pradesh. *International Journal of Fisheries and Aquatic Studies*, *8*(2), 157–162.

Sinansari, S., & Priono, B. (2019). Opportunity and business challenge of marine ornamental fishes in Indonesia as a potential commodity of fisheries. In *International Conference on Green Agro-industry and Bioeconomy. IOP Conference Series: Earth and Environmental Science*, *230*, 012067. https://doi.org/10.1088/1755-1315/230/1/012067 (accessed on 10th July 2025).

Tlusty, M. (2002). The benefits and risks of aquacultural production for the aquarium trade. *Aquaculture*, *205*(3–4), 203–219.

Weir, M., Rajić, A., Dutil, L., Cernicchiaro, N., Uhland, F. C., Mercier, B., & Tuševljak, N. (2012). Zoonotic bacteria, antimicrobial use and antimicrobial resistance in ornamental fish: A systematic review of the existing research and survey of aquaculture allied professionals. *Epidemiology & Infection*, *140*(2), 192–206.

Willis, S., & Bakuwel, P. (2018). Update on the ornamental aquatic export industry 2016: Trade data and conservation. *OFI Journal*, *86* (February).

Yue, G. (2019). The ornamental fish industry in Singapore. *Journal of Fishery Sciences of China*, *43*(1), 116–127. https://doi.org/10.11964/jfc.20180911442 (accessed on 10th July 2025).

CHAPTER 3

Potential and Future Prospects of Sea Cage Farming in India: A Promising Technology for Doubling Farmers' Income

ISHITA J. BAMBHANIYA,[1,*] V. C. BAJANIYA,[1] and BAMANIYA PINAK KAMLESHBHAI[2]

[1]*Department of Aquaculture, College of Fisheries Science, Kamdhenu University, Veraval, Gujarat, India*

[2]*Department of Aquatic Environment Management, College of Fisheries Science, Kamdhenu University, Veraval, Gujarat, India*

*Corresponding author

ABSTRACT

India ranks second in total aquaculture production in the world, with most of its output coming from inland culture. Mariculture in India, however, is still in its infancy. Sea cage farming represents one of the most promising technologies to enhance marine fish production. The Central Marine Fisheries Research Institute (CMFRI) has identified 134 ideal sites, covering an area of 46,958.2 ha, for sea cage farming. Andhra Pradesh has the largest coastal area among all states, with 11,792 hectares, followed by Gujarat with 11,572.2 hectares and Tamil Nadu with 7,673 hectares. Choosing the right location is vital for the success of cage farming. Numerous parameters must be considered to ensure its success. This chapter addresses site selection, illustrates cage sites at different coordinates in India, and discusses cage materials and a mooring system suitable for Indian seas. It also highlights species selection based on market value and cultural

Sustainable Aquaculture: Recent Advances and Techniques. Dhanaji W. Patil, Sandip S. Markad (Eds.)

DOI: 10.1201/9781779641113-3

conditions across different regions of India. The goals of the Blue Growth Mission and the Pradhan Mantri Matsya Sampada Yojana (PMMSY) have encouraged fisheries stakeholders and authorities to explore open-sea cage farming in coastal states.

3.1 INTRODUCTION

The fisheries and aquaculture sector in India has evolved from a traditional livelihood into a modern, technology-driven commercial industry. India possesses vast marine fisheries resources, including an 8,118 km coastline, a territorial sea of 193,834 km^2 (extending 12 nautical miles or 22.2 km from the shoreline), and around 4 million marine fishermen residing in 3,432 marine fishing villages across 66 coastal districts in 9 maritime states and 2 union territories. Additionally, India includes 2 island territories: Andaman & Nicobar and Lakshadweep.

The country's infrastructure comprises 6 major fishing harbors, 40 minor fishing harbors, and 1,537 marine fish landing centers. The shallow inshore waters along the extensive coastline of the mainland and island territories offer opportunities for sea cage farming—particularly in sheltered areas such as bays, lagoons, and semi-exposed or exposed coasts with minimal wave action. The existing marine infrastructure and marine fisher population serve as valuable assets for the development of this sector.

Cage farming is believed to have originated in China during the Han Dynasty, which occurred between 100 and 200 BCE (Li, 1994). The practice of cage culture in Southeast Asia was first recorded in the late 1800s, particularly in Kampuchea. Japan pioneered saltwater fish cage farming in the 1950s with the commercial cultivation of yellowtail (*Seriola quinqueradiata*) at the Kinki University Fisheries Laboratory (Gopakumar, 2009).

In India, fish cage farming has undergone significant development. Air-breathing fishes were first cultivated in marshes; major carps were raised in the Jamuna and Ganga rivers in Allahabad; and various species, including carp, snakehead, and tilapia, were cultured in standing water bodies in Karnataka (Rao et al., 2013). Historically, cages were only slightly larger than modified fish traps or baskets. In ancient times, fish were housed in bamboo and wooden cages and fed with leftover food and discarded fish. Later, modern cages were constructed using synthetic materials. At present, fish and other aquatic organisms are kept in cages for extended periods during their growth stages.

3.2 SEA CAGE FARMING: AN OVERVIEW

Cage culture is a form of aquaculture production system in which cultured organisms are held within a floating net structure. These cages are used to cultivate various types of finfish or shellfish in diverse water systems, including freshwater, brackish water, and marine environments (Mojjada, 2015). Sea cage culture specifically refers to the cultivation of marine fishes in the sea.

Cage culture involves raising fish within an enclosure, which can take various shapes such as circular, rectangular, or square. These enclosures may be suspended or floating in the water, or fixed to the seabed near the shoreline in the littoral zone. This is considered a super-intensive culture system with a high unit production rate. The unit production from marine cage culture systems typically ranges from 20 to 40 kg/m^3, depending on the species being cultured (Figure 3.1).

FIGURE 3.1 Cage culture setup in the sea.

3.2.1 SALIENT FEATURES OF CAGE

The salient features of a cage system are presented in Table 3.1.

TABLE 3.1 Salient Features of a Sea Cage System

Sl. No.	Components	Features
1.	Shape	Round, square, or rectangular. Round cages are considered more suitable for harsh marine conditions with wave and wind-induced turbulence, while cube-, square-, and rectangular-shaped cages are preferred in reservoirs.
2.	Size	i. Circular Cage: Diameter – 6 m; Depth – 4 m; Volume = 113 m^3 ii. Rectangular Cage: 6 × 4 × 4 m; Volume = 96 m^3. iii. A battery consists of 6, 12, or 24 cages.
3.	Materials	i. Framing Materials: Bamboo, mild steel (MS), galvanized iron (GI), polyvinyl chloride (PVC), and virgin-grade high-density polyethylene (HDPE). ii. Base Collar: • Inner ring size: 6 m • Outer ring size: 7 m • Middle ring (catwalk ring) size: 6.5 m • Base supports: 8 nos. • Vertical supports: 8 nos. • Diagonal supports: 8 nos. iii. Handrail: A 6-meter HDPE cage with 90 mm pipes installed approx. 1 meter above the inner collar ring, connected to the base collar rings via vertical and diagonal supports. iv. Mooring System: • Mooring rope • Anker marker line • Buoys • Moring chain • D-shackles • Anchors/Gabion boxes
4.	Cage net	i. Outer Predator Net Cage: Braided UV-treated HDPE netting, diameter: 7 m; depth: 5 m (submerged). ii. Fish Rearing/Grow-Out Net Cage: Twisted HDPE netting, diameter: 6 m; depth: 5 m iii. Bird Net: Twisted HDPE, twine size: 1.25 mm; mesh size: 60–80 mm.
5.	Ballast	i. To maintain the cylindrical shape of the net cages, concrete blocks of appropriate weight are attached at regular intervals. ii. HDPE pipes with 1.5-inch diameter can be used, with 10 mm thick MS chains or wire ropes inserted into them as ballast.

3.2.2 *CAGE SETUP*

At the selected site, one can choose to install either a circular cage with a minimum diameter of 6 meters and a depth of 4 meters (resulting in a volume of 113 cubic meters), or a rectangular cage measuring 6 meters in length, 4 meters in width, and 4 meters in height (yielding a volume of 96 cubic meters). These cages are enclosed by a larger outer net, referred to as a predator net, with a one-meter clearance on all sides. Additionally, a bird net is placed over the cage frame to prevent avian predation (Rao, 2012).

The cages offer flexibility, as they can be either free-floating or fixed in place. To secure the cages at the desired position and depth, a mooring system or assembly is used, incorporating mooring lines, chains, and anchors. With appropriate mesh sizes, these cages are suitable for rearing fish fry to fingerlings (nursery cages) and/or for growing table fish (Figure 3.2).

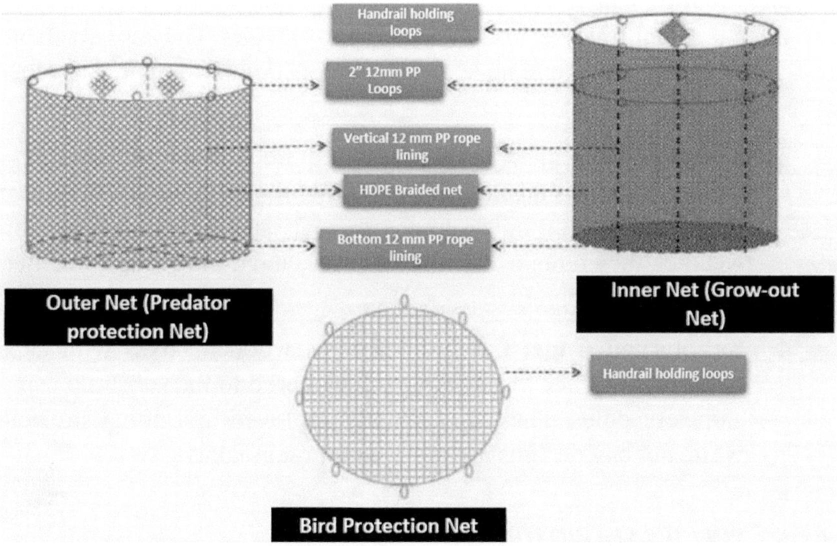

FIGURE 3.2 Outer predator protection net, inner grow-out net, and upper bird protection net of cage.

3.2.3 *SITE SELECTION*

Effective site selection plays a crucial role in determining the profitability and sustainability of cage farming. It significantly influences construction expenses, operational costs, fish growth and survival rates, and the durability of cage equipment. Before establishing cage farming operations, a thorough understanding of the site's environmental conditions—including topography and water quality parameters—is essential (Biswajit, 2009).

This process involves conducting detailed surveys, gathering secondary information from existing literature and government sources, and obtaining firsthand insights from local residents.

3.2.4 TYPES OF CAGES

Beveridge (1996) outlines four main types of cages used in aquaculture:

1. **Fixed Cage:** Cost-effective to establish, these cages consist of a net supported by pillars anchored to the lake or riverbed. They are typically used in protected shallow areas with adequate water depth.
2. **Floating Cage:** The most widely used type, floating cages feature a buoyant frame or collar that supports the net. They are available in various sizes and have fewer site constraints compared to other types.
3. **Submersible Cage:** These cages require a frame or rigging to maintain their shape. Their position in the water column can be adjusted to optimize environmental conditions. During calm weather, they remain above the surface, and during rough weather, they can be submerged.
4. **Submerged Cage:** Constructed from wooden boxes with gaps between the slats, these cages are anchored to the substrate using stones or poles. They are typically deployed in running streams, while net bag designs are more common in lakes.

3.2.5 SPECIES SELECTION

The selection of fish species is a critical aspect of cage culture operations. When choosing species for cultivation, it is essential to prioritize criteria such as biological characteristics, economic viability, and consumer preferences to ensure long-term economic sustainability (Jadhav, 2009).

Key considerations before selecting fish species include:

- Sufficient availability of fish seed;
- Availability of hatchery technology for seed production;

- Acceptance of artificial feeds;
- Tolerance to different environmental conditions;
- Compatibility with various culture systems;
- Resistance to disease and stress;
- Nutritional value;
- Consumer acceptance;
- Economic value in local and international markets;
- Regional preference.

List of marine species suitable for cage farming is given in Table 3.2.

TABLE 3.2 Marine Species Suitable for Cage Farming (Anon, 2018)

Species Cultured	Stocking Size (Length/Weight)	Stocking Density (nos./m^3)	Production per Cage (kg)
Pompano 	15 cm / 35 g	8–10	2,400 kg/7 months
Cobia 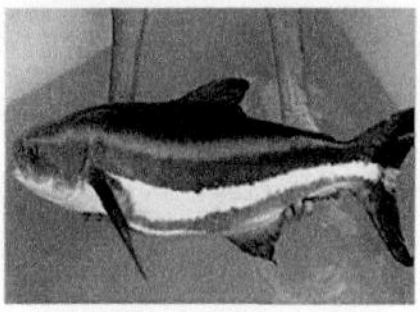	10 cm / 35 g	30–40	1,800 kg / 8 months
Seabass 	10 cm / 30 g	30–40	2,000 kg / 8 months
Grouper 	15 cm / 40 g	15–20	2,000 kg / 7 months

3.3 CURRENT STATUS OF SEA CAGE FARMING IN INDIA

In 2007, India set up its first sea cage fish farming system through the Indian Council of Agricultural Research–Central Marine Fisheries Research Institute (ICAR-CMFRI). They successfully grew *Asian seabass* in cages placed in the sea near Visakhapatnam, on the east coast at the Bay of Bengal (Mojjada, 2015; Rao, 2012). This has attracted many stakeholders and policymakers in the country. Sea cage farming is carried out in various coastal areas of India, such as Kerala (Imelda et al., 2010), Karnataka (Philipose et al., 2012, 2013), Gujarat (Damodaran et al., 2018; Mojjada et al., 2012a), Odisha (Mojjada et al., 2012b), and Andhra Pradesh (Ranjan et al., 2014). Along the Indian coast, over 3,900 cage farming units were operating with guidance from ICAR-CMFRI. Choosing the right locations for sea cage farming is crucial for success and for boosting sustainable seafood production. Tailoring site selection, design, and operations to each region is vital for successful cage culture. Selecting suitable sites can impact costs, fish growth, survival rates, and overall production (Beveridge, 2004; Loka et al., 2012). A study conducted by Divu et al. (2023) showed that there are 134 identified areas suitable for cage culture spread across 11 states and Union Territories (UTs), covering a total area of 46,958.2 hectares. Andhra Pradesh's nearshore waters stand out as the leading maritime state with 11,792 hectares for potential cage culture. On the west coast, Gujarat leads with 11,572.2 hectares available for the same purpose. The top three coastal states in India with the most potential area for sea cage farming are Andhra Pradesh, Gujarat, and Tamil Nadu (Table 3.3).

TABLE 3.3 Total Numbers of Potential Sites and Their Area for Sea Cage Farming in Maritime States of India (Divu et al., 2023)

State/UT	Number of Site	Total Area of Site (in ha)
West Bengal	1	85
Odisha	9	1,968
Andhra Pradesh	33	11,792
Tamil Nadu	45	7,673
Kerala	4	1,292
Karnataka	5	2,681
Goa	11	7,230
Maharashtra	5	1,915
Gujarat	16	11,572.2
Diu	2	615
Lakshadweep	3	135
Total Site	134	46958.2

3.3.1 CASE STUDY OF SEA CAGE FARMING IN INDIA

3.3.1.1 INITIATIVES OF OPEN SEA CAGE CULTURE FARMS IN INDIA

ICAR-CMFRI launched a significant project focused on cage aquaculture of finfish, in collaboration with NFDB, an organization dedicated to advancing cage aquaculture. NFDB approved funding for the establishment of open sea floating cages to be utilized in demonstration farms for open sea cage culture, spanning various maritime states across India (Table 3.4).

TABLE 3.4 Open Sea Cage Culture Initiatives by ICAR-CMFRI Across Maritime States of India, Including Site Locations, Number of Cages, and Species Cultured

Maritime States	Cage Culture Site	Number of Cages	Species Cultured
Maharashtra	Kalimba, Mumbai	1	Lobster *P. polyphagus*
Andhra Pradesh	Kakinada	6	Seabass
	Srikakulam		*Lates calcarifer*
	Visakhapatnam		Pearl spot
	Nellore		Pompano
	Antarvedi, Narsapur (Godavari estuary)		
	Nagaya Lanka (Krishna estuary)		
Kerala	Cochin (open sea and backwater)	2	Seabass
	Vizhinjam		Pearl spot
			Grey mullet
			Red snapper
Tamil Nadu	Chemmanchery, Chennai	3	Seabass
	Kanyakumari		Lobster *P. homarus*
	Mandapam		
Karnataka	Karwar	2	Cobia
	Byndoor, Mangalore		Red snapper
			Sea bream
			Pompano
Orissa	Balasore	1	Seabass
Gujarat	Sutrapada, Veraval	1	Lobster *P. polyphagus*

Since 2007, the Central Marine Fisheries Research Institute (CMFRI) has undertaken pilot-scale inshore marine cage culture experiments along the Visakhapatnam coast in the Bay of Bengal. These endeavors have been made possible through financial backing from the Ministry of Agriculture, Government of India. Notably, a significant achievement was attained with a successful harvest of 550 kg of seabass within a culture period of approximately four months. This outcome was achieved by stocking 1,400 fingerlings.

3.3.1.2 ASIAN SEABASS, LATES CALCARIFER CULTURE AT MUNAMBAM, COCHIN, INDIA IN OPEN SEA CAGE

A preliminary investigation was conducted to explore the cultivation of Asian seabass (*Lates calcarifer*) in an open sea cage located at Munambam, along the Cochin coast of the Arabian Sea. This pilot study aimed to assess both ecological and biological factors within a 120-day production cycle. The chosen site, situated approximately 2 km west of Cherai Beach, boasted wave heights ranging from 1 to 1.2 meters and a maximum tidal amplitude of 1.23 meters. The location, positioned at 10 meters depth at coordinates 10°08′083″ N; 076°08′915″ E, was deliberately selected to ensure minimal exposure to direct land-based pollution. For this study, a high-density polyethylene (HDPE) cage measuring 6 meters in diameter, equipped with a catwalk, handrails, and provisions for connecting various nets, was utilized. The Asian seabass (*Lates calcarifer*) was chosen as the local species for cultivation. Over the course of December 2008 to April 2009, the seabass were cultivated within the cage for the entire 120-day duration. Notably, the seabass exhibited significant growth, with an initial weight averaging 3.5 ± 1.5 grams, escalating to an average weight of 315.5 grams by the study's conclusion. The fish production rate within the cage was determined to be 4 kilograms per cubic meter.

3.3.1.3 CULTURE OF SPINY LOBSTERS, PANULIRUS POLYPHAGUS IN OPEN SEA CAGES AT GUJARAT, INDIA

Following a twelve-week acclimatization period in 8,000-liter FRP tanks, 2,256 juveniles of mud spiny lobsters (*Panulirus polyphagus*), with an average carapace length of 40.44 mm and a mean body weight of 67.83 g, were introduced into an open sea cage near Veraval on the coast of Gujarat, India, at a density of 12.4 m^{-3}. The lobsters were provided with a diet consisting of fresh fish and mollusc meat twice daily, equivalent to 10% of the lobster biomass per day. Throughout the study, water quality parameters in the vicinity of the cage were continuously monitored and found to be well within favorable ranges for shellfish culture, indicating minimal environmental impact from the cage. After 120 days, the lobsters were harvested from the cage, exhibiting a mean wet weight of 212.65 g, with a commendable survival rate of 86.1%. These initial findings suggest promising commercial potential for the advancement of sea cage culture of *Panulirus polyphagus* in Gujarat (Mohamed et al., 2010).

3.3.1.4 CULTURE OF SEABASS IN SEA CAGE AT VIZHINJAM BAY

A study conducted in Vizhinjam Bay, situated on the southwest coast of India, evaluated the growth potential of Asian seabass (*Lates calcarifer*) in a floating sea cage. Initially, seabass seedlings, with an average weight of 2.5 grams and total length of 53 millimeters, were nurtured in hapas in December 2008. Over a period of 60 days, they were fed pellet feed and exhibited notable growth, reaching an average size of 28 grams and 135 millimeters, with a survival rate of 60%. After this nursery phase, the juveniles were transferred to the cages in February 2009, with a stocking density of 60 individuals per cubic meter. During the subsequent 112-day rearing period, the seabass flourished, attaining an average size of 540 grams and 328 millimeters. The rate of weight gain per day experienced a significant increase, rising from 0.2 grams in December 2008 to 7.71 grams in May 2009, while the specific growth rate (SGR) decreased from 5.88 to 2.47. Throughout the study, hydrological parameters such as temperature, pH, salinity, dissolved oxygen, and microbial load were consistently maintained at optimal levels conducive to the normal growth of seabass. The findings suggest that cage culture of seabass in the marine environment offers substantial advantages, including accelerated growth rates and efficient utilization of water volume (Anil et al., 2010).

3.3.1.5 SEABASS CAGE CULTURE IN KARNATAKA

Two cages, stationed offshore near Mangalore approximately 3 kilometers from the coast at a depth of 10 meters, were established in collaboration with the Traditional Fishermen Society of Byndoor, near Byndoor (Upunda). The stocking of the cages commenced with *Penaeus indicus* (150,000 individuals, 35 mm TL) on 02-01-2009, followed by *Penaeus monodon* (200,000 individuals, 12 mm TL) on 19-02-2009. At the time of stocking, the average length and weight of *P. indicus* were recorded as 98 mm and 9 grams, respectively, while for *P. monodon*, these were 78 mm and 4 grams, respectively. *P. monodon* was provided with commercial feed (starter) four times daily during the first month, and this was reduced to three times thereafter. Throughout the study period, various parameters including hydrographic conditions, benthic fauna, and epifauna around the cages were closely monitored. Additionally, an underwater survey was conducted in the vicinity before installing the cages and again after one month of installation to assess the potential of the cages to serve as fish aggregating devices. Furthermore, the study included the monitoring of wild fish assemblages and natural entrants to the cages over the duration of the study (Dineshbabu et al., 2011).

3.3.1.6 SEABASS, LATES CALCARIFER CULTURE AT BALASORE, ODISHA IN SEA CAGE

In Balasore, a cage was deployed in January 2009 and initially stocked with 4,000 seabass averaging 55–85 grams in size. Remarkably, within a span of three months, these fish grew to around 300 grams, indicating successful cultivation. However, the first batch of stocked seabass was unfortunately lost due to social issues involving local fishermen. The nets were damaged, resulting in the escape of all fish into the sea. Subsequently, precautionary measures were implemented, and with the assistance of the state government, a second stocking was carried out, leading to the resumption of culture activities. In Srikakulam, despite the deployment of a sea cage, stocking was postponed due to the unavailability of seeds of appropriate size. Currently, the collected seeds are in the nursery phase, and they will be transferred to the cages once they reach the required size. Similarly, in Kakinada, the cage has been moored in the sea, awaiting the transfer of seeds from the nursery. Once the seeds attain the suitable size, they will be introduced into the cages for further cultivation (Mojjada et al., 2012).

3.4 ECONOMIC STATUS OF SEA CAGE FARMING

3.4.1 INITIAL INVESTMENT OF THE SEA CAGE CULTURE

Initial investment for sea cage culture is given in Table 3.5.

TABLE 3.5 Breakdown of Initial Investment for Sea Cage Culture

SL. No.	Item	Cost (in Lakhs)
1.	HDPE cage frame	1.91
2.	Ballast	0.08
3.	Floats	0.19
4.	Nets	0.58
5.	Chain, swivel, shackles	0.80
6.	Stone anchor	0.64
7.	One time launching charge	0.8
8.	Total	5.00

3.4.2 COSTS AND RETURNS

Costs and returns are given in Table 3.6.

TABLE 3.6 Estimated Costs and Returns from a Sea Cage Culture Unit (Cobia)

SL. No.	Item	Cost (in Lakhs)
1.	Setup cost of cage: Circular cage unit ($6 \times 4 = 113$ m^3)	₹5.00 lakhs
2.	Stocking density per cage	1,000 nos.
3.	Total culture duration	7 months
4.	Weight at final harvest (cobia, avg.)	3.0 kg
5.	Yield of crop (at 80% survival)	2,400 kg
6.	Estimated returns at end of 7 months (sale price @ ₹300/kg)	₹7.20 lakhs
7.	Net returns per cage per 7 months	₹2.20 lakhs
8.	Total Profit	₹2.20 lakhs

3.5 FINANCIAL ASSISTANCE IN SEA CAGE FARMING

The Pradhan Mantri Matsya Sampada Yojana (PMMSY), launched by the Central Government in September 2020, aims to boost mariculture. Managed by the Department of Fisheries under the Ministry of Fisheries,

Animal Husbandry, and Dairying, the initiative is responsible for its execution. Financial aid is available for sea cage installations through this scheme. Before setting up these cages, recipients must obtain necessary approvals from their respective State/Union Territory (UT) governments and other relevant authorities. Allocation of sea space for cage culture adheres to the leasing policy/guidelines of the State/UT.

Recipients are required to submit a comprehensive self-contained proposal (SCP), detailing technical specifications, cost estimates, targeted species, recurring expenses, permissions obtained, and financial resources. This includes confirmation from banks regarding loans covering the beneficiary's contribution or a declaration of personal investment. Governmental support is capped at five cages per individual beneficiary. Groups like fisher Self-Help Groups (SHGs), Joint Liability Groups (JLGs), Fisher Cooperatives, or those operating in clusters can receive assistance up to twice the number of group members, with a maximum of 50 cages per group. However, a cluster may consist of multiple groups or societies. Fish Farmer Producer Organizations (FFPOs/Cs) are subject to approval by the Competent Approval Committee (CAC) for implementation details and determining the total eligible units for support.

3.6 ADVANTAGES AND DISADVANTAGES OF SEA CAGE FARMING PRACTICES IN INDIA

3.6.1 ADVANTAGES

1. **Increased Production Efficiency:** Cage culture boasts a production per unit (m^3) that is 50 times higher than shore-based systems, maximizing yield potential.
2. **Cost Efficiency:** Cage farming entails lower recurring expenditure for infrastructure development and maintenance compared to shore-based farming practices, offering economic benefits to operators.
3. **Simplified Monitoring:** Stock monitoring in cage farming is straightforward, allowing for regular observation of behavior, feeding patterns, and growth rates. This simplifies management and helps mitigate issues related to stress and disease outbreaks.
4. **Convenient Harvesting:** Harvesting in cage farming is straightforward and can be planned according to market demand.

This enables producers to deliver high-quality products at competitive prices, enhancing profitability.

3.6.2 DISADVANTAGES

1. **Environmental Impact:** Sea cage farming can lead to environmental degradation due to waste accumulation, excess feed, and chemical usage. This can harm water quality, disrupt ecosystems, and contribute to algae blooms and habitat degradation.
2. **Escapes and Introductions of Non-Native Species:** There is a risk of fish escaping from sea cages, potentially introducing non-native species to local ecosystems, which can disrupt native biodiversity and ecosystems.
3. **Disease Spread:** Dense populations of fish in sea cages can increase the risk of disease outbreaks, which can spread rapidly and affect wild populations as well as other farmed fish in the area.
4. **Dependency on External Factors:** Sea cage farming is vulnerable to external factors such as weather events, rough seas, and natural disasters, which can damage or destroy cages, leading to financial losses and production disruptions.
5. **Social Conflicts:** Sea cage farming can sometimes lead to conflicts with local communities, fishermen, and other stakeholders due to issues such as competition for space, conflicts over resource use, and concerns about environmental impacts.
6. **Regulatory Challenges:** Sea cage farming often requires permits and approvals from various regulatory bodies, which can be time-consuming and costly. Compliance with regulations related to environmental protection, waste management, and fish health can also pose challenges for operators.
7. **High Initial Investment:** Setting up sea cage farming operations requires a significant initial investment in infrastructure, equipment, and technology. This can be a barrier to entry for small-scale farmers and new entrants to the industry.

8. **Limited Control Over Environmental Conditions:** Sea cage farmers have limited control over environmental conditions such as water temperature, currents, and salinity, which can impact fish growth and health.

3.7 CONCLUSION

Our nation boasts a vast coastline spanning 8,129 kilometers, presenting a significant opportunity for substantial marine fish production through mariculture within its territorial waters. A total of 134 prime locations covering an area of 46,958.2 hectares have been pinpointed for sea cage farming, particularly suitable for open sea cage operations. Among these, Andhra Pradesh leads with 11,792 hectares, followed closely by Gujarat with 11,572.2 hectares, and Tamil Nadu with 7,673 hectares. Although challenges like seed and feed availability may arise initially, addressing these hurdles can pave the way for advancements in hatchery and feed technologies, given the immense potential and resources at hand. Sea cage farming holds promise in significantly contributing to the nation's economy while satisfying the escalating demand for seafood.

KEYWORDS

- **cost efficiency**
- **financial aid**
- **financial assistance**
- **non-native species**
- **sea cage farming**
- **sea cage installations**
- **self-contained proposal**
- **self-help groups**

REFERENCES

Anil, M. K., Santhosh, B., Jasmine, S., Saleela, K. N., George, R. M., Kingsly, H. J., & Rao, G. S. (2010). Growth performance of the seabass *Lates calcarifer* (Blotch) in sea cage at Vizhinjam Bay along the south-west coast of India. *Indian Journal of Fisheries, 57*(4), 65–69.

Aquaculture technologies implemented by NFDB. (n.d.). National Fisheries Development Board, Department of Fisheries, Ministry of Fisheries, Animal Husbandry & Dairying, Govt. of India, Hyderabad–500052.

Beveridge, M. (2004). *Cage Aquaculture* (2nd ed.). Blackwell Publishing, Oxford, UK, 368 p.

Beveridge, M. (2004). *Cage Aquaculture* (2nd ed.). Wiley-Blackwell.

Damodaran, D., Mohammed Koya, K., Mojjada, S. K., Lalaji, C. D., Dash, G., Vase, V. K., & Sreenath, K. R. (2018). Optimization of the stocking parameters for mud spiny lobster *Panulirus polyphagus* (Herbst, 1793) capture-based aquaculture in tropical open sea floating net cages. *Aquaculture Research*, *49*(2), 1080–1086.

Dash, B., Kumar, S., Mojjada, G., & Rao, S. (2009). Integration of seaweed (*Kappaphycus alvarezii*) and pearl oyster (*Pinctada fucata*) along with Asian seabass (*Lates calcarifer*) in open sea floating cage off Andhra Pradesh coast.

Dineshbabu, A. P., Thomas, S., & Sasikumar, G. (2011). Successful demonstration of seabass cage culture in Karnataka. *Fishing Chimes*, *31*(1), 80–82.

Divu, D., Mojjada, S. K., Muktha, M., Azeez, P. A., Tade, M. S., Subramanian, A., & Gopalakrishnan, A. (2023). Mapping of potential sea-cage farming sites through spatial modelling: Preliminary operative suggestions to aid sustainable mariculture expansion in India. *Indian Journal of Fisheries*, *70*(4), 1–17.

Ghosh, S., Jeeva, J. C., Raju, S. S., Megarajan, S., Ranjan, R., Xavier, B., Edward, L., & Gopalakrishnan, A. (2024). Cage aquaculture of Indian pompano for livelihood diversification of artisanal fishers: Insights from Andhra Pradesh, India. *Indian Journal of Fisheries*, *71*(1). https://doi.org/10.21077/ijf.2024.71.1.146898-13.

Gopakumar, G. (2009). History of cage culture, cage culture operations, advantages and disadvantages of cages and current global status of cage farming. In: Course manual: National training on cage culture of seabass. CMFRI & NFDB, Kochi, pp. 8–12.

Imelda, J., Joseph, S., Ignatius, B., Rao, G. S., Sobhana, K. S., Prema, D., & Varghese, M. A. (2010). A pilot study on culture of Asian seabass *Lates calcarifer* (Bloch) in open sea cage at Munambam, Cochin coast, India. *Indian Journal of Fisheries*, *57*(3), 29–33.

Jadhav, U. (2009). *Aquaculture Technology and Environment* (p. 63). PHI Learning, India.

Li, S. F. (1994). Fish culture in cages and pens. In S. Li & J. Mathias (Eds.), *Freshwater Fish Cultures in China: Principles and Practices* (pp. 305–346). Elsevier, Amsterdam.

Loka, J., Vaidya, N. G., & Philipose, K. K. (2012). Site and species selection criteria for cage culture. In K. K. Philipose, J. Loka, S. R. Krupesha Sharma, & D. Divu (Eds.), *Handbook on Open-Sea Cage Culture* (pp. 27–36). ICAR-Central Marine Fisheries Research Institute, Karwar, Karnataka, India.

Mohamed, G., Rao, G. S., & Ghosh, S. (2010). Aquaculture of spiny lobsters in sea cages in Gujarat, India. *Journal of the Marine Biological Association of India*, *52*(2), 316–319.

Mojjada, S. K. (2015). *Studies on Technological Biological Innovations in Open Sea Cage Farming* (PhD thesis). Mangalore University, Mangalore, India.

Mojjada, S. K., Imelda, J., Koya, M., Sreenath, K. R., Dash, G., Dash, S. S., & Rao, G. S. (2012a). Capture based aquaculture of mud spiny lobster, *Panulirus polyphagus* (Herbst, 1793) in open sea floating net cages off Veraval, north-west coast of India. *Indian Journal of Fisheries*, *59*(4), 29–33.

Mojjada, S. K., Imelda, J., Maheswarudu, G., Ranjan, R., Dash, B., Ghosh, S., & Rao, G. S. (2012b). Open sea mariculture of Asian seabass *Lates calcarifer* (Bloch, 1790) in marine floating cage at Balasore, Odisha, north-east coast of India. *Indian Journal of Fisheries*, *59*(3), 89–93.

Philipose, K. K., Loka, J., Sharma, S. R., & Divu, D. (2012). *Handbook on Open Sea Cage Culture* (144 p.). ICAR-Central Marine Fisheries Research Institute, Karwar, India.

Philipose, K. K., Loka, J., Sharma, S. R., Sadhu, N., Dube, P., Gopakumar, G., & Rao, G. S. (2013). Farming of cobia, *Rachycentron canadum* (Linnaeus, 1766) in open sea floating cages in India. *Indian Journal of Fisheries*.

Ranjan, R., Xavier, B., Dash, E. L., Maheswarudu, G., & Rao, G. S. (2014). Domestication and brood stock development of the orange spotted grouper, *Epinephelus coioides* (Hamilton, 1822) in open sea cage off Visakhapatnam coast. *Indian Journal of Fisheries*, *61*(1), 21–25.

Rao, G. S. (2012). Cage culture: Mariculture technology of the Millennium in India. In: *Handbook on Opensea Cage Culture.* Central Marine Fisheries Research Institute, pp. 1–14.

Rao, G. S., Imelda, J., Philipose, K. K., & Mojjada, S. K. (2013). *Cage Aquaculture in India*. Central Marine Fisheries Research Institute.

CHAPTER 4

Biofloc Technology: A Solution for Global Food Security Challenges

SANA AZIZ,[1,*] MUHAMMAD SARFRAZ ALI,[2] NISAR AHMAD BHATTI,[1] SHAKEELA MANZOOR,[3] and SAJID ABDULLAH[3]

[1]*Department of Zoology, University of Jhang, Jhang, Punjab, Pakistan*

[2]*Department of Biosciences, COMSATS University Islamabad (CUI), Islamabad, Pakistan*

[3]*Department of Zoology, Wildlife and Fisheries, University of Agriculture, Faisalabad, Punjab, Pakistan*

*Corresponding author

ABSTRACT

Aquaculture practice is rapidly growing and is extensively being applied due to increased food demand. Depending on the existing and suitable conditions of water and land for aquaculture, intensive practices are also being implemented in various culture systems. This intensification may cause pathological conditions and deterioration of existing environmental parameters, leading to degradation due to increased population density in the system. Therefore, to mitigate the impacts associated with aquaculture, emerging biofloc technology supports sustainable intensification in the system. This technology uses a no-water-exchange approach, a widely adopted practice where health security and environmental integrity are concerned. The use of probiotics in the biofloc system is a viable alternative to ensure better health status of cultured organisms and to manage stress associated with intensification, thereby reducing environmental

Sustainable Aquaculture: Recent Advances and Techniques. Dhanaji W. Patil, Sandip S. Markad (Eds.)

DOI: 10.1201/9781779641113-4

deterioration more effectively. This cumulative effect of favorable water quality factors, along with the activity of heterotrophic bacteria, maintains the carbon–nitrogen (C:N) ratio, ultimately promoting the reduction of inorganic toxic substances. This chapter focuses on the efficacy and impact of probiotics on water quality parameters in the biofloc system.

4.1 INTRODUCTION

Aquaculture is a promising food production sector that can effectively utilize natural resources to combat hunger, malnutrition, and poverty (Food and Agriculture Organization, 2017; Hambrey, 2017). Fish and other aquatic animals are a good source of amino acids and fatty acids, which are crucial components of a healthy diet (Khanjani, 2015). With the world population predicted to reach 9.6 billion (El-Sayed, 2017), there is a growing demand for sustainable protein-rich food sources, making aquaculture an important solution. However, natural water resources are being overfished globally, leading to a significant gap between demand and production capacity (Avnimelech, 2012). Fish aquaculture faces challenges such as declining animal welfare standards, frequent disease outbreaks, and antibiotic resistance due to overuse of antimicrobial agents (Kim et al., 2020; Santos and Ramos, 2018).

To address these challenges, the development of a sustainable aquaculture industry should prioritize designing systems that achieve high productivity and profitability (Avnimelech and Kochba, 2009; Khanjani, 2015) while utilizing fewer resources and minimizing ecological impact. This is essential to meet the growing need for aquatic products (FAO, 2017; Khanjani and Sharifinia, 2021). Environmentally sustainable and intelligent technologies are therefore essential to build a resilient aquaculture ecosystem.

Biofloc technology (BFT) is an innovative and promising alternative for sustainable aquaculture, offering several benefits such as efficient use of limited water resources, reduced feed input, and natural development of microbial biomass. It employs microorganisms to absorb harmful waste compounds and promotes a self-nitrification process without the need for water exchange. This environmentally friendly approach helps reduce the overall ecological impact of aquaculture (Ekasari et al., 2015). The BFT method minimizes water exchange, maintains biofloc stability, and enhances fish productivity, making it economically attractive to aquapreneurs. This chapter summarizes studies on BFT systems addressing global food security challenges and highlights their principles, types, applications, economic aspects, challenges, and future prospects.

4.2 PRINCIPLE OF BIOFLOC TECHNOLOGY

Biofloc technology (BFT) is a sustainable method of farming aquatic organisms. It involves the production of masses of microorganisms in the water that serve as a food source for the farmed species, thereby reducing the need for external feed. The BFT concept was first developed at the French Research Institute for Exploration of the Sea in the early 1970s, and has since been used to culture a variety of aquatic species including *F. merguiensis*, *L. stylirostris*, *L. vannamei*, and *P. monodon* (Avnimelech, 1999). Nutrient recycling is the core principle in BFT (Ray et al., 2010), achieved by enhancing the carbon/nitrogen (C:N) ratio, which increases microbial communities such as algae and heterotrophic bacteria within a closed structure (Avnimelech, 2006). This technology follows the principle of flocculation.

4.3 BASIC TYPES AND CONCEPT OF BFT

There are two types of biofloc technology (BFT) systems. The first type is called green water biofloc, which is an outdoor system exposed to light. This system is mainly dominated by autotrophic microorganisms. The second type is known as brown water biofloc, which is an indoor system devoid of natural light exposure and is dominated by heterotrophic or chemotrophic microorganisms (El-Sayed, 2021) (Figure 4.1).

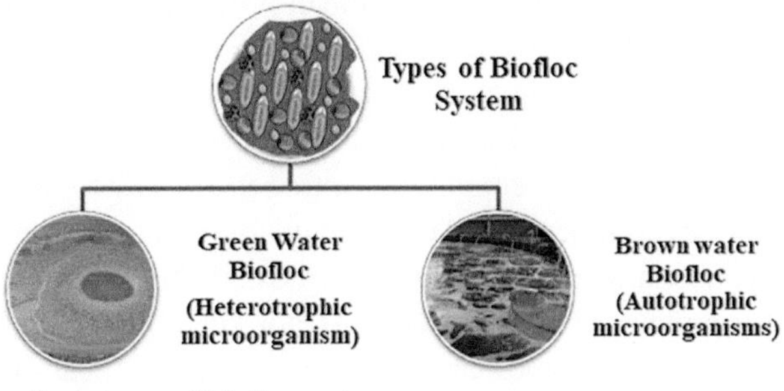

FIGURE 4.1 Basic types of biofloc systems.

It is crucial to maintain carbon/nitrogen supplementation in pond water in both systems (Nisar et al., 2022) because nitrogen is vital for pond ecosystem productivity, and nitrogen-cycle bacteria mineralize organic nitrogen into inorganic forms. This process affects fish survival and health. At a pH below 8.0, ammonia (toxic) is transformed into ammonium ions (non-toxic). Therefore, the use of biofloc technology has evolved to prevent the buildup of organic nitrogenous substances, which are toxic to aquatic life.

4.4 PROCESS OF BIOFLOC FORMATION

For large-scale biofloc-based farming, polyethylene-lined ponds of high density with central drainage and well-constructed dikes are mostly chosen (Figure 4.2).

FIGURE 4.2 Polyethylene-lined ponds for biofloc system.

This system can be used in recirculating aquaculture or raceway structures. It can be added directly to the system or produced through an activated sludge system and then added to the production system as biofloc. A small amount of nitrogenous source, such as fish feed or urea, combined with a carbon source (e.g., molasses, starch, or wheat flour), promotes biofloc development. Wastewater containing nitrogenous waste from multiple farms can serve as an inoculum. Adding clay (20 g), ammonium sulfate (10 mg), and organic matter from a carbon source (200 mg) such as molasses to 1 L of water stimulates biofloc formation (Gaona et al., 2011). If the solution has a balanced ratio of carbon and nitrogen, ammonium can be converted into bacterial biomass along with organic nitrogenous waste (Schneider et al., 2005). Adding a uniform distribution of sugars stimulates heterotrophic bacterial growth, which, in turn, facilitates the uptake of nitrogen through the synthesis of various microbial proteins (Khanjani et al., 2022).

Biofloc is a unique mixture of suspended particles and microorganisms that are held together by extracellular polymeric substances. It contains a diverse range of microorganisms, including bacteria, fungi, algae, invertebrates, and detritus (El-Sayed et al., 2021). This natural food is created by transforming fish waste and unutilized feed through a mixture of microorganisms and particles. This process occurs in a system with proper aeration and ample sunlight. The microorganisms and particles are held together by mucus released by bacteria and are finally bound by microorganisms with filamentous structures or through electrostatic attraction. Although some flocs are visible to the naked eye, most are microscopic, with sizes ranging from 50–200 microns. Biofloc is a rich source of protein, with a content of 25–50%, and fat content ranging from 0.5–15%, as well as vitamins and minerals—particularly phosphorus. Biofloc can be efficiently used in feed for fish and shrimp, replacing fishmeal as a protein source, and the unused portion of this feed can be easily converted into biofloc to develop high-quality feed (Avnimelech, 2009; Hargreaves, 2013; Khanjani et al., 2020a; Kuhn et al., 2016). The preparation of biofloc technology is shown in Figure 4.3.

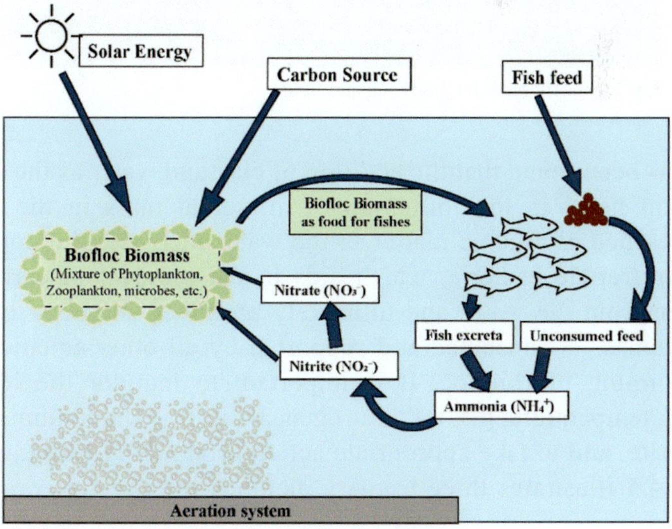

FIGURE 4.3 Schematics of biofloc fish culture system.

To improve the formation of microbial mass, the primary inoculum plays a crucial role in the biofloc production cycle. It is important to exercise caution before using any commercially available inoculum for this purpose. In order to maintain the floc in suspension, significant aeration is necessary. The composition of biofloc can be checked using Imhoff cones (Figure 4.4).

FIGURE 4.4 Imhoff cones for floc measurement.

It has been found that the addition of clay and water as the primary inoculum helps in the formation of microbial mass in the system. The presence of organic matter in the water increases the population of heterotrophic bacteria, which aids in the removal of carbon and nitrogen from the water and ultimately produces microbial biomass. This biomass is consumed and supported by all other aquatic organisms forming the biofloc. It is important to monitor the levels of oxygen, temperature, pH, total nitrogen, alkalinity, ammonium, nitrate, and nitrite, and to take appropriate action promptly (Khanjani, 2015). Figure 4.5 illustrates three primary methods for effective production of biofloc.

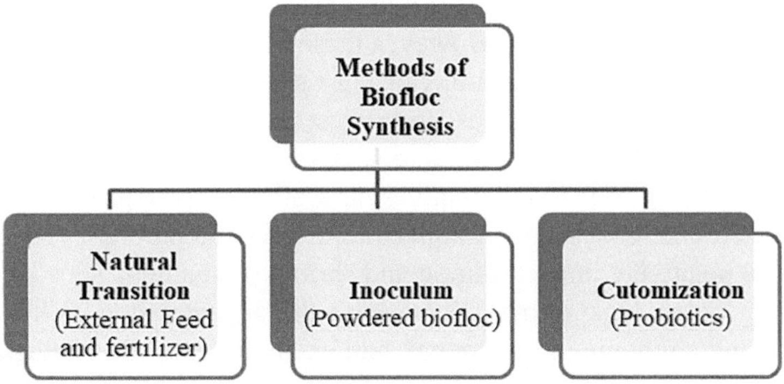

FIGURE 4.5 Three primary methods for effective production of biofloc system.

4.4.1 NATURAL TRANSITION METHOD

In this method, autotrophs are produced by using fertilizers, fish or shrimp feed, and other ingredients in a culture system. These autotrophs are then transformed into heterotrophs by maintaining the carbon-to-nitrogen (C:N) ratio between 12:1 and 15:1. The level and duration of biofloc formation are influenced by the type of carbon used and the amount of salinity (Khanjani et al., 2020b). However, this approach is time-consuming.

4.4.2 INOCULUM METHOD

This technique involves assessing the compatibility of nutrient availability and water quality from the earlier crop before inoculating new biofloc-based culture water. First, fermented carbon sources like rice bran and molasses are aerated in water for 24–48 hours to generate biofloc (Zemor et al., 2019). The cultured biofloc mass is then dried, converted into powder form, and stored. The powder is later dissolved with a carbon source and added as a fermented product. This is a time-saving process.

4.4.3 CUSTOMIZATION METHOD

This biofloc production method blends probiotics to create a healthier environment and increase aquacultural output without causing adverse effects on health. Previously, the excessive use of antibiotics and various chemical compounds was common in controlling disease outbreaks in aquaculture (Carbone and Faggio, 2016). However, these compounds have been proven to be harmful to human health, causing risks such as immune suppression. Microorganisms such as probiotics and microalgae can enhance aquaculture production by modulating immunity. Therefore, natural and environmentally friendly alternatives, as well as growth promoters, have become crucial for sustainable aquaculture. Bacterial probiotics have been found to positively impact salmon and shrimp production. They do so by breaking down organic matter, reducing the accumulation of hydrogen sulfide and ammonia in the system, and excluding harmful pathogens by promoting the growth of desirable bacterial species (Crab et al., 2010).

4.5 FACTORS AFFECTING BIOFLOC ORGANISMS (BFOS) COMPOSITION IN THE BFT SYSTEM

Biofloc communities have the ability to respond quickly to environmental changes by activating or deactivating pathways and altering their composition, structure, and functionality. However, monitoring these fluctuations can be challenging (Bentzon et al., 2016). Changes in environmental parameters—such as biotic factors like predation and abiotic factors like salinity—can disturb the synthesis, abundance, and biodegradation actions of BFOs (Avnimelech, 2007). The production of BFOs can also be affected by the type of cultured species, environmental conditions, as well as the food and carbon sources available (Monroy-Dosta et al., 2013).

4.5.1 CARBON/NITROGEN RATIO

Maintaining the correct carbon-to-nitrogen (C:N) ratio is important for controlling organic nitrogen and promoting microbial communities in water systems (Emerenciano, 2012). Under aerobic conditions, the C:N ratio should be higher than the bacterial body composition (Emerenciano, 2013). Different carbon sources affect bacterial abundance in BFT systems. However, balancing BFO densities with appropriate dissolved oxygen (DO) levels requires constant monitoring (Emerenciano et al., 2012).

Glucose can absorb around 10 mg NH_4^+-N/L with a C:N ratio of 10:110. Low-cost carbon sources are used to maintain a C:N ratio higher than 10:1 (Crab, 2010). Nitrogen not consumed by organisms can be used to produce microbial protein instead of forming toxic compounds. The supplemented carbon captures excreted ammonia, which is then converted into bacterial biomass (Schneider et al., 2005). The C:N ratio plays a crucial role in aquaculture. Manipulating the ratio can reduce toxic ammonia levels, influence BFO communities, alter ammonia pathways, and affect the growth and welfare of cultivated organisms (Khanjani et al., 2020a, b). Studies have shown that a C:N ratio of 15 is ideal for effective nutrient removal. Increasing the C:N ratio has been found to boost the biomass of heterotrophic bacteria and change the abundance of different bacterial species (Abu Bakar et al., 2015; Panigrahi et al., 2018). Differences in water quality and nitrogen dynamics are most likely due to the responses of the bacterial community in water bioflocs to different C:N ratios.

4.5.2 SALINITY

The salinity of water has a significant impact on fish growth in aquaculture. Therefore, maintaining an optimum salinity range is crucial. With the rise in salinity, the aggregation of particles and floc size also increase (Hakanson, 2005). It can influence the mass of heterotrophic bacteria (Khanjani et al., 2020a, b; Maica et al., 2012). In a study conducted by Maica et al. (2012), *Chlorophytes* dominated at low salinities, while diatoms were more prevalent at higher salinity levels. Salinity is one of the most important factors controlling ciliate density, and increasing salinity levels can lead to a reduction in ciliate biomass (Decamp et al., 2003).

4.5.3 TOTAL SUSPENDED SOLIDS (TSS)

Suspended solids can directly impact water quality in BFT systems and affect the variety and concentration of BFOs (Ray et al., 2010). Some studies have shown that removing suspended solids from water can affect the abundance of certain types of microorganisms. Nematodes, cyanobacteria, rotifers, and bacteria tend to decrease, while *Chlorophytes*, diatoms, and dinoflagellates remain largely unaffected (Ray et al., 2010). Microbial biomass often increases with rising TSS levels in BFT due to the utilization of nitrogen and carbon from fertilizers and animal waste by microorganisms (MOs).

4.5.4 TOTAL DISSOLVED SOLIDS (TDS)

Total dissolved solids (TDS) in water provide a qualitative measure of water quality based on the concentration of dissolved ions. However, TDS measurement does not offer specific insights into certain water quality issues, such as increased hardness, corrosiveness, or salty taste. During the maturation period, the concentration of TDS in reused water decreased, which could be attributed to the growth of microalgae that utilize inorganic nutrients (Natrah et al., 2013).

4.5.5 AERATION

In aquaculture, aeration affects the concentration and composition of BFO communities (Krummenauer et al., 2021). Adequate dissolved oxygen (DO) levels enhance the activity of heterotrophic and chemoautotrophic bacteria, enabling them to digest organic and inorganic carbon sources and transform nitrogen (Hargreaves, 2013). Crab et al. (2012) suggest that a DO concentration of at least 5 mg/L is necessary for the survival of both BFOs and cultivated species. One of the primary goals of aeration is to keep BFOs suspended, preventing particles from settling and creating anaerobic zones at the bottom of the medium. Regardless of the aeration layout, adjustments are made to ensure concentrations remain above 5 mg/L (Hargreaves, 2013).

In semi-biofloc photoautotrophic systems, oxygen can be naturally supplemented by phytoplankton. However, this oxygen is unstable and may affect pH levels and the toxicity of nitrogen compounds (Hargreaves, 2013). There is an increased demand for dissolved oxygen in aquaculture systems, not only due to fish respiration but also due to the growing number of microorganisms that aggregate and decompose accumulated organic material (Jiang and Pan, 2005).

4.5.6 LIGHT

Light is a crucial parameter in aquaculture that affects the structure and performance of the BFO community (Tao et al., 2021). Light conditions influence phytoplankton abundance, causing variations in the nutritional value of biofloc. Algae are a significant food source for cultivated aquatic species, promoting their development. Outdoor cultivation has more positive effects on photosynthetic organisms and *Litopenaeus vannamei* than indoor cultivation (Neal et al., 2010). The type and source of light can also impact the *L. vannamei* BFT bacterial community. Light exposure has the ability to inhibit the growth and function of both nitrifying communities and heterotrophic microorganisms (Khanjani and Sharifinia, 2021). The BFT system is generally more efficient under minimal light exposure. Therefore, further research is needed to understand how light intensity, wavelength, type, and photoperiod regimes impact BFT systems.

4.5.7 STOCKING DENSITY

The population of microbes in a BFT (biofloc technology) system is influenced by several factors, one of which is stocking density. Studies by Deng et al. (2019) and Sandoval-Vargas et al. (2020) indicate that stocking density is closely related to feeding rates and nutrient inputs, which in turn affect the microbial population. Therefore, maintaining an appropriate stocking density is crucial for the success of a BFT system. At high densities, excess organic matter can be removed via settling (Ray et al., 2010) and siphoning procedures, impacting the floc microbial community (Sandoval-Vargas et al., 2020).

Research studies by Deng et al. (2019) and Sandoval-Vargas et al. (2020) have shown that the water microbial community in BFT can be categorized into different groups based on stocking density. This suggests that density can significantly alter the microbial community in BFT, potentially affecting its functionality. These changes may be related to varying nutrient input ratios in the culture system, which impact the abiotic environment of the water. High-density biofloc systems can increase bacterial diversity by promoting microbial reactions through the accumulation of minerals and other compounds (Deng et al., 2019).

4.5.8 TEMPERATURE

Temperature is a measure of hotness or coldness. Fluctuations in water temperature can affect the body temperature of aquatic organisms, and extreme temperatures can impact their metabolism, reproduction, and survival. High water temperatures can increase toxicity levels.

Temperature also influences the diversity and density of flocs by altering the metabolic processes of microbes in biofloc technology (BFT).

Studies have found that in warmer water (40°C), there is more de-flocculation of stimulated sludge floc compared to water at cooler temperatures (18–20°C), possibly due to decreased microbial activity at lower temperatures (Wilen et al., 2000). Conversely, at higher temperatures (30–35°C), sludge accumulation occurs due to excessive production of extracellular polymeric substances. Additionally, high temperatures can lead to sludge formation, causing a decrease in floc volume as flocs lose floatability and sink to the bottom. According to further studies, the optimal temperature range for the mineralization of organic nutrients into usable forms is between 23.2°C and 31.8°C (Harini et al., 2016). However, adjusting the temperature in BFT ponds can be challenging due to high costs and dependence on climatic conditions.

4.5.9 PHOSPHATES

Phosphate is a crucial factor in maintaining fishpond fertility, and it exists in various forms, including ortho-phosphate and organic phosphate. Its availability depends on environmental factors such as fertilizer runoff, rainfall, and the pond bottom area. When phosphate concentrations increase, weed growth accelerates, leading to scum formation, which can ultimately enhance fish growth. In BFT systems, elevated phosphate concentrations are predicted to result in cyanobacterial blooms. However, *Bacillus* spp. and *Pseudomonas* spp. are dominant and effective in reducing phosphate in BFT systems by solubilizing it through intracellular accumulation in the form of polyphosphate (Krishnaswamy et al., 2009; Zhao et al., 2012). Probiotics have been found to significantly reduce phosphorus levels in pond sediments.

4.6 THE POTENTIAL ROLE OF THE BIOFLOC SYSTEM

4.6.1 INCREASED GROWTH OF FISHES

Microbial biotechnology involves the use of microorganisms to ensure food safety, improve food quality, enhance human nutrition, and prevent infections caused by flora and fauna. This biotechnology is especially important in sustainable aquaculture, where microorganisms are used to maintain and manage ecosystems by recycling nutrients and regulating water quality parameters. Various microbial techniques have been developed to create eco-friendly ecosystems that support fish growth and maximize production in aquaculture.

Microorganisms in biofloc stimulate fish digestive enzymes, enhancing the feed conversion ratio (FCR) and improving fish growth. The growth and FCR of fishes cultured in biofloc aquaculture systems are represented in Table 4.1. Elayaraja et al. (2020) reported that jaggery-based BFT provided a suitable ecosystem for healthier bacterial activities like nitrification and assimilation, resulting in improved water quality parameters. Likewise, Abakari et al. (2020) showed that biochar-based BFT affects water quality through the assimilation and nitrification by heterotrophic bacteria, resulting in increased levels of nitrogen and NO_3^-.

A study conducted by Van Doen et al. (2020) demonstrated that polyphenols obtained from chestnut (*Castanea sativa*) were used in BFT to culture Nile tilapia, resulting in higher mucosal and serum resistance against pathogenic *Streptococcus agalactiae*. Menaga et al. (2020) reported that growth parameters such as fork length, total length, final weight, and immunity of *Oreochromis niloticus* cultured in FRP tanks were improved when associated with BFT. Azim and Little (2008) showed that the growth of *O. niloticus* cultured in biofloc was enhanced by 35–38% compared to the control group, highlighting the importance of BFT in fish nutrition. The growth of *O. niloticus* raised in biofloc peaked at salinities of 5 and 6 g/L but declined at salinities of 9 and 12 g/L, suggesting that growth in biofloc is strongly dependent on salinity.

Kishawy et al. (2020) reported that growth parameters of *O. niloticus* raised in biofloc systems improved with the use of mannan oligosaccharides and glycerol compared to the control group, indicating growth enhancement depending on carbon sources. Wang et al. (2015) observed an improvement in the growth performance of crucian carp (*Carassius auratus*) cultured in biofloc systems as the C:N ratio increased. The findings of this study suggest that even a slight increase in the C:N ratio can boost biofloc development, thereby enhancing fish growth.

TABLE 4.1 Summary of Studies Conducted in Different Fish Species Related to Biofloc Systems

Fish Species Cultured	Technology Used	Effect on Cultured Water	Effect on Fish	References
Nile tilapia (*Oreochromis niloticus*)	Jaggery-based BFT	More assimilation and nitrification of bacteria	Higher survival, growth, immunity, and improved antioxidant ability	Elayaraja et al. (2020)
Red tilapia (*Oreochromis* sp.)	Cornmeal + molasses (120 and 240 fish/m^3)	–	Improvement in feed efficiency	Banuelos-Vargas et al. (2021)
Nile tilapia (*O. niloticus*)	Indoor biofloc system	Enabled more heterotrophic bacteria to convert NH_4^+ into bacterial biomass	Fish weight at harvest increased up to 22%	Luo et al. (2014)
Labeo rohita	Carbon sources (wheat, tapioca, sugar bagasse, corn)	–	Better growth	Ahmad et al. (2016)
Labeo rohita	Wheat flour	Good nutritional quality	Better growth	Mahanand et al. (2013)
Nile tilapia (*O. niloticus*)	Glucose-added biofloc groups	Mature biofloc system with good water quality	Higher final body weight	Liu et al. (2018)
Nile tilapia (*O. niloticus*)	BFT with salinized water	Environmentally viable	Better growth performance	De Alvarenga et al. (2018)
Common carp (*Cyprinus carpio* L.)	Molasses	–	Highest weight gain	Najdegerami et al. (2016)
Common carp (*C. carpio* L.)	Carbon sources (sugar beet molasses, sugar)	–	No differences in growth performance	Bakhshi et al. (2018)
Common carp (*C. carpio* L.)	BFT at two stocking densities	–	Higher fish growth, lower feed conversion ratio	Adineh et al. (2019)
Common carp (*C. carpio* L.)	20% and 30% crude protein in sugarcane molasses and rice bran	Better water quality	Better growth performance	Ebrahimi et al. (2020)

TABLE 4.1 *(Continued)*

Fish Species Cultured	Technology Used	Effect on Cultured Water	Effect on Fish	References
Common carp (*C. carpio* L.)	–	Better water quality	Better growing efficiency	Tabarrok et al. (2020)
Nile tilapia (*O. niloticus*)	Microbial proteins (MP) and poly-β-hydroxybutyrate (PHB)	Better water quality	No effect on survival during early stages	Klanian et al. (2020)
Nile tilapia (*O. niloticus*)	Rice bran and molasses (1:1)	Better water quality	Better growth performance	Hwihy et al. (2021)
Nile tilapia (*O. niloticus*)	Biochar-based BFT	Fewer suspended solids; more assimilation and nitrification	No significant effect on growth or physiology	Abakari et al. (2020)
Nile tilapia (*O. niloticus*)	Isolated probiotics (*B. infantis*, *E. profundum*, *B. subtilis*, *B. megaterium*)	Increased probiotics in water	Better growth and immunity	Menaga et al. (2020)
Nile tilapia (*O. niloticus*)	Chestnut polyphenols-based BFT	–	Better growth, survival, and immunity against *S. agalactiae*	Van Doan et al. (2020)
Nile tilapia (*O. niloticus*) juveniles	BFT with prebiotics and probiotics	Lower nitrite concentration	Higher growth, weight, and improved blood parameters	Laice et al. (2021)
Indian major carps	BFT for polyculture	NH_4^+-N, NO_2^--N, and NO_3^--N within optimal range	Higher specific growth	Deb et al. (2020)
Nile tilapia (*O. niloticus*)	BFT	Changes in salinity	Improved growth performance	Azim and Little (2008)

TABLE 4.1 *(Continued)*

Fish Species Cultured	Technology Used	Effect on Cultured Water	Effect on Fish	References
Nile tilapia (*O. niloticus*)	Carbon sources (glycerol and mannan oligosaccharides)	–	Better growth performance	Kishawy et al. (2020)
Crucian carp (*Carassius auratus*)	BFT	–	Enhanced growth performance	Wang et al. (2015)
African catfish (*Clarias gariepinus*)	Sucrose, rice bran, glycerol at C:N ratio of 15:1	Improved water quality; higher biofloc production	–	Dauda et al. (2017)
African catfish (*C. gariepinus*) and lemon fin barb hybrid (LFBH) (*Hypsibarbus wetmorei* ♂ × *Barbodes gonionotus* ♀)	Glycerol	–	Better growth performance of both species	Dauda et al. (2018)
Asian stinging catfish (*Heteropneustes fossilis*)	Sugarcane molasses	–	Better growth rate	Zafar et al. (2021)
Grey mullet (*Mugil cephalus*)	BFT	–	Better growth performance	Haridas et al. (2021)
Flatfish (*Paralichthys olivaceus*)	Starch	Better water quality	Better growth performance	Kim et al. (2020)
Bluegill (*Lepomis macrochirus*)	Corn starch, sucrose starch	Better water quality	–	Fischer et al. (2020)

4.6.2 INCREASED BACTERIAL COMMUNITY AND MAINTENANCE OF MICROBES

Biofloc refers to organic particles that are suspended in the water column. These particles contain a large number of heterotrophic bacteria, ciliates, flagellates, microalgae, small metazoans, cyanobacteria, and fungi (Avnimelech, 2012). In a culture system, the addition of carbohydrates facilitates the growth of heterotrophic bacterial loads. These bacteria use the added carbohydrate as a food source and synthesize microbial protein by assimilating nitrogen from the system. The microbial protein, in turn, can be utilized as a food source by other organisms such as shrimp and fish (Rahmatulla and Beveridge, 1993).

In BFT (biofloc technology) units, microorganisms play three primary roles. First, they help maintain water quality. Second, they provide nutrients that reduce feed costs and improve the feed conversion ratio (FCR). Third, they compete with pathogens, ensuring biosecurity. Additionally, microorganisms also appropriate greenhouse gases (GHGs), making the system eco-friendly. Overall, the biological functions of microorganisms in BFT units contribute to increased fish yield and production success.

4.6.3 WATER QUALITY IMPROVEMENT AND WASTE UTILIZATION

Water quality is a critical factor in ensuring the environmental and economic sustainability of farming systems, especially in intensive systems (Emerenciano et al., 2013). The BFT method is a composite trophic technology that incorporates autotrophic and heterotrophic microorganisms in the process of bioremediation. Autotrophic microbes are excellent bioremediators because they can synthesize their food from inorganic substances; this group includes nitrite- and ammonia-oxidizing bacteria.

Heterotrophic bacteria play a central role in BFT by breaking down non-living organic material. They use carbon to build their cells and act as electron donors during the oxidation of harmful contaminants. Unlike autotrophs, heterotrophic bacteria have a smaller role in nitrification and denitrification but are primarily responsible for decomposing organic waste such as feces, uneaten feed, and dead biomass. These bacteria transform nitrogenous ammonia into microbial aggregates or flocs, which are non-harmful. The combined processes of nitrification, photosynthesis, and assimilation play a vital role in improving and maintaining water quality. This nutrient recycling is especially important in systems with high nutrient content and low water exchange, such as BFT (Emerenciano et al., 2022a; Samocha et al., 2007).

4.6.4 CONTROL OF PATHOGENS

Biofloc technology (BFT) is an effective method for controlling pathogens in aquaculture. Unlike traditional disease management methods, which can lead to the development of resistance in pathogenic species, BFT promotes a balanced microbial community comprising both beneficial and pathogenic microorganisms. Bioflocs have probiotic properties that provide a favorable substrate for potential probiotic bacteria, thereby enhancing digestion, metabolism, immunity, and overall development in cultured animals (Daniel and Nageswari, 2017).

In BFT systems with optimal water quality, probiotic bacteria tend to dominate and reduce the prevalence of pathogens in both the water and the intestinal tracts of cultured species. This occurs through the secretion of a wide array of exoenzymes and bioactive polymers (Cienfuegos-Martínez et al., 2022). The natural probiotics present in BFT systems can significantly reduce water treatment costs—by up to 30%—as these systems typically require only 0.5–1% water exchange per day (Crab et al., 2010).

4.6.5 APPLICATION OF BFOS AS NUTRITIONAL REQUIREMENT

Aquaculture producers face a significant challenge in the form of high feed costs. Aquatic organisms require optimal protein levels and a balanced amino acid profile to maintain growth and overall well-being. However, these nutrients are expensive and directly affect the market price of feeds (Hisano et al., 2020). In the case of tilapia, feeding expenses can account for up to 50% of operational costs, particularly when diets contain high or inadequate protein content (El-Sayed, 1999). Therefore, alternative feed sources such as phytoplankton, zooplankton, and algae are used to reduce feeding costs in tilapia culture. These natural sources enhance the growth, survival, and productivity of fish, as they contain essential nutrients (Narimbi et al., 2018).

Studies have shown that tilapia can uptake up to 240 mg of nitrogen per kilogram of biofloc—equivalent to about 25% of the nitrogen found in fish protein diets (Avnimelech and Kochba, 2009). Biofloc-integrated diets have improved growth and digestibility by up to 75% in common carp (*Cyprinus carpio*) (Najdegerami et al., 2016). Among BFT's nutritional benefits is its ability to replace expensive ingredients such as fishmeal and soybean meal with protein-rich alternatives. Improvements in feed conversion ratio (FCR) and a reduced dietary protein requirement are considered particularly important.

Moreover, BFOs, in conjunction with commercial feeds, can fulfill dietary requirements and promote growth performance by supplementing amino acids, minerals, and unsaturated fatty acids in cultured species (Bakhshi et al., 2018; Izquierdo et al., 2006; Khanjani et al., 2022a). Several studies on biofloc systems with external carbon sources have reported the significant role of BFOs in providing essential nutrients (Deng et al., 2018; Khanjani and Sharifinia, 2022b). However, biofloc particles have variable compositions, containing 12–49% protein, 0.5–12.5% lipids, and 13–46% ash (Bakhshi et al., 2018; Khanjani et al., 2022a).

Furthermore, biofloc also contain bioactive compounds such as bromophenols, carotenoids, chlorophylls, and phytosterols (Zhao et al., 2016). Various studies have reported biofloc content ranging from 12.8–29.65 g/kg calcium (Ca), 11.25–22.6 g/kg phosphorus (P), 2.41–24.4 g/kg sodium (Na), 1.68–13.7 g/kg potassium (K), and 3.28–18.1 g/kg magnesium (Mg) (Kuhn et al., 2016).

The use of BFO biomass as a feed additive has been shown to improve FCR, growth, and survival rates in carps. BFOs help reduce feed costs by lowering the feed conversion rate. Additionally, BFOs are efficient at assimilating nitrogen from excreta and uneaten feed, which can then be consumed by prawns and shrimp. The aforementioned studies highlight the nutritional and water quality benefits of limited-exchange biofloc technology (BFT) systems. BFOs contribute to recycling toxic nitrogenous compounds and reduce the need for exogenous nitrogen input in aquafeeds.

4.6.6 ECONOMIC BENEFITS

Biofloc technology (BFT) systems offer better growth performance and improved water quality for aquatic species compared to conventional aquaculture systems. The aquaculture industry aims to reduce production costs while maximizing profits, and survival rate parameters significantly impact cost-effectiveness and profitability (Khanjani and Sharifinia, 2020).

The strength of BFT lies in its lower investment cost and higher protein utilization efficiency compared to conventional fish production systems. BFT systems require only sunlight, a carbon source, and aeration, which makes them an effective replacement for conventional feed-based systems, owing to their positive effects on fish survival and growth performance. The per-kilogram production cost of *Penaeus semisulcatus* (green tiger shrimp) and tilapia in BFT systems can be reduced by approximately 33% and 10%, respectively, depending on species, feed, and the quantity of consumed biofloc (Khanjani and Sharifinia, 2020; Megahed, 2010).

Additionally, BFT systems use zero water exchange, which reduces operational costs by about 30%, shortens the cultivation cycle, and improves growth and survival rates compared to traditional methods. Therefore, BFT systems represent a feasible and economically advantageous approach for aquaculture production.

4.7 CHALLENGES IN BIOFLOC RESEARCH SYSTEMS

The biofloc system faces several challenges that need to be addressed. These include selecting and correctly positioning appropriate aerators in tanks, integrating biofloc technology into existing systems such as raceways and polyculture systems, and identifying microorganisms that produce bioflocs with beneficial properties such as improved nutritional quality and biocontrol effects. Other challenges include developing effective monitoring methods for floc characteristics, optimizing the nutritional profile of bioflocs (e.g., fatty acid and amino acid composition, vitamin content), and determining the impact of various carbon source types on biofloc quality and formation.

4.8 CONCLUSION AND FUTURE PROSPECTS

Aquaculture must become more sustainable to meet the demands of a growing global population. Biofloc technology can improve water quality, increase productivity, and reduce environmental impact. By integrating biofloc systems with other food production methods, it is possible to generate more food and feed using fewer inputs while enhancing biosecurity. This technology is both environmentally and economically sustainable. However, further research and development are essential for establishing BFT as a cornerstone of future sustainable aquaculture.

KEYWORDS

- **aquaculture**
- **bioflocs**
- **carotenoids**
- **chlorophylls**
- **feed conversion rate**
- **global food security**
- **nitrogen recycling**
- **phytosterols**

REFERENCES

Abakari, G., Luo, G., Meng, H., Yang, Z., Owusu-Afriyie, G., Kombat, E. O., & Alhassan, E. H. (2020). The use of biochar in the production of tilapia (*Oreochromis niloticus*) in a biofloc technology system-BFT. *Aquaculture Engineering*, *91*, 102–123.

Abu Bakar, N. S., Nasir, N. M., Lananan, F., Abdul Hamid, Lam, S. S., & Jusoh, A. (2015). Optimization of C/N ratios for nutrient removal in aquaculture system culturing African catfish (*Clarias gariepinus*) utilizing bioflocks technology. *International Biodeterioration and Biodegradation*, *102*, 100–106.

Adineh, H., Naderi, M., Hamidi, M. K., & Harsij, M. (2019). Biofloc technology improves growth, innate immune responses, oxidative status, and resistance to acute stress in common carp (*Cyprinus carpio*) under high stocking density. *Fish and Shellfish Immunology*, *95*, 440–448.

Ahmad, H., Verma, A. K., Babitha Rani, A. M., Rathore, G., Saharan, N., & Gora, A. H. (2016). Growth, non-specific immunity and disease resistance of rohu (*Labeo rohita*) against *Aeromonas hydrophila* in biofloc systems using different carbon sources. *Aquaculture*, *457*, 61–67.

Avnimelech, Y. (1999). Carbon/nitrogen ratio as a control element in aquaculture systems. *Aquaculture*, *176*, 227–235.

Avnimelech, Y. (2006). Bio-filters: The need for a new comprehensive approach. *Aquaculture Engineering*, *34*, 172–178.

Avnimelech, Y. (2007). Feeding with microbial floc by tilapia in minimal discharge biofloc technology ponds. *Aquaculture*, *264*, 140–147.

Avnimelech, Y. (2012). *Biofloc Technology: A Practical Guide Book* (p. 272). The World Aquaculture Society, Baton Rouge.

Avnimelech, Y., & Kochba, M. (2009). Evaluation of nitrogen uptake and excretion by tilapia in biofloc tanks, using 15N tracing. *Aquaculture*, *287*, 163–168.

Azim, M. E., & Little, D. C. (2008). The biofloc technology (BFT) in indoor tanks: Water quality, biofloc composition, and growth and welfare of Nile tilapia (*Oreochromis niloticus*). *Aquaculture*, *283*, 29–35.

Bakhshi, F., Najdegerami, E. H., Manaffar, R., Tokmechi, A., Rahmani Farah, K., & Shalizar Jalali, A. (2018). Growth performance, hematology, antioxidant status, immune response and histology of common carp (*Cyprinus carpio* L.) fed biofloc grown on different carbon sources. *Aquaculture Research*, *49*, 393–403.

Banuelos-Vargas, I., de Oca, G. A. R. M., Martinez-Montano, E., Perez-Jimenez, A., Mendoza-Gamboa, O. A., Estrada-Godínez, J. A., & Hernandez, C. (2021). Antioxidant and immune response of juvenile red tilapia (*Oreochromis* sp.) cultured at different densities in seawater with biofloc plus probiotics. *Aquaculture*, *544*, 737112.

Carbone, D., & Faggio, C. (2016). Importance of prebiotics in aquaculture as immunostimulants: Effects on immune system of *Sparus aurata* and *Dicentrarchus labrax*. *Fish and Shellfish Immunology*, *54*, 172–178.

Cienfuegos-Martínez, K., Monroy-Dost, M. D. C., Hamdan-Partid, A., Hernández-Vergara, M. P., Aguirre-Garrido, J. F., & Bustos-Martínez, J. (2022). Effect of the probiotic *Lactococcus lactis* on the microbial composition in the water and the gut of freshwater prawn (*Macrobrachium rosenbergii*) cultivated in biofloc. *Aquaculture Research*, *53*, 3877–3889.

Crab, R., Chielens, B., Wille, M., Bossier, P., & Verstraete, W. (2010). The effect of different carbon sources on the nutritional value of bioflocs, a feed for *Macrobrachium rosenbergii* post larvae. *Aquaculture Research*, *41*, 559–567.

Crab, R., Defoirdt, T., Bossier, P., & Verstraete, W. (2012). Biofloc technology in aquaculture: Beneficial effects and future challenges. *Aquaculture*, *356*, 351–356.

Daniel, N., & Nageswari, P. (2017). Exogenous probiotics on biofloc-based aquaculture: A review. *Current Agriculture Research Journal*, *5*, 88–107.

Dauda, A. B., Romano, N., Chen, W. W., Natrah, I., & Kamarudin, M. S. (2018). Differences in feeding habits influence the growth performance and feeding efficiencies of African catfish (*Clarias gariepinus*) and lemon fin barb hybrid (*Hypsibarbus wetmorei* ♂ × *Barboides gonionotus* ♀) in a glycerol-based biofloc technology system versus a recirculating system. *Aquaculture Engineering*, *82*, 31–37.

Dauda, A. B., Romano, N., Ebrahimi, M., Karim, M., Natrah, I., Kamarudin, M. S., & Ekasari, J. (2017). Different carbon sources affect biofloc volume, water quality and the survival and physiology of African catfish *Clarias gariepinus* fingerlings reared in an intensive biofloc technology system. *Fisheries Science*, *83*, 1037–1048.

De Alvarenga, E. R., Alves, G. F. D. O., Fernandes, A. F. A., Costa, G. R., da Silva, M. A., Teixeira, E. D. A., & Turra, E. M. (2018). Moderate salinities enhance growth performance of Nile tilapia (*Oreochromis niloticus*) fingerlings in the biofloc system. *Aquaculture Research*, *49*, 2919–2926.

Deb, S., Noori, M. T., & Rao, P. S. (2020). Application of biofloc technology for Indian major carp culture (polyculture) along with water quality management. *Aquaculture Engineering*, *91*, 102–106.

Decamp, O., Cody, J., Conquest, L., Delanoy, G., & Tacon, A. G. (2003). Effect of salinity on natural community and production of (*Litopenaeus vannamei*) (Boon) within experimental zero-water exchange culture system. *Aquaculture Research*, *34*, 345–355.

Deng, M., Chen, J., Gou, J., Hou, J., Li, D., & He, X. (2018). The effect of different carbon sources on water quality, microbial community and structure of biofloc systems. *Aquaculture*, *482*, 103–110.

Deng, Y., Xu, X., Yin, X., Lu, H., Chen, G., Yu, J., & Ruan, Y. (2019). Effect of stock density on the microbial community in biofloc water and Pacific white shrimp (*Litopenaeus vannamei*) gut microbiota. *Applied Microbiology and Biotechnology*, *103*, 4241–4252.

Ebrahimi, A., Akrami, R., Najdegerami, E. H., Ghiasvand, Z., & Koohsari, H. (2020). Effects of different protein levels and carbon sources on water quality, antioxidant status and performance of common carp (*Cyprinus carpio*) juveniles raised in biofloc-based system. *Aquaculture*, *516*, 734639.

Ekasari, J., Rivandi, D. R., Firdausi, A. P., Surawidjaja, E. H., Zairin, M., Jr., Bossier, P., & De Schryver, P. (2015). Biofloc technology positively affects Nile tilapia (*Oreochromis niloticus*) larvae performance. *Aquaculture*, *441*, 72–77.

Elayaraja, S., Mabrok, M., Algammal, A., Sabitha, E., Rajeswari, M. V., Zágoršek, K., Ye, Z., Zhu, S., & Rodkhum, S. (2020). Potential influence of jaggery-based biofloc technology at different C:N ratios on water quality, growth performance, innate immunity, immune-related gene expression profiles, and disease resistance against *Aeromonas hydrophila* in Nile tilapia (*Oreochromis niloticus*). *Fish and Shellfish Immunology*, *107*, 118–128.

El-Sayed, A. (2017). Preeclampsia: A review of the pathogenesis and possible management strategies based on its pathophysiological derangements. *Taiwanese Journal of Obstetrics and Gynecology*, *56*, 593–598.

El-Sayed, A. F. M. (1999). Alternative dietary protein sources for farmed fish. *Aquaculture*, *179*, 149–168.

El-Sayed, A. F. M. (2021). Use of biofloc technology in shrimp aquaculture: A comprehensive review, with emphasis on the last decade. *Reviews in Aquaculture*, *13*, 676–705.

Emerenciano, M., Ballester, E. L., Cavalli, R. O., & Wasielesky, W. (2012). Biofloc technology application as a food source in a limited water exchange nursery system for pink shrimp *Farfantepenaeus brasiliensis* (Latreille, 1817). *Latin American Journal of Aquatic Research*, *43*, 447–457.

Emerenciano, M., Cuzon, G., Arévalo, M., Miquelajauregui, M. M., & Gaxiola, G. (2013). Effect of short-term fresh food supplementation on reproductive performance, biochemical composition, and fatty acid profile of *Litopenaeus vannamei* (Boone) reared under biofloc conditions. *Aquaculture International*, *21*, 987–1007.

FAO. (2017). *The Future of Food and Agriculture–Trends and Challenges*. Rome: Food and Agriculture Organization of the United Nations.

Fischer, H., Romano, N., Renukdas, N., Egnew, N., Sinha, A. K., & Ray, A. J. (2020). The potential of rearing juveniles of bluegill, *Lepomis macrochirus*, in a biofloc system. *Aquaculture Reports*, *17*, 100398.

Gaona, C., Poersch, L., Krummenauer, D., Foes, G., & Wasielesky, W. (2011). The effect of solids removal on water quality, growth and survival of *Litopenaeus vannamei* in a biofloc technology culture system. *International Journal of Recirculating Aquaculture*, *12*, 54–73.

Hakanson, L. (2005). The relationship between salinity, suspended particulate matter and water clarity in aquatic systems. *Ecological Research*, *21*, 75–90.

Hambrey, J. (2017). The 2030 Agenda and the Sustainable Development Goals: The challenge for aquaculture development and management. In *FAO Fisheries and Aquaculture Circular*. FAO: Rome, Italy.

Hargreaves, J. A. (2013). *Biofloc Production Systems for Aquaculture* (SRAC Publication No. 4503, pp. 1–11).

Haridas, H., Chadha, N. K., Sawant, P. B., Deo, A. D., Ande, M. P., Syamala, K., Sontakke, R., & Lingam, S. S. (2021). Growth performance, digestive enzyme activity, non-specific immune response and stress enzyme status in early stages of grey mullet reared in a biofloc system. *Aquaculture Research*, *52*, 4923–4933.

Harini, C., Rajagopalasamy, C., Kumar, J., & Santhakumar, R. (2016). Role of biofloc in the growth and survival of blue morph, *Pseudotropheus saulosi. Indian Journal of Science and Technology*, *9*, 1–7.

Hisano, H., Parisi, J., Cardoso, I. L., Ferri, G. H., & Ferreira, M. F. (2020). Dietary protein reduction for Nile tilapia fingerlings reared in biofloc technology. *Journal of the World Aquaculture Society*, *51*, 452–462.

Hwihy, H., Zeina, A., Abu Husien, M., & El-Damhougy, K. (2021). Impact of biofloc technology on growth performance and biochemical parameters of *Oreochromis niloticus*. *Egyptian Journal of Aquatic Biology and Fisheries*, *25*, 761–774.

Izquierdo, M., Forster, I., Divakaran, S., Conquest, I., Decamp, O., & Tacon, A. (2006). Effect of green and clear water and lipid source on survival, growth and biochemical composition of Pacific white shrimp (*Litopenaeus vannamei*). *Aquaculture Nutrition*, *12*, 192–202.

Jiang, L., & Pan, L. (2005). Effect of dissolved oxygen on immune parameters of the white shrimp (*Litopenaeus vannamei*). *Fish and Shellfish Immunology*, *18*, 185–188.

Khanjani, M. (2015). *The Effect of Different Feeding Levels in Biofloc System on Water Quality, Growth Performance and Carcass Composition of Pacific white shrimp (Litopenaeus vannamei)* (Ph.D. thesis, Hormozgan University, Iran), *165*, 23.

Khanjani, M. H., & Sharifinia, M. (2021). Production of Nile tilapia *Oreochromis niloticus* reared in a limited water exchange system: The effect of different light levels. *Aquaculture*, *542*, 736912.

Khanjani, M. H., Alizadeh, M., & Sharifinia, M. (2020a). Rearing of the Pacific white shrimp, *Litopenaeus vannamei*, in a biofloc system: The effects of different food sources and salinity levels. *Aquaculture Nutrition*, *26*, 328–337.

Khanjani, M. H., Sharifinia, M., & Hajirezaee, S. (2020b). Effects of different salinity levels on water quality, growth performance and body composition of Pacific white shrimp (*Litopenaeus vannamei*) cultured in a zero-water exchange heterotrophic system. *Annals of Animal Science*, *20*, 1471–1486.

Khanjani, M. H., Sharifinia, M., & Hajirezaee, S. (2022a). Recent progress towards the application of biofloc technology for tilapia farming. *Aquaculture*, *552*, 738021.

Kim, J. H., Sohn, S., Kim, S. K., & Hur, Y. B. (2020). Effects on hematological parameters, antioxidant and immune responses, AChE, and stress indicators of olive flounders, *Paralichthys olivaceus*, raised in biofloc and seawater challenged by *Edwardsiella tarda*. *Fish and Shellfish Immunology*, *97*, 194–203.

Kishawy, A. T. Y., Sewid, A. H., Nada, H. S., Kamel, M. A., El-Mandrawy, S. A. M., Abdelhakim, T. M. N., El-Murr, A. E., El Nahhas, N., Hozzein, W. N., & Ibrahim, D. (2020). Mannanoligosaccharides as a carbon source in biofloc boost dietary plant protein and water quality, growth, immunity and *Aeromonas hydrophila* resistance in Nile tilapia (*Oreochromis niloticus*). *Animals*, *10*, 1724.

Klanian, M. G., Díaz, M. D., Solís, M. J. S., Aranda, J., & Moral, P. M. (2020). Effect of the content of microbial proteins and the poly β-hydroxybutyric acid in biofloc on the performance and health of Nile tilapia (*Oreochromis niloticus*) fingerlings fed on a protein-restricted diet. *Aquaculture*, *519*, 734872.

Krishnaswamy, U., Muthusamy, M., & Perumalsamy, L. (2009). Studies on the efficiency of the removal of phosphate using bacterial consortium for the bio-treatment of phosphate wastewater. *European Journal of Applied Sciences*, *1*, 6–15.

Krummenauer, D., Reis, W. G. D., Abreu, P. C., Bezerra, A., Jr, W. W., & Advent, B. (2021). Effect of aeration on the development of the microbial community in shrimp biofloc technology systems. *Global Seafood Alliance*.

Kuhn, D. D., Lawrence, A. L., Crockett, J., & Taylor, D. P. (2016). Evaluation of bioflocs derived from confectionary food effluent water as a replacement feed ingredient for fishmeal or soy meal for shrimp. *Aquaculture*, *454*, 66–71.

Laice, L. M., Filho, R. A. C. C., Ventura, A. S., Farias, K. N. N., Silva, A. L. D. N. N., Fernandes, S. E., Silva, A. C. F., Barbosa, P. T. L., Souza, A. I. D., Emerenciano, M. G. C., & Povh, J. A. (2021). Use of symbiotics in biofloc (BFT)-based Nile tilapia culture: Production performance, intestinal morphometry and hematological parameters. *Aquaculture*, *530*, 735715.

Liu, G., Ye, Z., Liu, D., Zhao, J., Sivaramasamy, E., Deng, Y., & Zhu, S. (2018). Influence of stocking density on growth, digestive enzyme activities, immune responses, antioxidant of *Oreochromis niloticus* fingerlings in biofloc systems. *Fish and Shellfish Immunology*, *81*, 416–422.

Luo, G., Gao, Q., Wang, C., Liu, W., Sun, D., Li, L., & Tan, H. (2014). Growth, digestive activity, welfare, and partial cost-effectiveness of genetically improved farmed tilapia (*Oreochromis niloticus*) cultured in a recirculating aquaculture system and an indoor biofloc system. *Aquaculture*, *422*, 1–7.

Mahanand, S. S., Moulick, S., & Rao, P. S. (2013). Optimum formulation of feed for rohu, *Labeo rohita* (Hamilton), with biofloc as a component. *Aquaculture International*, *21*, 347–360.

Maica, P. F., de Borba, M. R., & Wasielesky, W. Jr. (2012). Effect of low salinity on microbial floc composition and performance of *Litopenaeus vannamei* (Boone) juveniles reared in a zero-water-exchange super-intensive system. *Aquaculture Research*, *43*, 361–370.

Megahed, M. E. (2010). The effect of microbial biofloc on water quality, survival and growth of the green tiger shrimp (*Penaeus semisulcatus*) fed with different crude protein levels. *Journal of the Arab Aquaculture Society*, *5*, 119–142.

Menaga, M., Felix, S., Charulatha, M., Gopalakannan, A., Mohanasundari, C., & Boda, S. (2020). *In vivo* efficiency of *Bacillus* sp. isolated from biofloc system on growth, hematological, immunological and antioxidant status of genetically improved farmed tilapia (GIFT). *Indian Journal of Experimental Biology*, *58*, 714–721.

Monroy-Dosta, M. D. C., De Lara-Andrade, R., Castro-Mejia, J., Castro Mejia, G., & Coelho-Emerenciano, M. G. (2013). Microbiology community composition and abundance associated to biofloc in tilapia aquaculture. *Revita de Biology Marina y Oceanography*, *48*, 511–520.

Najdegerami, E. H., Bakhshi, F., & Lakani, F. B. (2016). Effects of biofloc on growth performance, digestive enzyme activities, and liver histology of common carp (*Cyprinus carpio* L.) fingerlings in zero-water exchange system. *Fish Physiology and Biochemistry*, *42*, 457–465.

Narimbi, J., Mazumder, D., & Sammut, J. (2018). Stable isotope analysis to quantify contributions of supplementary feed in Nile tilapia *Oreochromis niloticus* (GIFT strain) aquaculture. *Aquaculture Research*, *49*, 1866–1874.

Natrah, F. M. I., Bossier, P., Sorgeloos, P., Yusoff, F. M., & Defoirdt, T. (2013). Significance of microalgal-bacterial interactions for aquaculture. *Reviews in Aquaculture*, *6*, 48–61.

Neal, R. S., Coyle, S. D., & Tidwell, J. H. (2010). Evaluation of stocking density and light level on the growth and survival of the Pacific white shrimp (*Litopenaeus vannamei*), reared in zero-exchange systems. *Journal of the World Aquaculture Society*, *41*, 533–544.

Nisar, U., Peng, D., Mu, Y., & Sun, Y. (2022). A solution for sustainable utilization of aquaculture waste: A comprehensive review of biofloc technology and aquamimicry. *Frontiers in Nutrition*, *8*, 791738.

Panigrahi, A., Saranya, C., Sundaram, M., Vinoth Kannan, S. R., Das, R. R., Satish Kumar, R., Rajesh, P., & Otta, S. K. (2018). Carbon: Nitrogen (C:N) ratio level variation influences microbial community of the system and growth as well as immunity of shrimp (*Litopenaeus vannamei*) in biofloc based culture system. *Fish and Shellfish Immunology*, *81*, 329–337.

Rahmatullah, & Beveridge, M. C. M. (1993). Ingestion of bacteria in suspension by Indian major carps (*Catla catla*, *Labeo rohita*) and Chinese carps (*Hypophthalmichthys molitrix*, *Aristichthys nobilis*). *Hydrobiologia*, *264*, 79–84.

Ray, A. J., Lewis, B. L., Browdy, C. L., & Leffler, J. W. (2010). Suspended solids removal to improve shrimp (*Litopenaeus vannamei*) production and an evaluation of a plant-based feed in minimal-exchange, superintensive culture systems. *Aquaculture*, *299*, 89–98.

Samocha, T. M., Patnaik, S., Speed, M., Ali, A. M., Burger, J. M., & Almeida, R. V. (2007). Use of molasses as carbon source in limited discharge nursery and grow-out systems for *Litopenaeus vannamei*. *Aquaculture Engineering*, *36*(2), 184–191.

Sandoval-Vargas, L. Y., Jiménez-Amaya, M. N., Rodríguez-Pulido, J., Guaje-Ramírez, D. N., Ramírez-Merlano, J. A., & Medina-Robles, V. M. (2020). Applying biofloc technology in the culture of juvenile *Piaractus brachypomus* (Cuvier, 1818): Effects on zootechnical performance and water quality. *Aquaculture Research*, *51*, 3865–3878.

Schneider, O., Sereti, V., Eding, E. H., & Verreth, J. A. J. (2005). Analysis of nutrient flows in integrated intensive aquaculture systems. *Aquaculture Engineering*, *32*, 379–401.

Tabarrok, M., Seyfabadi, J., Salehi Jouzani, G., & Younesi, H. (2020). Comparison between recirculating aquaculture and biofloc systems for rearing juvenile common carp

(*Cyprinus carpio*): Growth performance, haemato-immunological indices, water quality and microbial communities. *Aquaculture Research*, *51*, 4881–4892.

Tao, C. T., Tran, Khoa, T. N. D., Truyen, P. M., Hoa, N. V., An, H. C. M., & Hai, T. N. (2021). Effects of light intensity on growth and survival rate of freshwater prawn (*Macrobrachium rosenbergii*) at larvae and postlarvae stages in biofloc system. *AACL Bioflux*, *14*, 3556–3565.

Van Doan, H., Hung, T. Q., Lumsangkul, C., Jaturasitha, S., El-Haroun, E., & Paolucci, M. (2020). Dietary inclusion of chestnut (*Castanea sativa*) polyphenols to Nile tilapia reared in biofloc technology: Impacts on growth, immunity, and disease resistance against *Streptococcus agalactiae*. *Fish and Shellfish Immunology*, *105*, 319–326.

Wang, G., Yu, E., Xie, J., Yu, D., Li, Z., Luo, W., Qiu, L., & Zheng, Z. (2015). Effect of C/N ratio on water quality in zero-water exchange tanks and the biofloc supplementation in feed on the growth performance of crucian carp (*Carassius auratus*). *Aquaculture*, *443*, 98–104.

Wilen, B. M., Neilsen, J. L., Keiding, K., & Nielsen, P. H. (2000). Influence of microbial activity on the stability of activated sludge flocs. *Colloids and Surfaces B: Bio Interfaces*, *18*, 145–156.

Zafar, M. A., Talha, M. A., & Rana, M. M. (2021). Effect of biofloc technology on growth performance, digestive enzyme activity, proximate composition, and hematological parameters of Asian stinging catfish (*Heteropneustes fossilis*). *Journal of Applied Aquaculture*, *34*, 1–19.

Zemor, J., Wasielesky, W., Foes, G., & Poersch, L. (2019). The use of clarifiers to remove and control the total suspended solids in large-scale ponds for production of *Litopenaeus vannamei* in a biofloc system. *Aquaculture Engineering*, *85*, 74–79.

Zhao, D., Pan, L., Huang, F., Wang, C., & Xu, W. (2016). Effects of different carbon sources on bioactive compound production of biofloc, immune response, antioxidant level, and growth performance of *Litopenaeus vannamei* in zero-water exchange culture tanks. *Journal of the World Aquaculture Society*, *47*, 566–576.

Zhao, P., Huang, J., Wang, X. H., Song, X. L., Yang, C., & Zhang, X. G. (2012). The application of bioflocs technology in high-intensive, zero exchange farming systems of *Marsupenaeus japonicus*. *Aquaculture*, *354–355*, 97–106.

CHAPTER 5

Aquaponics: A Novel Agri-Aqua Technology

M. S. VISHNUPRIYA,[1,*] SUTANU KARMAKAR,[2] and A. K. SINGH[1]

[1]*Department of Fisheries Engineering and Technology, College of Fisheries Science, Birsa Agricultural University, Ranchi, Jharkhand, India*

[2]*Department of Aquatic Environment Management, Faculty of Fishery Science, West Bengal University of Animal and Fishery Sciences, Kolkata, West Bengal, India*

*Corresponding author

ABSTRACT

Aquaponics is an integrated fish and plant production technology, essentially comprising two sub-systems: 'aquaculture' and 'hydroponics.' Aquaculture involves raising aquatic animals in tanks, while hydroponics involves cultivating plants in water without soil. Aquaponic systems come in various sizes, from small indoor setups to large commercial operations, and can be operated in both freshwater and brackish water. There is a symbiotic relationship between plants and fish, where fish provide nutrients for plant growth, and plants, in turn, treat and purify the wastewater by utilizing those nutrients. The underlying principle of aquaponics is to efficiently use water to produce two crops instead of one, sharing nutrient resources between fish and plants. In low- and middle-income nations, aquaponics can offer a sustainable food source depending on the conditions, particularly in areas with a temperate environment. A unique advantage of an aquaponic system is its ability to conserve water more effectively

Sustainable Aquaculture: Recent Advances and Techniques. Dhanaji W. Patil, Sandip S. Markad (Eds.)

DOI: 10.1201/9781779641113-5

than traditional irrigation systems. The development of new, sustainable aquaculture techniques is essential, and aquaponic systems play a vital role in achieving environmental, social, and financial sustainability.

5.1 INTRODUCTION

Aquaponics is an integrated system that combines both fish and plant production technologies. It has two main subsystems: 'aquaculture' and 'hydroponics.' The word "aqua" in aquaponics refers to the aquaculture component of the system, which involves the rearing and cultivation of aquatic animals, while "ponics" is derived from hydroponics—the process of growing plants without soil by cultivating them in water (Eck et al., 2019).

Aquaponics is a type of recirculating aquaculture system that incorporates soilless plant production, in which water is continuously reused. The high level of fish production generates nutrient-rich water, which is utilized by the plants. In turn, the plants purify the wastewater by removing ammonia, nitrites, nitrates, phosphates, potassium, and organic carbon from fish waste. These substances are essential nutrients for plant growth, and once absorbed, the water can be returned to the fish tanks.

Since aquaponics uses less water than traditional aquaculture systems, it mimics natural ecosystems and is promoted as a sustainable agricultural method (Masabni and Niu, 2022; Mullins et al., 2016; Patil and Markad, 2022; Preethi and Devarshi, 2023; Soumyadip and Sutanu, 2018).

Aquaponic systems vary in size and application: they may be used for personal or hobby purposes, community and economic development, science education and teaching, or to enhance food production in urban areas where space and environmental conditions restrict conventional agriculture. In certain situations, particularly where land and water are scarce, aquaponics may prove more productive and economically viable. While the most visible outcomes of aquaponic systems are the production of fish and vegetables, it is essential to understand that aquaponics is fundamentally the management of an entire ecosystem comprising three main groups of organisms: bacteria, plants, and fish (Gosh and Chowdhury, 2019).

5.2 PRINCIPLE OF AQUAPONICS

The key principles behind aquaponics are as follows:

- Waste products from one biological system can be used as nutrients for another biological system;
- Integrating fish and plant culture can result in the production of multiple products at once;
- Water can be recycled through biological filtration and recirculation processes; and
- A local food production system can improve the local economy and provide access to healthy foods (Diver, 2006).

The most important biological components in aquaponics are discussed in subsections (Figure 5.1).

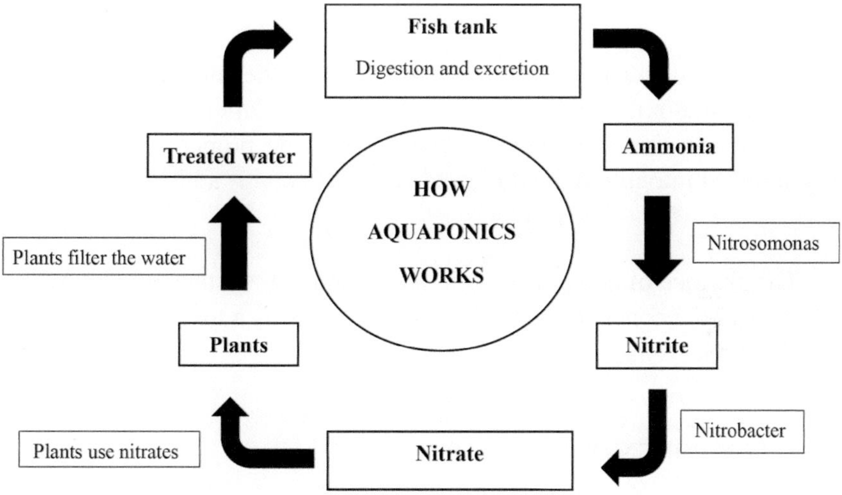

FIGURE 5.1 Aquaponics cycle.

5.2.1 NITROGEN CYCLE

Nitrogen is the most important and essential inorganic nutrient for plants. As it cannot be consumed directly, it must be converted into a usable form through the nitrogen fixation process. The health of the aquaponics system strongly depends on the efficiency of the bacteria involved. In the initial stage, efficiency is limited until the bacterial colony develops.

The nitrogen cycle generally takes about 30 days to complete, but since every pond and aquaponics system is unique, there is no fixed duration. The cycle's timing depends on factors such as fish quantity, tank size, and the number of plants.

Bacteria involved in nitrogen fixation chemically alter nitrogen (N_2) by adding hydrogen and oxygen compounds, resulting in new compounds such as ammonia (NH_3) and nitrate (NO_3), which can be directly absorbed by plants. Fish excretion primarily contains ammonia, which is then converted by fungi and bacteria (Schmautz et al., 2022; Wongkiew et al., 2017).

Nitrifying bacteria play a vital role in the aquaponics system by breaking down harmful ammonia into more useful compounds. These bacteria first convert ammonia into nitrite, and then into nitrate.

5.2.2 BIO-FILTER

The two main nitrifying bacteria employed in the nitrification process are ammonia-oxidation bacteria and nitrite-oxidation bacteria, which are needed in the aquaponics system to function properly. Ammonia is metabolized into nitrate by the bacteria *Nitrosomonas* and *Nitrobacter*.

$$NH_3 \rightarrow NO_3^- \rightarrow NO_2^-$$

The absence of these bacteria in the aquaponics system may lead to an increase in the concentration of ammonia, which is a toxic substance that affects and kills fish (Kasozi et al., 2016).

5.3 COMPONENTS OF AQUAPONICS

5.3.1 PRIMARY COMPONENTS

Three main components in the aquaponics system (Lennard and Goddek, 2019):

- Fish;
- Planta;
- Bacteria.

5.3.2 SECONDARY COMPONENTS

- Fish tank;
- Plant growing unit;
- Water pumps;
- Air pumps;
- Water heater;
- Filtration – mechanical and biofilters;
- Aeration.

5.4 DIFFERENT TYPES OF AQUAPONICS SYSTEM

There are primarily three different designs for aquaponics systems:

- Nutrient film technique (NFT);
- Media bed; and
- Deep water culture (DWC).

5.4.1 NUTRIENT FILM TECHNIQUE (NFT)

The technology known as the Nutrient Film Technique (NFT) originated in the hydroponics sector. Plants are grown in shallow horizontal pipes using this technique, and the water is circulated through the pipe, forming a thin layer (1–2 cm) of water that interacts with plant roots, which use those nutrients to thrive (Lennard and Goddek, 2019) (Figure 5.2).

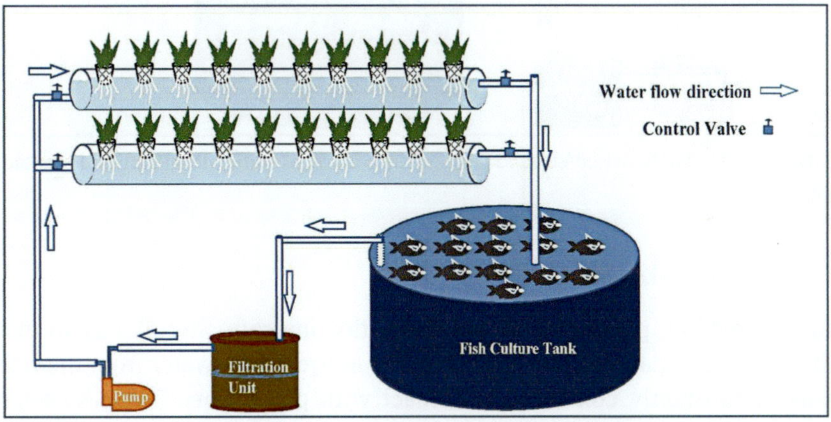

FIGURE 5.2 Nutrient film technique Source: Modified from Patil and Markad, 2022.

5.4.2 MEDIA BED

Media-based systems, often known as flood-and-drain systems, are relatively simple to build. Water from the fish tank is periodically poured into a plant-containing substrate. The water then flows back into the fish tank or sump, allowing bacteria to nitrate and draw oxygen into the substrate for plant roots. The media bed acts as a biological and solids filter while providing support for the plant during its growth. These systems are popular among amateurs and in underdeveloped areas because of their ease of building and operation. However, because media beds are less productive than other varieties covered below, it is rare to find commercial production employing simply media beds (Soumyadip and Sutanu, 2018).

5.4.2.1 TYPES OF SUBSTRATES

Peat, coconut fiber, organic substrates, perlite, expanded clay, stone wool, expanded polystyrene, polyurethane foam (Figure 5.3).

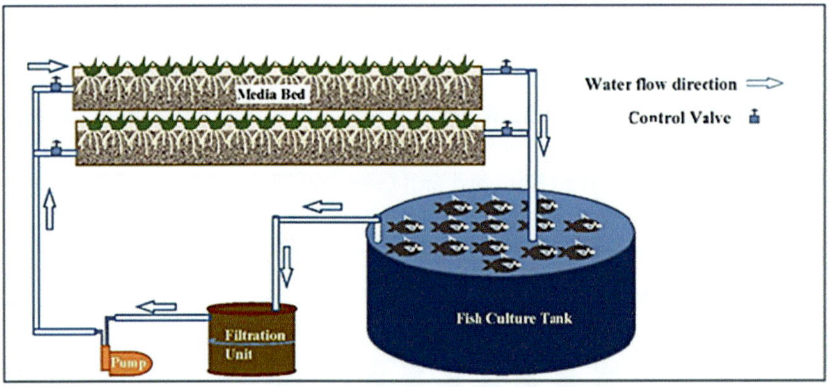

FIGURE 5.3 Media-based/ebb and flow system Source: Modified from Patil and Markad, 2022.

5.4.3 DEEP WATER CULTURE (DWC)

In this method, plants are attached to a floating raft (Styrofoam) so that their roots can dangle into the water. The nutrient-rich water from the fish tank is constantly coming into contact with plant roots. These systems must have effective solids filtration in order to keep sediments out of the plant bed and prevent blockage of the plant roots. To maintain adequate oxygen levels in the plant troughs for both the roots and beneficial bacteria, aeration is also necessary. The rafts' undersides and the troughs' linings not only provide a large water-holding capacity that helps stabilize water quality metrics, but also offer sufficient space for nitrifying bacteria to colonize. Because the design keeps roots submerged in water even when water is lost, it also acts as a buffer against power outages (Preethi and Devarshi, 2023) (Figure 5.4).

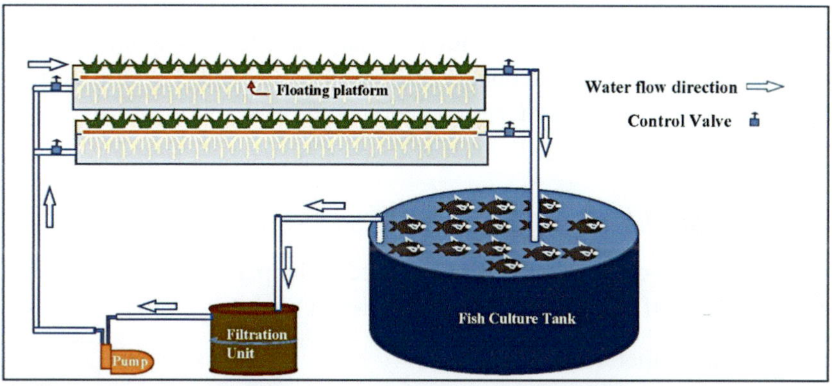

FIGURE 5.4 Deep water culture/Floating raft System. Source: Modified from Patil and Markad, 2022.

5.5 SELECTION OF FISH AND PLANTS FOR THE SYSTEM

5.5.1 FISH IN AQUAPONICS

In aquaponic systems, a number of fish species have been successfully grown. Nile tilapia (*Oreochromis niloticus*), common carp (*Cyprinus carpio*), channel catfish (*Ictalurus punctatus*), largemouth bass (*Micropterus salmoides*), rainbow trout (*Oncorhynchus mykiss*), barramundi (*Lates calcarifer*), and some of the various ornamental fish species, including goldfish and tetra, are more suitable and commonly used in aquaponics systems (Hager et al., 2021; Patil et al., 2023; Rakocy, 2012).

5.5.2 PLANTS IN AQUAPONICS

Typically, over 150 types of different crops, flowers, and herbs have been successfully cultivated in aquaponics systems. Leafy green crops with a low value per unit and large production are grown by aquaponic farmers. Herbs such as lettuce, Swiss chard, kale, basil, cabbage, peppers, mint, spinach, coriander, and others can be harvested in three to five weeks after transplanting (six to eight weeks from seed), providing a consistent source of revenue. Fruiting plants, such as tomatoes, cucumbers, and peppers, require longer growing seasons and lower yields to harvest, but their individual worth is higher. Harvesting takes place between ten and sixteen weeks.

5.5.3 BACTERIA IN AQUAPONICS

Bacteria in aquaponics act as an intermediary between fish waste and plant fertilizer. Where fish waste contains ammonia, it is converted into nitrates (plant fertilizer) by the nitrifying bacteria, and the water is returned to the fish tank.

5.6 MAINTENANCE OF AQUAPONICS SYSTEM

The main production parameters that need to be perfectly set to meet the optimal needs of plants and fish are (Table 5.1):

- Air temperature;
- Water temperature;
- The concentration of macro- and micronutrients;
- Dissolved oxygen in air and water—which depends on the filtration method used;
- CO_2 concentrations in air and in the water;
- pH;
- Light.

TABLE 5.1 Ideal Parameters for Aquaponics

Temperature (°C)	pH	DO (mg/liter)	Ammonia (mg/liter)	Nitrite (mg/liter)	Nitrate (mg/liter)
18–30	6–7	>5	<1	<1	5–150

5.7 ADVANTAGES AND DISADVANTAGES

➢ **Advantages:**

- It has high water efficiency: uses 90% less water than the traditional farming system.

- Low maintenance: it is easy and inexpensive to construct.
- Plant growth is significantly faster, with high production yield and quality.
- No additional nutrients are required.
- Helps reduce external contamination and disease.
- Waste reduction: waste from the fish is taken up by plants for growth.
- Nutrient efficiency is high in aquaponics due to circulation of the same water in the system.
- Two culture products are obtained from one system (fish and plants).
- Flexibility – it requires less space and water, and is adaptable to all types of locations.

➢ **Disadvantages:**

- Daily maintenance is required.
- Initial start-up expense is high compared to hydroponics and soil-based vegetable cultivation.
- The system can fail in different ways.
- Sometimes, production (fish and plants) is not considered sustainable.
- Technical knowledge is needed from two different fields for successful farming.

5.8 CONCLUSION

Aquaponics is a novel technology for both agriculture and aquaculture. These technologies use less water to produce nutritious food, and they also help reduce wastewater in aquaculture by recirculating it. It is a new, rapidly emerging, eco-friendly agricultural technology that can integrate RAS with hydroponics. It may have the potential to increase not only sustainability but also the productivity of fish and plants and ensure food and nutrition security.

KEYWORDS

- **aquaculture**
- **aquaponics**
- **dual culture products**
- **eco-friendly technology**
- **food and nutrition security**
- **hydroponics**
- **nutrient efficiency**

REFERENCES

Diver, S. (2006). Aquaponics—integration of hydroponics with aquaculture. Amy Smith, Production. A publication of ATTRA—National Sustainable Agriculture Information Service. Retrieved from http://www.attra.ncat.org (accessed on 10th July 2025).

Eck, M., Körner, O., & Jijakli, M. H. (2019). Nutrient cycling in aquaponics systems. In *Aquaponics Food Production Systems: Combined Aquaculture and Hydroponic Production Technologies for the Future* (pp. 231–246).

Gosh, K., & Chowdhury, S. (2019). Review of aquaponics system: Searching for a technically feasible and economically profitable aquaponics system. *Journal of Agricultural, Environmental and Consumer Sciences*, *19*, 5–13.

Greenfish Aquaponics. (2015). The nitrogen cycle–Understanding the nitrogen cycle within aquaponics. http://aquaponics.ie/wordpress/index.php/what-is-aquaponics/the-nitrogen-cycle/ (accessed on 10th July 2025).

Hager, J., Bright, L., Tidwell, J., & Dusci, J. (2021). *A Practical Handbook for Growers—Aquaponics Production Manual*. Kentucky State University, Kentucky, USA.

Kasozi, N., Abraham, B., Kaiser, H., & Wilhelmi, B. (2021). The complex microbiome in aquaponics: Significance of the bacterial ecosystem. *Annals of Microbiology*, *71*(1), 1–13.

Lennard, W., & Goddek, S. (2019). Aquaponics: The basics. In *Aquaponics Food Production Systems* (pp. 113).

Masabni, J., & Niu, G. (2022). Aquaponics. In *Plant Factory Basics, Applications and Advances* (pp. 167–180). Academic Press.

Maurya, P., & Ranjan, D. (2023). Aquaponics. In *Traditional and Recent Aquaculture Practices* (pp. 91–106).

Mullins, C., Nerrie, B., & Sink, T. D. (2016). *Principles of Small-Scale Aquaponics*. Oklahoma Cooperative Extension Service.

Patil, D. W. and Markad, S. S. (2022). Aquaponics. In Gore S. K. (Eds) *Environmental Pollution and Solution* Dnyanopasak Publication, Jintur, Maharashtra, India pp.102–113.

Patil, D. W., Gaikwad, J. M., & Markad, S. S. (2023). Growth performance and survival of tilapia, *Oreochromis mossambicus* reared in aquatic system. *BIOINFOLET*, *20*(2B), 351–354.

Patil, D., & Markad, S. (2022). Aquaponics. In S. Gore (Ed.), *Environmental Pollution and Solution* (pp. 102–114). ISBN: 978-81-957387-0-0. https://doi.org/10.5281/zenodo.6673622

Purkait, S., & Karmakar, S. (2018). Aquaponics: A promising technology for future food production. In *Aqua Star* (pp. 45–49).

Rakocy, J. E. (2012). Aquaponics—Integrating fish and plant culture. In *Aquaculture Production Systems* (pp. 344–386).

Schmautz, Z., Walser, J. C., Espinal, C. A., Gartmann, F., Scott, B., Pothier, J. F., & Smits, T. H. (2022). Microbial diversity across compartments in an aquaponic system and its connection to the nitrogen cycle. *Science of the Total Environment*, *852*, 158426.

Wongkiew, S., Hu, Z., Chandran, K., Lee, J. W., & Khanal, S. K. (2017). Nitrogen transformations in aquaponic systems: A review. *Aquacultural Engineering*, *76*, 9–19.

CHAPTER 6

Advancing *In Vivo* Biomedical Research: Design and Implementation of an Affordable Zebrafish Housing Unit

B. MUKILARASI,[1] S. B. LAKSHMI SWARUBHAA,[1] A. NATARAJAN,[1] S. SWATHI,[1] S. RAMAKRISHNAN,[1] S. SARAVANAN,[2] and S. VIMALRAJ[3,*]

[1]*Department of Applied Mechanics and Biomedical Engineering, Indian Institute of Technology Madras, Chennai, Tamil Nadu, India*

[2]*Department of Prosthodontics, Saveetha Dental College and Hospital, Saveetha Institute of Medical and Technical Sciences (SIMATS), Saveetha University, Chennai, Tamil Nadu, India*

[3]*Department of Biotechnology, Faculty of Biomedical Sciences and Technology, Sri Ramachandra Institute of Higher Education and Research (SRIHER) (Deemed to be University), Porur, Chennai, Tamil Nadu, India*

*Corresponding author

ABSTRACT

In vivo studies are essential for advancing findings from *in vitro* research. While *in vitro* studies offer affordability and simplicity, extrapolating data accurately presents challenges. Zebrafish (*Danio rerio*) emerges as a promising model due to its genomic similarity with humans and practical advantages. This chapter presents a cost-effective zebrafish housing and breeding unit with a recycling system, suitable for small to medium research institutions. The setup involves a reservoir tank, filtration system, inlet and exhaust pipelines, and fish tanks interconnected for water circulation. The filtration system employs mechanical, chemical, and biological

Sustainable Aquaculture: Recent Advances and Techniques. Dhanaji W. Patil, Sandip S. Markad (Eds.)

DOI: 10.1201/9781779641113-6

filtration. Compared to commercial facilities, this unit offers significant cost and time savings, simplified maintenance, and efficient space utilization. It ensures optimal environmental conditions for zebrafish growth and facilitates *in vivo* studies with reduced complexity and expense. Such affordable and straightforward zebrafish facilities cater to the increasing demand for *in vivo* research opportunities in life sciences and technology.

6.1 INTRODUCTION

As it pertains to the field of scientific research, the move from *in vitro* to *in vivo* studies represents a pivotal crossroads in the progression of knowledge and the implementation of discoveries into practical applications. Even though *in vitro* research has several distinct advantages, including cost-effectiveness, time efficiency, and ease of use, it also presents difficulties when it comes to effectively extrapolating findings to situations that occur in the real world (Lee and Seki, 2023). When it comes to the interpretation of *in vitro* results, the employment of mathematical modeling and systems biology methodologies is frequently required. This is especially true as the complexity of the studies increases. The need to perform clinical trials to validate findings and assure repeatability is highlighted by the fact that there are differences in the behavior of organs and tissues *in vitro* compared to their behavior *in vivo* in pharmacological research (Fosse et al., 2023). Within the framework of this discussion, animal models play a crucial part, since they provide valuable insights into the toxicity and effectiveness of medications in a physiological setting that is more complex. Using animal models that are comparable to those used in testing gives a more reliable platform for assessing the efficacy and toxicity of various medications. When conducting *in vivo* research, it is essential to choose a suitable model, as well as to provide the animals with the right housing and to ensure that they are well maintained (Mukherjee et al., 2022). In the fields of pharmacology, pathology, and the treatment of disorders, the zebrafish has been receiving an increasing amount of attention as an animal model (Shin et al., 2002). Among these models, the zebrafish (*Danio rerio*) has emerged as a particularly effective tool in the field of pharmacology and disease research due to its multiple advantages (Verma et al., 2024). These advantages include the ability to produce a greater number of embryos in a single clutch, the ease with which embryos and larvae can be studied under a microscope, the fact that they require fewer complex experimental procedures, and, most importantly, the fact that they have a high degree of genomic similarity to humans (Choi et al., 2021; Howe et al., 2013). However, the establishment and maintenance of zebrafish facilities can create obstacles in terms of both logistics and finances. This is especially true for academic or research organizations that are small to medium in size and have limited resources (Ericsson et al., 2021; Kim et al., 2009; Lawrence et al., 2012; Lipscomb et al., 2023; Tarasco et al., 2019; Teoh et al., 2020). A detailed guide on constructing a zebrafish housing and breeding unit that is both affordable and standardized, together with a recycling system, is provided in this chapter. This page covers the issues that have been presented. Assembling a reservoir tank and filtration system, arranging inlet and exhaust pipelines, and setting up individual fish tanks within a multi-tiered rack system are all components of the setup that are covered in this book chapter. As a result of the painstaking design of each component, which is intended to promote the continuous circulation and filtration of water, an environment that is ideal for zebrafish breeding and study has been created. This zebrafish unit provides a cost-effective alternative to commercial facilities by utilizing materials that are easily available and construction procedures that are straightforward. This is accomplished without sacrificing the unit's functionality or operational efficiency. The integration of the filtration system with the reservoir tank is particularly noteworthy, since it simplifies the procedures for maintenance and decreases the amount of additional equipment that is required. In addition to significant cost reductions and simplified construction, this strategy offers several other benefits. This tiny unit, in contrast to larger and more complicated facilities, takes up very little space and can be constructed in a single day. As a result, it is an excellent choice for establishments that have limited infrastructure or financial constraints. Additionally, its modular design simplifies the process of expanding or customizing it to meet the requirements of a wide range of research projects. To summarize, the establishment of zebrafish facilities that are both economical and easily accessible has a great deal of potential for the advancement of *in vivo* research and the improvement of our understanding of the biological processes and disease mechanisms involved. This effort seeks to promote creativity and collaboration in the field of life sciences by democratizing access to critical research tools, such as animal models. The ultimate goal of this endeavor is to drive forward progress toward improved healthcare and scientific discoveries.

6.2 MATERIALS

A list of materials required for the setup is given in Table 6.1.

TABLE 6.1 List of Materials Required for the Setup

Materials	Quantity Required (Approximate)
½″ 90? Elbow (soc × soc)	20
½″ 90? Elbow (soc × FPT)	5
½″ Tee (soc × soc × soc)	20
Nylon mesh	1 m
¾″ diameter PVC pipe, 6″ long	18
¾″ diameter PVC pipe, 12″ long	2
¾″ diameter PVC pipe, 18″ long	2
½″ diameter PVC pipe, 8″ long	1
Bio balls	300
Soft sponge	10
Hard sponge	10
Ceramic rings	300
Carbon filters	200
Motor pump	1
Aeration device	1
5 mm acrylic quarantine tanks (15″ × 8″ × 8″)	18
Glass reservoir tank (36″ × 15″ × 18″)	1

6.3 ZEBRAFISH FACILITY SETUP

1. Obtain the materials listed in Table 6.1.
2. Set up the 4-shelf rack. The number of shelves can be customized as per the requirement.
3. Reservoir tank and filtration system setup: Using Figures 6.1(A) and 6.2 as references, assemble the 36″ × 15″ × 18″ reservoir tank.
   - Attach a compartment wall, leaving a 2″ gap from the floor of the reservoir tank, such that the compartment holds one-fifth of the reservoir tank.
   - Attach another compartment wall, leaving a 5″ gap from the ceiling of the reservoir tank.
   - Fill these compartments with layers of bio sponge, bio balls, filter mat, and ceramic rings, which will act as the filtration system.

- Install the pump and aerator into the reservoir tank.

4. Inlet line setup:
    - Extend a ½″ PVC pipeline from the main water source to the reservoir tank.
    - Measure and cut ¾″ PVC pipe pieces extending from the pump to the upper shelf of the rack as the primary inlet pipe, as shown in Figures 6.1(B) and 6.2.
    - Join these ¾″ PVC pipes with Tee connectors directed toward each shelf.
    - Add an on–off valve to the bottom Tee connector to serve as a drain for maintenance flushing of the system.
    - Measure and cut 0.3″ reverse osmosis pipe extending from the Tee connectors of the primary inlet line to the fish tanks on each shelf as the secondary inlet line.
    - Join these 0.3″ reverse osmosis pipes with Tee connectors directed toward the fish tanks.
    - Attach on–off valves to the Tee connectors in the secondary inlet line.

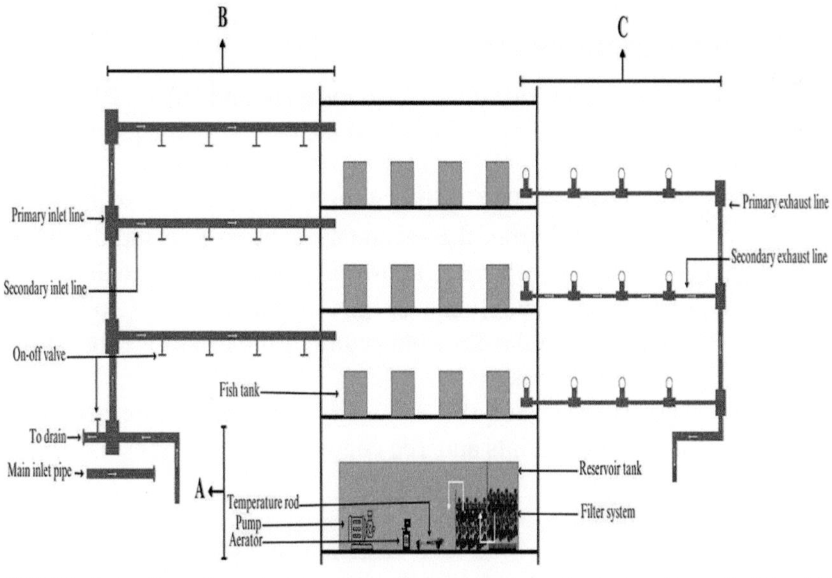

FIGURE 6.1 One-dimensional schematization of the zebrafish culture facility. (A) Reservoir setup; (B) inlet pipe framework; and (C) outlet pipe framework.

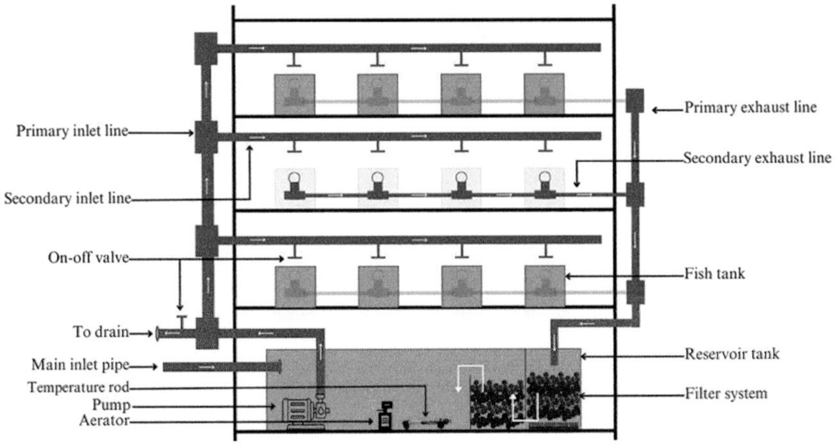

FIGURE 6.2 Schematic representation of the zebrafish facility.
Note: This figure presents a comprehensive schematic representation of the zebrafish facility, offering a detailed visual overview of its structural layout and key components. It provides valuable insights into the organization and interconnection of various elements within the facility, including the reservoir tank, filtration system, inlet and outlet pipes, and the arrangement of zebrafish housing units. By illustrating the spatial arrangement and functional relationships between different parts of the facility, this schematic aids in understanding the circulation of water, filtration processes, and overall workflow within the zebrafish research environment.

5. Exhaust pipeline setup:
   - For the primary exhaust line, measure and cut ¾″ PVC pipes to fit along the opposite side of the rack from the primary inlet line.
   - Join these ¾″ PVC pipes with Tee connectors so that they collect water from the secondary exhaust line and drain into the filtration system, as shown in Figures 6.1(C)–6.3(B).
   - For the secondary exhaust line, extend a ¾″ PVC pipe horizontally from the Tee connectors of the primary exhaust line to each fish tank.
   - The fish tank and the secondary exhaust line are connected using elbow joints and Tee connectors.
6. Fish tank setup:
   - Assemble a 15″ × 8″ × 8″ fish tank and drill a circular hole at the back of the tank 1″ below the upper edge using a hole saw, as shown in Figure 6.3.

- Attach a ½″ pipe to the hole at the back of the tank. Connect the pipe extending from the fish tank to the secondary exhaust line using an elbow joint.
- Attach a piece of fine mesh to the inner side of the back wall of the tank to prevent the fish from escaping.

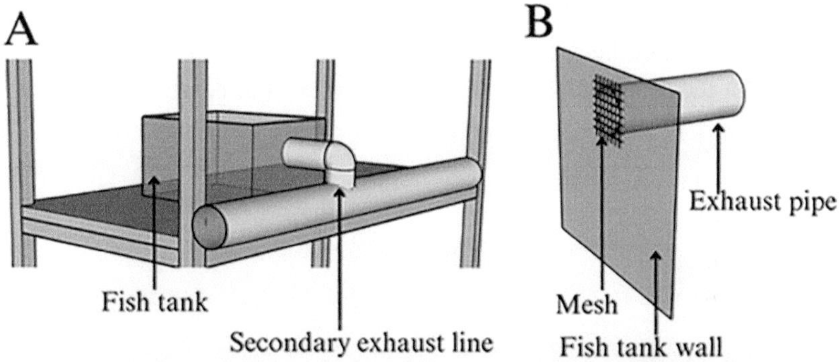

FIGURE 6.3 Schematic representation of a single fish tank of the culture unit. (A) Single fish tank; and (B) exhaust connection.
Note: This figure depicts a detailed schematic representation of a single fish tank within the culture unit. It provides an in-depth visual portrayal of the essential components and features of an individual tank, including its dimensions, structural design, and key elements such as the water inlet and outlet systems. This figure offers valuable insights into the configuration and setup of the fish tanks within the larger zebrafish culture unit, aiding in understanding their functionality and role within the overall facility.

6.4 OVERVIEW OF SETUP

The zebrafish facility comprises an eighteen-tank housing unit for the animals, connected to the reservoir tank and filtration system via inlet and outlet lines, respectively. Positioned on the bottom layer of the rack, the reservoir tank and filtration system support the upper layers of fish tanks (Figure 6.4). Initially, the reservoir tank is filled with water from the main water source through the main inlet line. Water is then pumped upward via the primary inlet line, which subsequently branches into secondary inlet lines, delivering water to each fish tank. The secondary inlet line, supplying the fish tanks, is equipped with an on–off valve to enable the isolation of individual tanks if necessary.

Water from the fish tanks is directed to the secondary exhaust line, which connects to the backside openings of the tanks. The secondary exhaust line collects water from each tank and channels it to the primary exhaust line. The openings on the back side of the tanks are fitted with fine mesh material on the inner surface of the wall to prevent fish from escaping.

The water collected by the primary exhaust line is then drained into the filtration system, where it undergoes a series of filtration processes, including mechanical, biological, and chemical means. The filtered water ultimately returns to the reservoir tank, completing the circulation cycle. The primary inlet line extending from the reservoir tank is also equipped with an on–off valve, allowing for the draining of water from the reservoir tank when necessary.

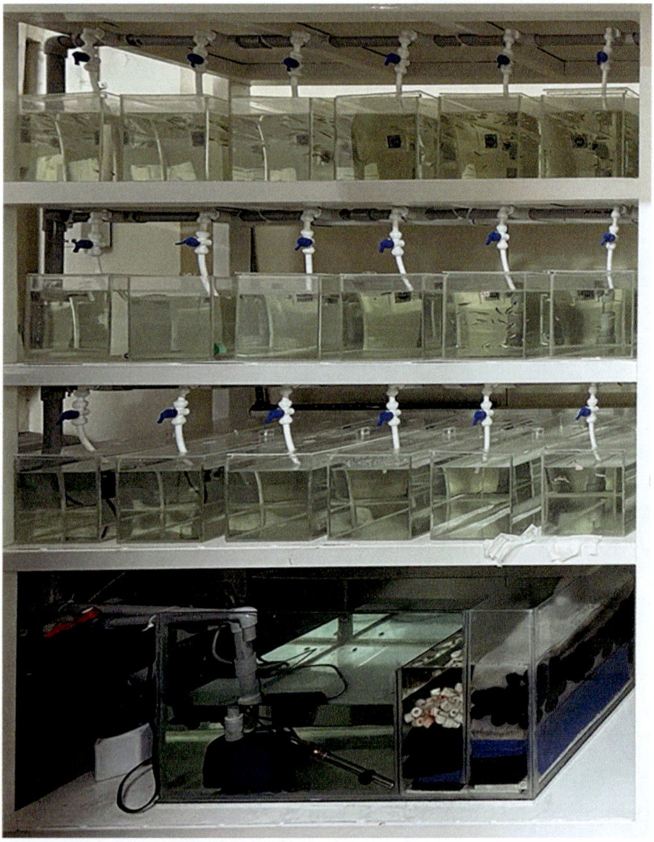

FIGURE 6.4 Overview of the setup of the zebrafish facility.

Note: This figure illustrates the arrangement and connections of the various components involved in housing and maintaining zebrafish, including the reservoir tank, filtration system, inlet and outlet pipes, and the layout of the housing units for the fish. It serves as a visual guide to understanding the overall structure and functioning of the zebrafish facility.

6.5 WORKING OF THE UNIT

The zebrafish facility comprises an eighteen-tank housing unit for the animals, which is connected to the water reservoir and filtration unit through inlet and outlet pipes, respectively. The entire facility is arranged in different layers of a rack, with the reservoir tank and filtration unit situated on the bottom layer and the eighteen tanks positioned on the upper layers. These three distinct units of the zebrafish facility are interconnected through inlet and outlet pipes. Various sections of the pipeline are joined using elbow joints or Tee joints to adjust and control the water flow through different pipe streams. This setup enables the water to be recycled and recirculated throughout the different units of the facility.

6.5.1 RESERVOIR TANK

The main inlet pipe, constructed of PVC with an inner diameter of 0.7 inches and an outer diameter of 0.8 inches, connects from the water source to the reservoir tank. Subsequently, water is pumped and lifted to the housing unit via primary and secondary inlet pipes, facilitated by a motor with the capacity to uplift 1,200 liters per hour. The primary inlet pipe, also made of PVC, has an inner diameter of 0.75 inches and an outer diameter of 1 inch, extending through the three layers of the rack where the housing tanks are positioned.

A reverse osmosis pipe, with an inner diameter of 0.3 inches and an outer diameter of 0.4 inches, acts as the secondary pipe branching from the primary pipe to each tank separately. It is connected ventrally to the tanks and incorporates a valve system that, when opened, delivers water to the tank. The secondary exhaust pipe, composed of PVC, is dorsally connected to the tanks, with each one opening into a common primary exhaust pipe.

The secondary exhaust pipe drains water from the tanks at a constant flow rate into the primary exhaust pipeline, which eventually drains it into the filtration system. The exhaust water undergoes filtration through various layers of filters before reaching the reservoir tank again. The recycled and filtered exhaust water from the reservoir is recirculated through the system in the same manner. Periodically, the reservoir tank can be cleaned to prevent contamination, facilitated by a valve system to exhaust water from the tank. Subsequently, the tank is refilled through the major inlet pipeline. Additionally, the reservoir tank is equipped with an aeration system to maintain the oxygen saturation level in the water.

6.5.2 HOUSING UNIT

The housing unit is specifically designed to accommodate eighteen tanks, each with uniform dimensions measuring 15″ × 8″ × 8″ and a water-holding capacity of approximately 4 gallons per tank (Figure 6.5). These acrylic quarantine tanks provide an optimal environment for the fish to acclimate to their new surroundings. Positioned in the upper rows of the rack, each row includes a primary inlet pipe responsible for carrying water from the reservoir. The primary pipeline bifurcates into secondary inlet pipes, delivering water ventrally to each tank continuously when in operation. This inlet pipe operates via a valve system that, when opened, fills the tank with water.

Utilizing a reverse osmosis pipe for the inlet line proves to be a cost-efficient and adaptable alternative to the commercial pipelines commonly found in larger-scale facilities. At the top end of each tank, a secondary exhaust pipe is positioned dorsally to facilitate continuous drainage of water from the tank. The opening of this pipe is covered with mesh to prevent fish from escaping. Waste- and debris-laden water from the quarantine tanks is collectively drained into the filtration system via the primary exhaust pipe. This constant influx and efflux of water from the tanks ensure cleaning efficiency and prevent the accumulation of waste materials within the animal shelters.

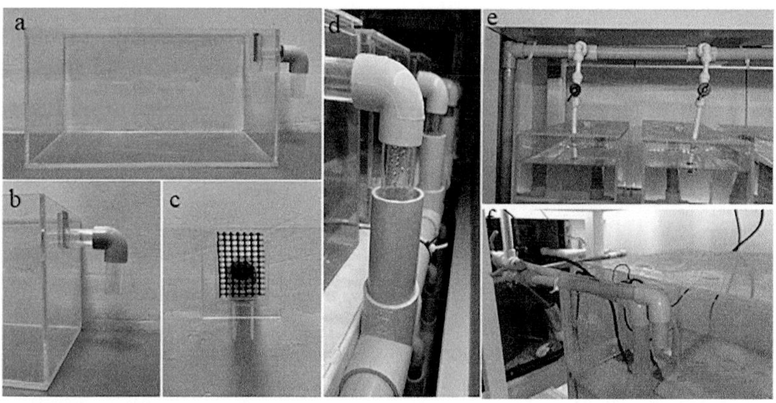

FIGURE 6.5 Components of the zebrafish culture unit. (a) Fish tank; (b) outlet valve; (c) mesh in the outlet valve; (d) exhaust pipe; (e) inlet valve; and (f) inlet and outlet pipes in the reservoir tank.

Note: This photograph illustrates the key elements of the zebrafish culture unit, including the tank for housing the fish, valves for controlling water flow (both inlet and outlet), and the exhaust pipe for removing water from the system. The inlet and outlet pipes connect the fish tanks to the reservoir tank, facilitating water circulation within the unit.

6.5.3 FILTRATION SYSTEM

The filtration unit is integrated into the reservoir tank, with internal boundaries to eliminate the need for additional pipeline connections. The system consists of mechanical, chemical, and biological means of filtration combined. Broadly, the filtration unit is divided into two chambers (Figure 6.6). The first chamber on the right is completely separated from the reservoir tank, whereas the second chamber on the left connects to the reservoir at the top. Water from the primary exhaust pipe drains into the right chamber and flows from right to left.

This exhaust water passes through a series of filters, each providing additional filtration benefits. The first layer of the filtration unit consists of an absorbent sponge that captures large to medium-sized dust particles. This sponge can be reused up to three times before being replaced. The next layer consists of a collection of bio balls. These bio balls offer a large surface area that supports beneficial bacteria, primarily nitrifying bacteria. These bacteria assist in removing nitrites and ammonia from the water.

Another layer of absorbent sponge is placed below the bio balls in a sandwich-like arrangement. This is followed by a layer of bio sponge, which provides mechanical filtration by trapping non-dissolved waste materials, sediments, and floc through its interstitial pores. These hard sponges have a broad surface area and open pores that filter unnecessary particles. They can be reused multiple times before replacement. This unit includes both blue and black hard bio sponges.

An under-gravel filter mat is placed at the bottom of the system, extending between both chambers. It prevents water stagnation by pulling water and impurities downward through the gravel and facilitating upward water flow to the second chamber.

The next filtration compartment contains three filter layers. Bio sponges are placed just above the filter mat and topped with a double layer of carbon filters. These activated carbon layers remove finer dust particles and can eliminate brownish or yellowish discoloration from the water. Activated carbon is especially effective in removing toxins and odors.

The final layer of the unit consists of ceramic rings, also known as bio rings. These are capable of hosting both nitrifying and denitrifying bacteria. Shaped like hollow cylinders, they contain small pores on both the outer and inner surfaces.

The filtered water falls into the reservoir tank from the top of the left chamber, from where it is recirculated throughout the system. The entire filtration system should be regularly inspected for clogging and waste buildup. Filter components should be replaced as needed.

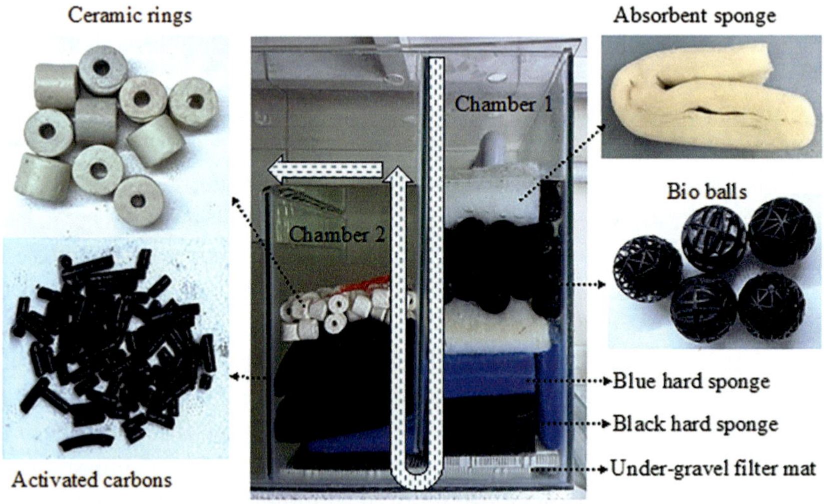

FIGURE 6.6 Filtration unit adapted for the zebrafish culture unit.

Note: The filtration unit is divided into two chambers. Water flows through the system as indicated by the arrows. Chamber 1 contains an absorbent sponge, bio balls, blue and black hard sponges, and an under-gravel filter mat. Chamber 2, from bottom to top, includes a hard sponge, activated carbon, and ceramic rings.

6.6 ADVANTAGES

This zebrafish unit, in comparison with commercial zebrafish facilities, proves to be highly cost-effective and time-efficient. The cost of the zebrafish facility is around 1 lakh, which is tenfold less than the cost of a commercial unit (Ericsson et al., 2021; Kim et al., 2009; Lawrence et al., 2012; Lipscomb et al., 2023; Tarasco et al., 2019; Teoh et al., 2020). The construction of a commercial unit may be labor-intensive and time-consuming, whereas this affordable and easy-to-build unit eliminates the need for high labor input and reduces labor costs. Additionally, the setup of the unit can be completed in a single day.

Commercial facilities, which are installed with large equipment and systems, occupy both necessary and unnecessary space. In contrast, this small-scale unit avoids occupying excess space, as the tanks and instruments are compact. However, the reduced size and space do not compromise the unit's quality or functional performance. The need for sophisticated architecture and structural complexity can be eliminated with simple zebrafish units.

In commercial units, a separate reverse osmosis system is typically installed to filter the source water, even though the water is already being filtered through the recirculation system. This extra system is avoided in the simpler unit. Some commercial—and even simpler—units use ultraviolet (UV) filtration in each tank, which is cost-intensive and may, at times, have harmful effects on fish growth and embryo production. The current system does not contain UV filtration; instead, it relies on enhanced biological, chemical, and physical filtration.

Larger commercial units may also contain fluidized bioreactors for filtration, which increase the overall cost and require separate maintenance. In this unit, the filtration system is partially integrated with the reservoir tank, eliminating the need for separate maintenance and connections. The unit is easy and feasible for manual cleaning whenever necessary, a feature that most larger facilities do not support.

Moreover, installation, repair, and renovation of the unit can be self-performed in simply built setups. Larger units may require trained personnel for installation and repair. The replacement of sophisticated components adds additional cost in commercial units. However, smaller units, built with simple and cost-efficient components, offer cheaper and more accessible replacements. Since these components are locally available, both purchase and transportation are easy.

Temperature and pH are the most important parameters in a housing unit. This zebrafish facility provides ideal temperature and pH conditions through the use of thermal rods and a pH meter, thereby maintaining a macroenvironment favorable for fish growth. The cost-effective nature of the unit does not compromise environmental maintenance.

6.7 CONCLUSION

The zebrafish unit, designed to accommodate approximately 1,000 to 1,500 fish, stands as a testament to simplicity without compromising effectiveness. Its unique aesthetic properties complement its functionality, offering a harmonious blend of form and function. The unit's operational capabilities rival those of commercial counterparts, achieving comparable efficacy at a fraction of the cost and time.

As scientific focus increasingly shifts toward *in vivo* studies and zebrafish emerges as a prominent animal model, the demand for accessible and efficient zebrafish facilities continues to grow. This unit, with its affordable price tag and straightforward construction, addresses this need with precision. By providing a platform for *in vivo* studies at a reasonable cost and with minimal maintenance requirements, it opens doors to myriad opportunities in the fields of life science and technology.

Overall, the zebrafish facility represents more than just a housing unit for aquatic life; it embodies the intersection of innovation and accessibility in scientific research. Its streamlined design and cost-effective nature make it a valuable asset for researchers seeking to conduct *in vivo* studies without breaking the bank. With its potential to facilitate discoveries across various disciplines, this zebrafish facility paves the way for advancements in understanding and addressing complex biological phenomena.

ACKNOWLEDGMENT

This work was supported by the Department of Science and Technology, Science and Engineering Research Board, Research Scientist Scheme, Government of India, through a research grant awarded to S. Vimalraj (Grant no. BS/SRS/2022-23/109/LS).

KEYWORDS

- **fish facility**
- **integrated filtration system**
- **recirculating system**
- **reverse osmosis system**
- **UV filtration**
- **water filtration**
- **zebrafish maintenance**
- **zebrafish unit**

REFERENCES

Choi, T. Y., Choi, T. I., Lee, Y. R., Choe, S. K., & Kim, C. H. (2021). Zebrafish as an animal model for biomedical research. *Experimental & Molecular Medicine*, *53*(3), 310–317. https://doi.org/10.1038/s12276-021-00571-5 (accessed on 10th July 2025).

Ericsson, A. C., Busi, S. B., Davis, D. J., Nabli, H., Eckhoff, D. C., Dorfmeyer, R. A., Turner, G., Oswalt, P. S., Crim, M. J., & Bryda, E. C. (2021). Molecular and culture-based assessment of the microbiome in a zebrafish (*Danio rerio*) housing system during set-up and equilibration. *Animal Microbiome*, *3*(1), 55. https://doi.org/10.1186/s42523-021-00116-1 (accessed on 10th July 2025).

Fosse, V., Oldoni, E., Bietrix, F., Budillon, A., Daskalopoulos, E. P., Fratelli, M., Gerlach, B., Groenen, P. M. A., Hölter, S. M., Menon, J. M. L., Mobasheri, A., Osborne, N., Ritskes-Hoitinga, M., Ryll, B., Schmitt, E., Ussi, A., Andreu, A. L., McCormack, E., & PERMIT group. (2023). Recommendations for robust and reproducible preclinical research in personalized medicine. *BMC Medicine*, *21*(1), 14. https://doi.org/10.1186/s12916-022-02719-0 (accessed on 10th July 2025).

Howe, K., Clark, M. D., Torroja, C. F., Torrance, J., Berthelot, C., Muffato, M., Collins, J. E., Humphray, S., McLaren, K., Matthews, L., McLaren, S., Sealy, I., Caccamo, M., Churcher, C., Scott, C., Barrett, J. C., Koch, R., Rauch, G. J., White, S., & Stemple, D. L. (2013). The zebrafish reference genome sequence and its relationship to the human genome. *Nature*, *496*(7446), 498–503. https://doi.org/10.1038/nature12111 (accessed on 10th July 2025).

Kim, S., Carlson, R., Zafreen, L., Rajpurohit, S. K., & Jagadeeswaran, P. (2009). Modular, easy-to-assemble, low-cost zebrafish facility. *Zebrafish*, *6*(3), 269–274. https://doi.org/10.1089/zeb.2009.0587 (accessed on 10th July 2025).

Lawrence, C., & Mason, T. (2012). Zebrafish housing systems: A review of basic operating principles and considerations for design and functionality. *ILAR Journal*, *53*(2), 179–191. https://doi.org/10.1093/ilar.53.2.179 (accessed on 10th July 2025).

Lee, Y. S., & Seki, E. (2023). *In vivo* and *in vitro* models to study liver fibrosis: Mechanisms and limitations. *Cellular and Molecular Gastroenterology and Hepatology*, *16*(3), 355–367. https://doi.org/10.1016/j.jcmgh.2023.05.010 (accessed on 10th July 2025).

Lipscomb, B., Seymour, N., Lewis, J., & LeBert, D. C. (2023). An affordable and easy-to-construct zebrafish housing system for stable long-term laboratory research. *Zebrafish*, *20*(6), 260–270. https://doi.org/10.1089/zeb.2023.0061 (accessed on 10th July 2025).

Mukherjee, P., Roy, S., Ghosh, D., & Nandi, S. K. (2022). Role of animal models in biomedical research: A review. *Laboratory Animal Research*, *38*(1), 18. https://doi.org/10.1186/s42826-022-00128-1 (accessed on 10th July 2025).

Shin, J. T., & Fishman, M. C. (2002). From zebrafish to human: Modular medical models. *Annual Review of Genomics and Human Genetics*, *3*, 311–340. https://doi.org/10.1146/annurev.genom.3.031402.131506 (accessed on 10th July 2025).

Tarasco, M., Martins, G., Gavaia, P. J., Bebianno, M. J., Cancela, M. L., & Laizé, V. (2020). ZEB316: A small stand-alone housing system to study microplastics in small teleosts. *Zebrafish*, *17*(1), 18–26. https://doi.org/10.1089/zeb.2019.1801 (accessed on 10th July 2025).

Teoh, S. L., Sapri, S. R. B., Yusof, M. R. B. M., Yahaya, M. F., & Das, S. (2020). Construction of an affordable open-design recirculating zebrafish housing system. *Journal of the American Association for Laboratory Animal Science*, *59*(5), 512–518. https://doi.org/10.30802/AALAS-JAALAS-19-000167 (accessed on 10th July 2025).

Verma, S. K., Nandi, A., Sinha, A., Patel, P., Mohanty, S., Jha, E., Jena, S., Kumari, P., Ghosh, A., Jerman, I., Chouhan, R. S., Dutt, A., Samal, S. K., Mishra, Y. K., Varma, R. S., Panda, P. K., Kaushik, N. K., Singh, D., & Suar, M. (2024). The posterity of zebrafish in paradigm of *in vivo* molecular toxicological profiling. *Biomedicine & Pharmacotherapy*, *171*, 116160. https://doi.org/10.1016/j.biopha.2024.116160 (accessed on 10th July 2025).

CHAPTER 7

Probiotics and Other Microbial Approaches in Aquaculture

PHILOMINAL PATHINATHAN* and K. T. HAFEEF ROSHAN

College of Fishery Science, Andhra Pradesh Fisheries University, Narasapuram, West Godavari, Andhra Pradesh, India

*Corresponding author

ABSTRACT

Aquaculture plays a vital role in global food production but faces challenges like disease outbreaks and limited antibiotic use. Probiotics have emerged as effective tools to enhance fish health and sustainability. These beneficial microorganisms—mainly bacteria and yeast—help maintain gut microbiota balance, improve nutrient absorption, regulate immune responses, and reduce disease risk. Their use can lower dependency on antibiotics, improve feed conversion, and reduce environmental impacts. Similarly, microbial protein is gaining attention as a sustainable feed alternative. Rich in essential amino acids, it supports the growth of aquatic species and promotes a circular economy by utilizing organic waste for production. Microbial proteins can also aid in gut microbiota modulation, enhancing digestive health and immunity. While promising, their broader adoption is limited by cost, scalability, and species-specific acceptance. Overall, probiotics and microbial proteins represent innovative strategies to address key issues in aquaculture. This chapter highlights their benefits, challenges, and potential to support sustainable practices, reduce reliance on traditional feed sources, and promote healthier aquatic environments.

Physiological Disorders of Fruits and Vegetables: Causes and Remedies. Ghulam Khaliq, Sajid Ali, and Mohammed Wasim Siddiqui (Eds.)

DOI: 10.1201/9781779641113-7

7.1 INTRODUCTION

Aquaculture involves cultivating aquatic organisms such as finfish, shellfish, crustaceans, and plants, mainly for human consumption. It represents the livestock industry's most rapidly expanding sector and contributes to over half of the global fish supply consumed by people (FAO, 2020). According to the Food and Agriculture Organization (FAO) of the United Nations, a conservative projection suggests that by 2030, an extra 40 million tons of fish protein will be required to meet the dietary needs of the growing human population (FAO, 2020). Meeting this worldwide demand involves significantly boosting the cultivation of aquaculture species through advanced technological and practical means (Tuan et al., 2013).

Between 1961 and 2017, global food fish consumption rose by an average annual rate of 3.1%, nearly double the yearly world population growth of 1.6% during that period. This increase surpassed the growth rates of all other animal protein foods, such as meat and dairy, which grew by an average of 2.1% per year (Sofia, 2020). As demand from the expanding global population rises, aquaculture production methods have been significantly intensified through advanced technological and practical measures (Tuan et al., 2013).

As aquaculture intensifies and becomes more commercialized to fulfill rising demand, it encounters several challenges. These include combating diseases and epizootics, enhancing broodstock and domestication, devising suitable feed and feeding methods, refining hatchery and grow-out technology, and managing water quality amidst this intensification and commercialization (Subasinghe, 2003). While proper husbandry methods can alleviate stress in aquaculture, complete elimination of stress remains unattainable. Ultimately, prolonged stress escalates the susceptibility of cultured fish to diseases.

Among these challenges, disease outbreaks have emerged as the primary constraint in cultivating numerous species, significantly impacting aquaculture production. This issue not only affects economic growth but also hampers social development in various countries (Qi et al., 2009; Tuan et al., 2013). Diseases have resulted in substantial economic losses for global finfish aquaculture, as highlighted in various studies (Johnson et al., 2004; Monir et al., 2015; Mustafa et al., 2001; Patil and Markad, 2015; Sahoo et al., 2013). These losses are globally estimated to range from US $1.05 to US $9.58 billion annually (Shinn et al., 2015).

Over time, antibiotics have been conventionally employed in aquaculture as a strategy to control and prevent fish diseases. Additionally, they have been utilized to enhance growth and improve feed conversion efficiency. The utilization of antimicrobials in aquaculture traces back to Gutsell's work in 1946, where the potential use of antibiotics, specifically sulfonamides, was acknowledged for combating furunculosis. The 2017 joint FAO/OIE/WHO expert meeting on antimicrobial use and resistance in aquaculture highlighted two primary hazards: antimicrobial residues and antimicrobial resistance.

Antibiotic resistance develops as bacteria encounter antibiotics at levels below therapeutic doses over prolonged periods, leading bacteria to develop antibiotic-resistant genes or plasmids. These resistant bacteria can horizontally transfer their resistance genes and plasmids to other bacteria.

Indiscriminate antibiotic use leads to the rise of antibiotic-resistant microorganisms and disrupts the balance of gut microflora in fish, negatively impacting their health. Moreover, residual deposition in fish muscle poses a potential health risk to consumers (Stentiford et al., 2017). Given the concerns surrounding antibiotic use in aquaculture, an alternative method to maintain fish health within the industry involves the use of probiotics. This microbial intervention approach has shown promising results, enhancing not only fish health but also, in many cases, promoting fish growth (Stentiford et al., 2017).

Recent reports emphasize that aquaculture is facing a critical juncture regarding antimicrobial resistance, underscoring the need to implement essential corrective and mitigation measures to protect the health of fish, aquaculture systems, and humans. Over an extended period, substantial research efforts have aimed to identify safe, cost-effective, and environmentally sustainable feed supplements offering remarkable advantages for farmed fish—including phytogenics, essential oils, prebiotics, and probiotics.

As the aquaculture sector advances and refines its production methods, the demand for feeds and ingredients continues to rise. Forecasts reliably indicate that the most significant factor that could hinder the industry's growth is the potential limitation in accessing ingredients for aquaculture feeds (Olsen and Hasan, 2012; Tacon and Metian, 2008). Numerous nutritional studies have focused on experimenting with various nontraditional protein sources. Plant meals, a common replacement for fish meal, have been extensively utilized in these studies. Yet, it is worth noting that plant proteins might not be universally suitable for all aquatic species (Kumar et al., 2012).

This limitation, coupled with rapid advancements in various biotechnology fields, suggests that the upcoming era of dietary proteins will heavily feature those sourced from microorganisms. In 1966, C. L. Wilson at the Massachusetts Institute of Technology introduced the term "single-cell protein" (SCP). SCP denotes the processed biomass or extracted protein sourced from microorganism cultures, offering potential as ingredients for food or animal feeds. These microorganisms present significant advantages for SCP production.

Extensive demonstrations from studies (Becker, 2007; Dewapriya and Kim, 2014; Overland et al., 2010) highlight the notable advantages of microorganisms in SCP production when compared to traditional nutrient sources. These organisms exhibit rapid growth rates and generate high yields. Numerous established industrial methods have succeeded in consistently generating substantial microbial biomass from microalgae, yeasts, and bacteria. This biomass can serve either as the final output of a particular process or emerge as a by-product of the primary process, with versatile applications spanning various purposes.

Recent research has indicated that bacterial biomass obtained through natural gas fermentation—specifically from bacterial cultures of *Methylococcus capsulatus*—holds significant promise as a protein source. This is supported by factors such as its amino acid composition, digestibility, and observed positive effects on growth and health in both aquatic and terrestrial animals (Overland et al., 2010; Patil and Markad, 2015; Romarheim et al., 2011).

Recent explorations in animal nutrition have also delved into alternative sources of bacterial biomass. For instance, Zhao et al. (2012) investigated a bacterial meal—a by-product of lysine extraction—as a substitute for soybean meal in pig diets. Their findings showed that replacing up to 25 percent of soybean meal protein with bacterial meal had no significant impact on nitrogen metabolism. Similarly, Zhang et al. (2013) demonstrated that *Corynebacterium glutamicum*, a single-cell protein, could effectively replace 50 percent of fish meal in pig diets while maintaining comparable growth rates, nutrient digestibility, and intestinal morphology to diets solely comprising fish meal as the protein source.

Regarding commercial yeast-derived products intended for animal nutrition, research indicates that incorporating even small amounts of these yeast fractions (comprising peptides, nucleotides, and other cytoplasmic elements) promotes significant growth and health advantages in fish (Berto et al., 2015; Lunger et al., 2006).

7.2 PROBIOTICS

The term "probiotics" was initially coined by Lilley and Stillwell in 1965 to refer to substances released by one microorganism that fostered the growth of another (Fuller, 1992). This term stood in direct contrast to "antibiotics." Later, in 1971, Sperti expanded its usage to describe "tissue extracts that promoted microbial growth" (Fuller, 1992). Probiotics refer to live microorganisms introduced into the gastrointestinal tract (GIT) through food or water, aiming to enhance internal microbial balance and promote good health. They are also described as "live microorganisms ingested to offer health benefits," and alternatively defined as "live microorganisms consumed in sufficient quantities as part of food, providing health advantages to the host."

Probiotics emerged as a recommended environmentally friendly approach for disease prevention in aquaculture. Administered orally, they offer distinctive health advantages, particularly in managing bacterial fish diseases. Presently, various commercial probiotics are commonly available in powdered forms for ease of use (Castex et al., 2010; Chang and Liu, 2002; Shelby et al., 2006). In 1992, Havenaar and Huis outlined probiotics as "microbial cultures, whether single or mixed, administered to animals or humans to enhance the characteristics of the host's natural microflora, thereby benefiting their health" (FAO/WHO, 2001). According to Gatesoupe (1999), in the context of aquaculture, probiotics encompass bacteria that enhance water quality upon introduction into the water and/or actively deter pathogens within the aquatic environment (Rengpipat, 2005). The Food and Agriculture Organization of the United Nations/World Health Organization (FAO/WHO) articulated probiotics as living microorganisms that, when administered in appropriate quantities, confer health benefits upon the host (FAO/WHO, 2001).

In 1986, Kosaza pioneered the initial practical application of probiotics in aquaculture, acknowledging the advantages observed from probiotic use in humans and poultry. More recently, Lazado and Caipang (2014) suggested that within an aquaculture context, probiotics could be defined as "microorganisms, whether alive or inactive or even specific components thereof, employing various mechanisms to confer beneficial effects either to the host or its environment" (Lazado and Caipang, 2014). Choosing the right probiotics is crucial, as improper strains can cause adverse effects in the host. The main goal in selecting probiotics is to ensure their safety and their ability to provide beneficial outcomes. Moreover, they need to maintain their efficacy throughout production, manufacturing, distribution, and storage before reaching the end consumer (Shewale et al., 2014). Jahangiri and Esteban (2018) reported different routes, such as delivery through feeding of probiotic-supplemented pellets, live food organisms, direct addition of probiotics to the water body, or through injection, for administration of probiotics to the host body (Figure 7.1).

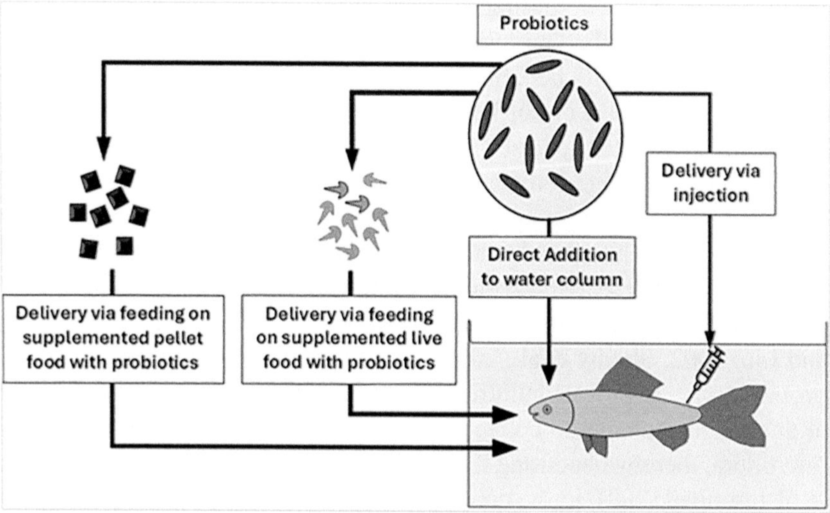

FIGURE 7.1 Different routes for administration of probiotics.

Source: Reprinted from Jahangiri and Esteban (2018). Copyright © 2018 by the authors. Licensee MDPI, Basel, Switzerland. http://creativecommons.org/licenses/by/4.0/

7.2.1 ESSENTIAL CHARACTERISTICS OF PROBIOTICS (MICHAEL, 2014)

i. The strain should have the ability to positively impact the host animal, such as promoting growth or bolstering disease resistance.
ii. It must be nonpathogenic and free from toxicity.

iii. It should exist in the form of live cells, preferably in significant quantities.
iv. The strain should demonstrate resilience and metabolic activity in the gut environment, including resistance to low pH levels and organic acids.
v. It needs to exhibit stability, maintaining viability over extended periods under both storage and field conditions.

The United Nations has outlined several specifications to be taken into account when selecting and approving a probiotic product, as detailed below (Aly, 2009).

- The probiotic's ability to survive passage through the gastrointestinal tract (GIT) should be demonstrated.
- Colonization within the GIT of the host should occur when the probiotic is present.
- The probiotic culture should competitively inhibit the attachment of pathogenic bacteria to the intestinal surface.
- Efficiency in inhibiting pathogenic bacteria according to *in vitro* tests should be demonstrated.
- Probiotics should exhibit resistance against other sanitation agents or disinfectants.
- The label of probiotic products should indicate the genus and species name according to international nomenclature.
- The label should include information regarding doses and expiration dates.
- It is desirable to have data ensuring that the product will not cause infections in immunocompromised animals.

7.2.2 SOURCES OF BACTERIAL PROBIOTICS

Microbes are naturally present in various environments, including humans, aquatic animals, snow, soils, sediments, groundwater, freshwater, and seawater. Biotic and abiotic environments have shown varying bacterial counts ranging from 10^2 to 10^{11} CFU/g. External bacteria, originating from sources like air, soil, or human contact, can enter water ecosystems, altering microbial populations. This influx can result in the colonization of diverse bacteria within the gastrointestinal tract (GIT) of aquatic animals.

This indicates that simply increasing the quantity of probiotics in feed does not necessarily lead to more significant growth. The dosage is a crucial factor when adding functional nutrients to aqua feeds. For instance, in pabda catfish (*Ompok pabda*), weight gain decreased when dietary commercial probiotics exceeded 0.2% (Chowdhury et al., 2020). This research highlighted that excessive concentrations might disrupt the overall body physiology of fish, potentially causing imbalances in carbohydrate and fat metabolism.

7.2.3 BENEFITS OF PROBIOTICS IN AQUACULTURE

7.2.3.1 IMPROVEMENTS IN FEED UTILIZATION

Numerous studies have consistently highlighted the positive impact of probiotics on enhancing feed utilization in various aquatic species. For instance, in white shrimp (*Litopenaeus vannamei*), the supplementation of *Lactobacillus pentosus* demonstrated significant improvements in feed utilization (Zheng and Wang, 2016). Additionally, research involving Nile tilapia showed promising results when incorporating heat-killed *Lactobacillus plantarum* at varying doses (50, 100, or 1,000 mg/kg) over 12 weeks. This supplementation notably boosted the activity of key enzymes like amylase, lipase, and protease (Dawood et al., 2019), indicating improved feed utilization in these fish. A study conducted on narrow-clawed crayfish, *Astacus leptodactylus*, revealed notable increases in the activity of essential enzymes—protease, amylase, alkaline phosphatase, and lipase—when fed with *Lactobacillus plantarum* at concentrations of 10^7, 10^8, and 10^9 CFU/g (Valipour et al., 2019).

7.2.3.2 IMPROVE WATER QUALITY

Probiotic bacteria can directly process organic matter and even break down harmful substances present in water, thereby enhancing its overall quality. These probiotics play a crucial role in decomposing fish or prawn waste, leftover food, plankton remnants, and other organic substances into compounds like carbon dioxide, nitrate, and phosphate. This process contributes significantly to improving the nutrient cycle within aquatic environments, creating and sustaining favorable water quality for cultured animals (Rao, 2010). According to Nimrat et al. (2012), the use of a combination of *Bacillus* probiotics notably improved crucial water quality parameters—including pH, ammonia levels, and nitrite concentrations—while rearing white shrimp. Additionally, when *Bacillus subtilis* was added to the water at concentrations ranging from 10^3 to 10^5 CFU/ml, it significantly decreased overall ammonia levels, thereby positively impacting water quality in shrimp culture, as observed by Subedi and Shrestha (2020).

7.2.3.3 STRESS TOLERANCE

The introduction of probiotics has shown substantial enhancements in stress tolerance among various species in aquaculture. In a study conducted by Nguyen Van Nguyen in 2018, it was observed that fish supplemented with *Lactobacillus plantarum* exhibited lower cortisol concentration increments when exposed to ammonia stress compared to those on a control diet. Cortisol, a stress hormone, remained at lower levels in fish fed with this probiotic. Furthermore, research by Dawood et al. (2019) demonstrated that *Aspergillus oryzae*, when used as a probiotic in Nile tilapia, significantly bolstered their defense mechanisms against hypoxia-induced stress.

Moreover, investigations in pabda catfish by Chowdhury et al. (2020) revealed that the supplementation of commercial probiotics at a 0.2% dosage notably improved tolerance to saline water stress. Interestingly, the study indicated that both higher and lower doses did not confer the same level of stress tolerance as the 0.2% supplementation in pabda catfish.

7.2.4 PATHOGENIC INHIBITIONS

Probiotics are believed to generate bacteriocins, siderophores, lysozymes, proteases, and hydrogen peroxide, which help deter pathogens (Nayak, 2010, 2021). One common method, known as the "spot on the lawn" technique, involves applying a pathogenic bacterium onto a tryptic soy agar (TSA) plate. After incubation, the potential probiotic candidate is then placed as spots on this agar plate (Vine et al., 2004). One rapid technique to assess bacterial isolates involves the double-layer method. This method begins by culturing a single colony of isolates on a TSA plate. Once colonies start growing, they are taken out, and semi-solid TSA containing the bacterial pathogen is poured over the plate. Using a well diffusion approach, fresh bacterial cells or bacterial supernatant are then introduced into holes on the plate and spread along with the pathogen (Hai et al., 2007).

One common procedure involves agar diffusion, where the potential probiotic is spread overnight on TSA agar, and then pathogenic testing is carried out by spotting cultures on the plate. An alternate method is the disc diffusion technique, involving a paper disc soaked in the cell-free supernatant of a bacterial broth from isolates. A dried agar plate hosting the pathogen is prepared, and subsequently, these paper discs are placed onto the plate (Balcázar et al., 2008). Quorum quenching showcases how probiotics can hinder the production of violacein by *Chromobacterium violaceum* (Villamil et al., 2014). Additionally, the "cross-streaking method" involves streaking isolated bacteria at the center of an agar plate, followed by removal and killing of the bacterial growth. Afterward, a pathogenic bacterial culture is applied to this plate (Hai et al., 2007). The emergence of a clear zone in these methods signifies the potential of the isolates to inhibit pathogens.

Bacterial pathogens like *Aeromonas* spp. and *Streptococcus* spp. are typically present in the gastrointestinal tract (GIT) of fish and harbor virulence genes like ***aemolysin*** and ***aerolysin***, which can lyse blood cells (El-Son et al., 2021). Their hemolytic activities are evaluated using various blood types such as human, horse, sheep, fish, and shrimp hemolymph (Muñoz-Atienza et al., 2013). Certain bacterial isolates, including *Bacillus* spp., *Citrobacter freundii*, *Lactobacillus plantarum*, and *Lactobacillus casei*, demonstrated no harmful effects in blood hemolysis tests.

7.2.5 BACTERIOCIN PRODUCTION

Microbes with beneficial properties have a direct way of hindering the growth of harmful bacteria by producing bacteriocins. These bacteriocins are essentially peptides known for their ability to combat specific targets by exhibiting antimicrobial activity (Dobson et al., 2012). Within the intricate environment of the gastrointestinal tract, bacteriocins serve various functions, such as signaling among microbes, aiding in colonization, and actively eliminating harmful organisms (Dobson et al., 2012; Fajardo and Martínez, 2008).

7.2.6 SUPPRESSION OF VIRULENCE EXPRESSION

Bacteria employ a mechanism called quorum sensing (QS) to communicate among themselves by releasing autoinducer molecules based on cell density (Defoirdt, 2014; Suga and Smith, 2003). QS is crucial in regulating various microbial traits, such as biofilm creation, expression of virulence, swarming behavior, and even control of bioluminescence (Waters and Bassler, 2005).

Acylated homoserine lactones (AHLs) play a pivotal role in quorum sensing. These compounds are synthesized by various significant pathogens in aquaculture, such as *Aeromonas salmonicida*, *A. hydrophila*, *Vibrio anguillarum*, *V. harveyi*, *V. salmonicida*, and *Yersinia ruckeri* (Defoirdt, 2014; Kastbjerg et al., 2007). Given that bacterial disease development is partly governed by QS, a potential approach for combating infections involves quorum quenching (QQ)—the disruption of QS. Notably, researchers have investigated multiple *Bacillus* strains associated with fish for their ability to quench quorum sensing and potentially combat aquatic pathogens (Kuebutornye et al., 2020).

7.2.7 FACTORS AFFECTING THE EFFICIENCY OF NOVEL PROBIOTICS FOR AQUACULTURE USE

In the realm of scientific exploration, preliminary trials conducted *in vitro* offer a promising avenue to diminish expenses and reduce the need for extensive live animal testing, especially in the context of *in vivo* studies. Pathogen antagonism tests stand out as a valuable initial screening method for assessing antagonistic capabilities (Balcázar et al., 2008; Chemlal-Kherraz et al., 2012). These assessments encompass a range of parameters such as blood hemolysis, evaluations of antibiotic resistance, adherence assays, pH and bile salt tolerances, and other relevant characteristics. These investigations serve as crucial benchmarks to evaluate the potential of probiotics within the confines of *in vitro* trials.

Numerous articles present varied approaches to selecting probiotics. For example, certain methodologies involve an initial assessment of pathogenic activities, followed by rigorous safety testing (Chemlal-Kherraz et al., 2012). This evaluation typically includes examinations for blood hemolysis and pathogenic inhibition. Conversely, some studies focus solely on bacterial aggregation as a selection criterion, while others emphasize a combination of pathogenic inhibition and adhesive capabilities. Additionally, there is a suggested correlation between cell surface hydrophobicity and auto-aggregation, proposed as a means to gauge the potential of probiotics (Wang and Han, 2007).

7.3 MICROBIAL PROTEIN AS AN ALTERNATIVE SOURCE FOR AQUACULTURE

The nutritional attributes of various microorganisms have garnered interest for integrating them into animal compound diets. Microbial biomass sourced from diverse origins can serve as an additive or ingredient, partially replacing other protein sources in animal feeds. In aquaculture, for instance, it can substitute portions of fish meal and soybean meal.

Newer research has shown that both the entire, untreated biomass and the extracted substances from yeast, bacteria, and microalgae provoke significant growth and immunostimulatory responses in the early stages of aquatic organism development (Biswas et al., 2012; Daniels et al., 2010; Macias-Sancho et al., 2014; Vidakovic et al., 2015). Studies have also shown successful production and utilization of yeast and photosynthetic bacteria for zooplankton culture and as substitutes for fish meal (Gamboa-Delgado et al., 2016a; Kim and Lee, 2000; Loo et al., 2013; Olvera-Novoa et al., 2002).

Advancements in aquaculture production technologies are clearly leading to increased animal densities. A strong immune response in farmed animals is vital for enhancing growth rates and stress resistance, especially in intensive aquaculture systems.

7.3.1 NUTRITIONAL VALUE OF MICROBIAL COMPONENTS

Typically, microalgae, yeasts, and bacteria exhibit elevated protein levels per dry weight (as shown in Table 7.1).

TABLE 7.1 General Nutritional Characteristics of Bacteria, Fish Meal, Microalgae, and Yeasts (Percent Composition on a Dry Weight Basis)

Component	Protein (%)	Lipids (%)	Ash (%)	Nucleic Acids (%)
Bacteria	50–65	1–3	3–7	8–12
Fish meal	59–74	8–14	10–22	0.2–0.6
Microalgae	6–62	3–45	8–43	3–8
Yeast	45–55	1–6	5–10	6–15

Various studies have detailed the protein content and amino acid profiles of these microorganisms, highlighting some as possessing significant biological value (Becker, 2007; FAO/WHO, 1973; Loosli and McDonald, 1968). Microalgae generally boast higher lipid content compared to bacteria and yeasts. Conversely, yeasts and bacteria showcase richness in protein, pigments, essential biological compounds, and vitamins. Furthermore, studies indicate that the nutritional composition of bacteria and microalgae is subject to considerable alteration through the application of distinct culture media and varied growth conditions (Table 7.1).

Yeast cell walls typically constitute 25 to 32% of the cell weight (Ferreira et al., 2010). Certain bacterial cell walls are noted to be more easily digestible compared to those present in microalgae and yeasts (Kobayashi and Kurata, 1978). While cell walls might pose indigestibility for certain organisms (Liu et al., 2016; Skrede et al., 1998), they remain a crucial nutrient source due to their primary composition of polysaccharides (85–88%) and protein (13%) (Nguyen et al., 1998). Although some organisms struggle to digest cell walls (Liu et al., 2016; Skrede et al., 1998), these structures remain vital as they predominantly consist of polysaccharides (85–88%) and protein (13%)—serving as a crucial nutrient source (Nguyen et al., 1998). While certain organisms face challenges in breaking down cell walls (Liu et al., 2016; Skrede et al., 1998), these structures retain their importance due to being primarily composed of polysaccharides (85–88%) and protein (13%), playing a pivotal role as a significant nutrient source (Nguyen et al., 1998).

7.3.2 CRITERIA FOR THE SELECTION OF MICROORGANISM

Swiftly multiplying microorganisms are essential for achieving substantial output, measured as the biomass weight generated within a given timeframe. Nevertheless, this heightened productivity leads to increased

RNA quantities within the cell. RNA production, despite being integral, poses a challenge as it can act as an undesirable factor in the final product due to its anti-nutritional properties. Biomass yield coefficient refers to the overall weight of newly formed cells per unit of substrate consumed (Waldron and Lacroute, 1975). This circumstance proves particularly valuable, particularly in instances where the substrate holds a high cost (Boehlke and Friesen, 1975). A slower growth rate necessitates a larger amount of substrate for consumption, resulting in reduced output. This decrement in output negatively impacts the yield coefficient (Leuenberger, 1972). Organisms exhibit varying levels of heat shock tolerance. Exposure to heat shock prompts cells from different microbes to develop resistance to diverse stresses. While initially lethal under normal conditions, cells adapt through a phenomenon known as acquired stress tolerance. This tolerance can be induced through various treatments such as exposure to chemicals or heat. Additionally, these treatments trigger the production of a specific class of proteins known as heat shock proteins (Tamura et al., 1998). Each organism possesses a unique capacity to manage temperature rises during exothermic fermentation, consequently decreasing the necessity for cooling. The capability of a microorganism to thrive optimally at specific temperatures eliminates the demand for extensive heating or cooling. A broader range of tolerable temperatures exhibited by microorganisms diminishes the imperative for stringent temperature control throughout the fermentation process.

7.3.3 MICROBIAL BIOMASS AS ADDITIVE AND FISHMEAL REPLACEMENT

7.3.3.1 MICROALGAE

Marine and freshwater microalgae have found extensive use in larviculture for fish, crustaceans, and mollusks. Established techniques enable continuous production of various microalgal species in hatcheries and nurseries. However, despite these advancements, the utilization of microalgae as dietary components in aquaculture feeds hasn't seen significant adoption. The focus primarily remains on cultivating the entire algal biomass rather than extracting and isolating the algal protein (Becker, 2007).

On the contrary, there's been notably more attention directed towards producing single-cell protein (SCP) from various microorganisms, particularly bacteria and yeast. This heightened interest stems largely from economic considerations, as current production methods often require substantial resources. Conversely, technical challenges persist in endeavors to achieve high microalgal biomass, contributing to this disparity in attention (Nasseri et al., 2011). Explorations into the feasibility of using SCP derived from microalgae in animal nutrition and the cultivation of algae as a protein source are limited in existing studies (Mahasneh, 1997).

However, considering the high digestibility and biological value of microalgal proteins—comparable to traditional vegetable proteins (Becker, 2007; Teimouri et al., 2013)—there's potential to anticipate notably increased dietary incorporation of microalgal biomass in aquaculture feeds. Much of this biomass is anticipated to be sourced from by-products of bio-refineries and industries engaged in producing valuable compounds like pigments.

Commonly examined microalgal species incorporated as additives in aquaculture diets include *Haematococcus pluvialis*, *Spirulina*, *Chlorella*, and *Schizochytrium*. Notably, species within the *Schizochytrium* genus are recognized as valuable sources of polyunsaturated fatty acids. Emerging methods for microalgae production hold significant promise for increasing future yields of microalgal biomass. Various microalgae species such as *Schizochytrium*, *Chlorella*, and *Pavlova* exhibit a shared physiological capability to be cultivated without light, using a carbon source for growth (heterotrophic growth) instead of relying on traditional light energy (autotrophic growth). Perez-Garcia et al. (2011) highlight that under specific heterotrophic growth conditions, these microalgae consistently yield reproducible biomass, achieving cell densities of 50–100 grams of dry biomass per liter. When compared, this production range significantly exceeds the maximum of 30 grams per liter of dry cell biomass in autotrophic cultures and matches the production of yeast dry biomass in commercial fermentors, which stands at 130 grams per liter. Findings from diverse studies investigating the utilization of microalgal biomass as an ingredient in aquaculture diets consistently indicate predominantly positive effects on the growth, survival, pigmentation, and immune response of aquatic organisms.

7.3.3.2 YEAST

Yeast was among the earliest recognized microorganisms that gained significance as a supplement in animal feed (Bekatorou et al., 2006; Nasseri et al., 2011). Known for their high protein content, yeasts and other single-celled fungi have long been incorporated into various diets for land animals. Moreover, they have been subject to experimental evaluations in diverse aquatic animal diets. *Saccharomyces cerevisiae* stands as the primary yeast strain employed in manufacturing compressed yeast, utilizing pure yeast culture and molasses as the key raw materials. The initial phase of yeast production involves cultivating the yeast from a pure culture in a sequence of fermentation vessels. After the final fermentation, the yeast is separated via centrifugation, followed by filtration of the solids to enhance the concentration of the yeast.

Following this stage, the yeast undergoes blending with additional components. Post-blending, the yeast cakes are extruded and subsequently either wrapped or dried to create dry yeast (EPA, 1995). Recent innovations in post-production methods have specifically targeted the removal of the yeast cell wall, resulting in the creation of a yeast extract possessing superior nutritional attributes compared to intact yeast cells, particularly beneficial for fish and crustaceans (Vidakovic et al., 2015; Zhao et al., 2015).

Brewer's yeast serves as a frequent addition to compound animal diets, valued for its supply of amino acids and vitamins, and believed to enhance diet palatability. This effect is attributed to the abundant presence of nucleotides within yeast cells. Studies have indicated that as much as 20% of the overall nitrogen content in *Saccharomyces cerevisiae* comprises the purine and pyrimidine bases found in nucleoproteins (Rumsey et al., 1992).

While yeast demonstrates high digestibility, it can't entirely substitute all animal protein in diets. Initial recommendations proposed inclusion rates of 5–10% for fish and 2–5% for shrimp (Hertrampf and Piedad-Pascual, 2000). However, elevated levels of yeast in diets have been linked to reduced palatability, feed intake, and growth in fish. Rumsey et al. (1991) observed decreased feed intake among rainbow trout when fed diets containing high yeast levels; diets surpassing 50% yeast became unappealing to the fish. Recent research indicates the possibility of significantly elevating dietary inclusion levels of yeast, granted that the diets maintain a well-balanced composition.

7.3.3.3 BACTERIAL BIOMASS AND DRY BIOFLOCCULATED MATERIAL

Various bacterial species have attracted attention due to their potential for large-scale production and utilization in aquaculture. Initial investigations have highlighted the successful production and application of photosynthetic bacteria like *Rhodopseudomonas palustris* in aquaculture, although studies have also delved into other species such as *Rhodobacter sphaeroides* and *Rhodovulum sulfidophilum* (Kim and Lee, 2000; Loo et al., 2013).

Extensive research has focused on methanotroph bacteria, cultivated using natural gas as a carbon source, for their potential as nutrient sources for both terrestrial and aquatic animals. *Methylococcus capsulatus*, among these bacteria, has undergone testing in various monogastric animals. Studies conducted on fish have demonstrated that the protein meal derived from this bacterial species performs effectively as an alternative protein source to fish meal in feed formulations for Atlantic salmon (Aas et al., 2006).

Numerous businesses have directed efforts toward generating bacterial biomass using agricultural and brewery waste. A primary challenge for these ventures has been achieving a nutritionally uniform end product suitable for substituting fish meals or serving as a dietary supplement. Nonetheless, initial experiments indicate that incorporating this bacterial biomass into diets can effectively enhance the survival, growth rate, and immune response of crustaceans.

Conversely, the expansion of production technologies for aquatic organisms has spurred the development of innovative methodologies that boost production yields within limited space. Microbial-based systems have emerged as a part of these techniques, particularly in the cultivation of shrimp and other species. The nutritional foundation of these systems hinges on the creation and sustained presence of aggregated material, primarily composed of microbial clusters or biofloc.

Biofloc systems come in two forms: autotrophic or heterotrophic, determined by the prevalence of microalgae, bacteria, or a combination of both. These systems often foster a diverse microbial community by employing particular substrates and inoculants. The resulting flocculated material serves as supplementary sustenance for the cultivated organisms and contributes positively to water quality. By the conclusion of production cycles in biofloc-based systems, the residual flocculated material could potentially serve as one of the generated by-products.

Likewise, water treatment facilities within certain food and beverage industries produce significant quantities of bioflocculated material. Research has delved into incorporating dry bioflocculated material into experimental diets for aquatic organisms, suggesting intriguing nutritional uses for various sustainable by-products generated by both the aquaculture and food industries.

7.3.4 EMERGING AND CURRENT MICROBIAL PRODUCTION METHODS

Alternatively, the successful development of harmless bacterial supplements utilizing inexpensive agricultural byproducts as a base (referenced in Glencross et al., 2014) suggests a potential rise in the production of microbial biomass for animal nutrition. In a separate nutritional context, Loo et al. (2013) cultivated non-sulfur bacteria (specifically *Rhodopseudomonas palustris*, *Rhodobacter sphaeroides*, and *Rhodovulum sulfidophilum*) in palm oil mill effluent. The resulting bacterial biomass was then utilized in the cultivation of rotifers for nutritional purposes.

7.4 CURRENT AND FUTURE CHALLENGES

Advancements in production methodologies and techniques aimed at boosting the nutritional quality of end products play crucial roles in determining the widespread adoption pace of microbial biomass within aquaculture feeds. Enhancing each stage of production will result in higher biomass yield and superior final products. However, any alterations made to microbial biomass after harvesting tend to raise production expenses. Further research is necessary to assess if various treatments applied to microbial biomass post-harvest yield positive nutritional benefits compared to the unaltered biomass, considering the associated costs.

7.5 CONCLUSION

Microbial biomass has emerged as a viable supplement and potential fishmeal replacement in aquafeeds. The nutritional value of these microbial components, coupled with advancements in production methodologies, presents a promising trajectory for enhancing aquaculture's efficiency and sustainability. However, the journey forward necessitates meticulous considerations and ongoing research. Evaluating the efficacy of probiotics, ensuring their safety and stability, and identifying optimal doses are imperative steps. Similarly, exploring microbial protein sources, refining production techniques, and assessing post-harvest treatments for microbial biomass is crucial to maximizing their nutritional benefits while managing production costs. In conclusion, the integration of probiotics and microbial biomass stands as a beacon of hope in navigating the challenges that loom over aquaculture. Through sustainable practices, innovative research, and strategic implementations, the aquaculture industry holds the potential not only to meet the escalating demand for seafood but also to foster a healthier ecosystem, ensuring the well-being of both aquatic life and consumers alike.

KEYWORDS

- **agricultural waste**
- **aquaculture**
- **aquatic animals**
- **bacterial biomass**
- **brewery waste**
- **gut health**
- **methanotroph bacteria**
- **microbial protein**
- **nutrient sources**
- **probiotics**

REFERENCES

Aas, T. S., Grisdale-Helland, B., Terjesen, B. F., & Helland, S. J. (2006). Improved growth and nutrient utilization in Atlantic salmon (*Salmo salar*) fed diets containing a bacterial protein meal. *Aquaculture*, *259*(1–4), 365–376.

Balcazar, J. L., de Blas, I., Ruiz-Zarzuela, I., Cunningham, D., Vendrell, D., & Múzquiz, J. L. (2008). Characterization of probiotic properties of lactic acid bacteria isolated from intestinal microbiota of fish. *Aquaculture*, *278*(1), 188–191.

Bekatorou, A., Psarianos, C., & Koutinas, A. A. (2006). Production of food grade yeasts. *Food Technology and Biotechnology*, *44*, 407–415.

Biswas, G., Korenaga, H., Takayama, H., & Sakai, M. (2012). Immune stimulant effects of a nucleotide-rich baker's yeast extract in the kuruma shrimp, *Marsupenaeus japonicus*. *Aquaculture*, *366–367*, 40–45.

Castex, M., Lemaire, P., Wabete, N., & Chim, L. (2010). Effect of probiotic *Pediococcus acidilactici* on antioxidant defences and oxidative stress of *Litopenaeus stylirostris* under *Vibrio nigripulchritudo* challenge. *Fish & Shellfish Immunology*, *28*(4), 622–631.

Chang, C. I., & Liu, W. Y. (2002). An evaluation of two probiotic bacterial strains, *Enterococcus faecium* SF68 and *Bacillus toyoi*, for reducing edwardsiellosis in cultured European eel, *Anguilla anguilla* L. *Journal of Fish Diseases*, *25*(5), 311–315.

Chemlal-Kherraz, D., Sahnouni, F., Matallah-Boutiba, A., & Boutiba, Z. (2012). The probiotic potential of lactobacilli isolated from Nile tilapia (*Oreochromis niloticus*) intestine. *African Journal of Biotechnology*, *11*(68), 13220–13227.

Chowdhury, G., Dey, S., & Nath, R. (2020). Effects of dietary probiotics on the growth, blood chemistry and stress response of Pabda catfish (*Ompok pabda*) juveniles. *AACL Bioflux*, *13*, 1595–1605.

Daniels, C. L., Merrifield, D. L., Boothroyd, D. P., Davies, S. J., Factor, J. R., & Arnold, K. E. (2010). Effect of dietary *Bacillus* spp. and mannan oligosaccharides (MOS) on European lobster (*Homarus gammarus* L.) larvae growth performance, gut morphology and gut microbiota. *Aquaculture*, *304*, 49–57.

Dawood, M., Koshio, S., Esteban, M. A., & Dossou, S. (2019). Modulation of digestive enzyme activity, blood health, oxidative responses and growth-related gene expression in GIFT by heat-killed *Lactobacillus plantarum* (L-137). *Aquaculture*, *505*. https://doi.org/10.1016/j.aquaculture.2019.02.053 (accessed on 10th July 2025).

El-Son, M. A. M., Nofal, M. I., & Abdel-Latif, H. M. R. (2021). Co-infection of *Aeromonas hydrophila* and *Vibrio parahaemolyticus* isolated from diseased farmed striped mullet (*Mugil cephalus*) in Manzala, Egypt–a case report. *Aquaculture*, *530*, 735–738.

EPA. (1995). *Food and Agricultural Industries in AP 42* (5th ed., Vol. I). Research Triangle Park, NC: EPA.

FAO. (2020). *Sustainability in Action: The State of World Fisheries and Aquaculture 2020*. Rome. https://doi.org/10.4060/ca9229en (accessed on 10th July 2025).

FAO. (2020). *The State of World Fisheries and Aquaculture 2020: Sustainability in Action*. Rome.

FAO/WHO. (2006). *Probiotics in Food: Health and Nutritional Properties and Guidelines for Evaluation*. Food and Agriculture Organization of the United Nations: Rome, Italy.

Fuller, R. (1989). A review: Probiotics in man and animals. *Journal of Applied Bacteriology*, *66*, 365–378.

Fuller, R. (1992). History and development of probiotics. In G. Sperti (Ed.), *Probiotics* (pp. 1–10). Springer: Dordrecht.

Gatesoupe, F. J. (1999). The use of probiotics in aquaculture. *Aquaculture*, *180*(1, 2), 147–165.

Hai, N. V., Fotedar, R., & Buller, N. (2007). Selection of probiotics by various inhibition test methods for use in the culture of western king prawns, *Penaeus latisulcatus* (Kishinouye). *Aquaculture*, *272*(1), 231–239.

Hertrampf, J. W., & Piedad-Pascual, F. (2000). *Handbook on Ingredients for Aquaculture Feeds*. Kluwer Academic Publishers: Dordrecht, the Netherlands, pp. 573.

Jahangiri, L., & Esteban, M. A. (2018). Administration of probiotics in the water in finfish aquaculture systems: A review. *Fishes*, *3*(3), 33. https://doi.org/10.3390/fishes3030033 (accessed on 10th July 2025).

Johnson, S. C., Treasurer, J. W., Bravo, S., Nagasawa, K., & Kabata, Z. (2004). A review of the impact of parasitic copepods on marine aquaculture. *Zoological Studies*, *43*, 229–243.

Kim, J. K., & Lee, B. K. (2000). Mass production of *Rhodopseudomonas palustris* as diet for aquaculture. *Aquacultural Engineering*, *23*, 281–293.

Kozasa, M. (1986). Toyocerin Z. *Bacillus toyoi* as growth promotor for animal feeding. *Microbiologie Aliments Nutrition*, *4*, 121–135.

Lazado, C. C., & Caipang, C. M. A. (2014). Mucosal immunity and probiotics in fish. *Fish & Shellfish Immunology*, *39*, 78–89.

Lilly, D. M., & Stillwell, R. H. (1965). Probiotics: Growth promoting factors produced by microorganisms. *Science*, *147*(3659), 747–748.

Liu, W., Taylor, M. H., & Pearce, C. M. (2016). Nutritional value of selected species of microalgae for larvae and early post-set juveniles of the Pacific geoduck clam, *Panopea generosa*. *Aquaculture*, *452*, 326–341.

Macias-Sancho, J., Nolasco, H., Marquez, L., Vargas-Albores, F., & Gollas-Galván, T. (2014). Fishmeal substitution with *Arthrospira* (*Spirulina platensis*) in a practical diet for *Litopenaeus vannamei*: Effects on growth and immunological parameters. *Aquaculture*, *426–427*, 120–125.

Mahasneh, I. A. (1997). Production of single cell protein from five strains of the microalga *Chlorella* sp. (Chlorophyta). *Cytobiosciences*, *90*, 153–161.

Michael, E. T., Amos, S. O., & Hussaini, L. T. (2014). A review on probiotics application in aquaculture. *Fish & Aquaculture Journal*, *5*, 111.

Monir, S., Bagum, N., Rahman, S., Ashaf-Ud-Doulah, M., Bhadra, A., & Borty, S. C. (2015). Parasitic diseases and estimation of loss due to infestation of parasites in Indian major carp culture ponds in Bangladesh. *International Journal of Fisheries and Aquatic Studies*, *2*, 118–122.

Muñoz-Atienza, E., Araújo, C., Martínez, B., Campanero, C., del Campo, R., Hernández, P. E., & Margolles, A. (2013). Antimicrobial activity, antibiotic susceptibility and virulence factors of lactic acid bacteria of aquatic origin intended for use as probiotics in aquaculture. *BMC Microbiology*, *13*(1), 15.

Mustafa, A., Rankaduwa, W., & Campbell, P. (2001). *Canadian Veterinary Journal*, *42*(1), 54–56.

Nasseri, A. T., Rasoul-Amini, S., Morowvat, M. H., & Ghasemi, Y. (2011). Single cell protein: Production and process. *American Journal of Food Technology*, *6*, 103–116.

Nayak, S. K. (2010). Probiotics and immunity: A fish perspective. *Fish & Shellfish Immunology*, *29*(1), 2–14.

Nayak, S. K. (2021). Multifaceted applications of probiotic *Bacillus* species in aquaculture with special reference to *Bacillus subtilis*. *Reviews in Aquaculture*, *13*(2), 862–906.

Nguyen, T. H., Fleet, G. H., & Rogers, P. L. (1998). Composition of the cell walls of several yeast species. *Applied Microbiology and Biotechnology*, *50*, 206–212.

Nimrat, S., Suksawat, S., Boonthai, T., & Vuthiphandchai, V. (2012). Potential *Bacillus* probiotics enhance bacterial numbers, water quality and growth during early development of white shrimp (*Litopenaeus vannamei*). *Veterinary Microbiology*, *159*, 443–450.

Patil, D. W., & Markad, S. S. (2015). Probiotics as an alternative source for antibiotics and chemotherapeutics in aquaculture. In *Proceedings of National Conference on "Current Trends in Aquaculture (CTIA-2015)"* (pp. 16–25). *International Journal of Advance Research in Basic & Applied Science, Special Issue*.

Perez-Garcia, O., Escalante, F. M. E., de-Bashan, L. E., & Bashan, Y. (2011). Heterotrophic cultures of microalgae: Metabolism and potential products. *Water Research*, *45*, 11–36.
Qi, Z. Z., Zhang, X. H., Boon, N., & Bossier, P. (2009). Probiotics in aquaculture of China–Current state, problems and prospect. *Aquaculture*, *290*, 15–21.
Rao. (n.d.). Bioremediation–The advanced biotechnology process to restore the health of aquaculture pond ecosystem.
Rumsey, G. L., Hughes, S. G., Smith, R. R., Kinsella, J. E., & Shetty, K. J. (1991). Digestibility and energy values of intact, disrupted and extracts from dried yeast fed to rainbow trout (*Oncorhynchus mykiss*). *Animal Feed Science and Technology*, *33*, 185–193.
Rumsey, G. L., Winfree, R. A., & Hughes, S. G. (1992). Nutritional value of dietary nucleic acids and purine bases to rainbow trout (*Oncorhynchus mykiss*). *Aquaculture*, *108*, 97–110.
Sahoo, P. K., Mohanty, J., Garnayak, S. K., Mohanty, B. R., Kar, B., Prasanth, H., & Jena, J. K. (2013). Estimation of loss due to argulosis in carp culture ponds in India. *Indian Journal of Fisheries*, *60*, 99–102.
Santos, L., & Ramos, F. (2018). Antimicrobial resistance in aquaculture: Current knowledge and alternatives to tackle the problem. *International Journal of Antimicrobial Agents*, *52*(2), 135–143.
Shelby, R. A., Lim, C., Yildirim-Aksoy, M., & Delaney, M. A. (2006). Effects of probiotic diet supplements on disease resistance and immune response of young Nile tilapia, *Oreochromis niloticus*. *Journal of Applied Aquaculture*, *18*(2), 23–34.
Shewale, R. N., Sawale, P., Khedkar, C., & Singh, A. (2014). Selection criteria for probiotics: A review. *International Journal of Probiotics & Prebiotics*, *9*, 17–22.
Stentiford, G. D., Sritunyalucksana, K., Flegel, T. W., Williams, B. A., Withyachumnarnkul, B., Itsathitphaisarn, O., & Bass, D. (2017). New paradigms to help solve the global aquaculture disease crisis. *PLoS Pathogens*, *13*(2), e1006160.
Subasinghe, R. P., Bondad-Reantaso, M., & McGladdery, S. (2003). Recent technological innovations in aquaculture. *FAO Fisheries Circular*, *886*, 85.
Subedi, B., & Shrestha, A. (2020). A review: Application of probiotics in aquaculture. *International Journal of Forestry, Animal and Fisheries Research*, *4*(5).
Teimouri, M., Amirkolaie, A. K., & Yeganeh, S. (2013). The effects of *Spirulina platensis* meal as a feed supplement on growth performance and pigmentation of rainbow trout (*Oncorhynchus mykiss*). *Aquaculture*, *396–399*, 14–19.
Tuan, T., Minh Duc, P., & Hatai, K. (2013). Overview of the use of probiotics in aquaculture. *International Journal of Research in Fisheries and Aquaculture*, *3*, 89–97.
Valipour, A., Falahatkar, B., & Akhavan, S. R. (2019). Dietary *Lactobacillus plantarum* affected on some immune parameters, air-exposure stress response, intestinal microbiota, digestive enzyme activity and performance of narrow clawed crayfish (*Astacus leptodactylus*, Eschscholtz). *Aquaculture*, *504*. https://doi.org/10.1016/j.aquaculture.2019.01.064 (accessed on 10th July 2025).
Vidakovic, A., Langeland, M., Sundh, H., Sundell, K., Olstorpe, M., & Vielma, J. (2015). Evaluation of growth performance and intestinal barrier function in Arctic charr (*Salvelinus alpinus*) fed yeast (*Saccharomyces cerevisiae*), fungi (*Rhizopus oryzae*) and blue mussel (*Mytilus edulis*). *Aquaculture Nutrition*. https://doi.org/10.1111/anu.12344 (accessed on 10th July 2025).

Villamil, L., Reyes, C., & Martínez-Silva, M. A. (2014). *In vivo* and *in vitro* assessment of *Lactobacillus acidophilus* as probiotic for tilapia (*Oreochromis niloticus*, Perciformes: Cichlidae) culture improvement. *Aquaculture Research*, *45*(7), 1116–1125.

Vine, N. G., Leukes, W. D., Kaiser, H., Daya, S., Baxter, J., & Hecht, T. (2004). Competition for attachment of aquaculture candidate probiotic and pathogenic bacteria on fish intestinal mucus. *Journal of Fish Diseases*, *27*(6), 319–326.

Wang, Y. B., & Han, J. Z. (2007). The role of probiotic cell wall hydrophobicity in bioremediation of aquaculture. *Aquaculture*, *269*(1), 349–354.

Zhao, L., Wang, W., Huang, X., Guo, T., Wen, W., & Feng, L. (2015). The effect of replacement of fish meal by yeast extract on the digestibility, growth and muscle composition of the shrimp *Litopenaeus vannamei*. *Aquaculture Research*. https://doi.org/10.1111/are.12883 (accessed on 10th July 2025).

Zheng, C., & Wang, W. (2016). Effects of *Lactobacillus pentosus* on the growth performance, digestive enzyme and disease resistance of white shrimp, *Litopenaeus vannamei* (Boone, 1931). *Aquaculture Research, 48.* https://doi.org/10.1111/are.13110 (accessed on 10th July 2025).

CHAPTER 8

Environmental Influence of Aerators in Aquaculture Operations Using Life Cycle Assessment (LCA) Approach

SUBHA M. ROY,[1,*] MIRZA MASUM BEG,[2] and TAEHO KIM[1]

[1]*Smart Aquaculture Research Centre, Chonnam National University, Yeosu, Jeonnam, Republic of Korea*

[2]*Department of Fisheries Science, Ramnagar College, Vidyasagar University, West Bengal, India*

*Corresponding author

ABSTRACT

Artificial aerators are extensively used in aquaculture operations for better fish production. During the use of aerators, the pond environment deteriorates and affects surrounding ecosystems. Therefore, it is necessary to determine the effects of artificial aerators on the environment using the life cycle assessment (LCA) approach. LCA measures the environmental influence of a product by considering the main stages of a product's life. These stages include raw material acquisition, processing, and product life. In the present study, environmental effects were assessed for various artificial aerators, i.e., paddle wheel aerator (PWA), spiral aerator (SA), propeller aerator (PA), circular stepped cascade aerator (CSCA), pooled circular stepped cascade aerator (PCSCA), and perforated tray aerator (PTA). The results showed that paddle wheel aerators provide higher efficiency but also result in greater emissions in terms of kg CO_2 equivalent due to higher energy use. Spiral aerators provide lower efficiency than paddle wheel aerators, but the environmental impact is also lower. In

Sustainable Aquaculture: Recent Advances and Techniques. Dhanaji W. Patil, Sandip S. Markad (Eds.)

DOI: 10.1201/9781779641113-8

small ponds, PCSCA, CSCA, and PA are generally used to achieve higher efficiency, remove ammonia and volatile organic matter, and reduce CO_2 emissions in the environment.

8.1 INTRODUCTION

Intensive and extensive aquaculture operations require aeration systems. These systems operate when dissolved oxygen (DO) falls to critical levels, during absence of sunlight, and in high stocking density ponds (Roy et al., 2022a, b). The primary function of artificial aeration is to supply DO to the fishpond. Moreover, the pond ecosystem may be affected by aeration systems (Cheatham et al., 2023; Roy et al., 2024). Aeration plays an important role in increasing DO levels in ponds. Various aerators are manufactured for use in aquaculture ponds. The popular aerators include paddle wheel aerator (PWA), spiral aerator (SA), propeller aerator (PA), circular stepped cascade aerator (CSCA), pooled circular stepped cascade aerator (PCSCA), and perforated tray aerator (PTA), which are used in both intensive and extensive aquaculture operations. The capital cost and maintenance expenses are reasonable for the above-mentioned aerators. These types of aerators are suitable for aquaculture (Beg et al., 2023; Boyd and McNevin, 2021; Roy et al., 2017).

Recently, scientists and researchers have developed many assessment tools to evaluate the ecological effects on food production systems (Roy et al., 2023), including risk investigation, environmental footprint, energy studies, and life cycle analysis (LCA) (Liu et al., 2016; Wilfart et al., 2013). LCA is regarded as one of the leading tools to estimate environmental impacts. It provides an inclusive analysis of pertinent environmental effects over a product's entire life cycle (Ghamkhar et al., 2020; Henriksson et al., 2012; Mungkung et al., 2013). LCA is a comprehensive process that considers inputs and outputs throughout the entire procedure to estimate the significant effects on a functional unit (d'Orbcastel et al., 2009; Hou et al., 2023). LCA consists of four steps, i.e., goal definition and scope, inventory, impact analysis, and interpretation (Ayer and Tyedmers, 2009; ISO, 2006). For goal definition and scope, a system boundary and functional unit should be defined. In the inventory phase, existing data, inputs, and outputs for various stages of the process are collected. Impact analysis is an essential step in LCA; this step includes evaluation methods, characterization, normalization, and weighting. The interpretation step measures the emission rate throughout the whole process (Haslawati et al., 2022; Vasquez-Mejia et al., 2023). Life cycle assessment (LCA) is now the most effective method for determining and contrasting the environmental effects of various industries.

Artificial aerators are extensively used in aquaculture operations to enhance fish production. However, during the use of aerators, the pond environment deteriorates, negatively affecting the surrounding ecosystems (Roy et al., 2022a). Therefore, it is essential to determine the environmental effects of artificial aerators using the LCA approach. LCA measures the environmental influence of a product by considering key stages in its life cycle, including raw material acquisition, processing, and product use. Very few studies have addressed the LCA of aerator manufacturing. In the present study, data for aerator manufacturing were collected from the literature, and LCA was employed to assess PWA, SA, PA, CSCA, PCSCA, and PTA in terms of their environmental impacts.

The outcomes might extend the life cycle of fundamental databases for aquaculture and enhance the environmental performance of aerator production operations (Arias et al., 2018; Bohnes and Laurent, 2021). Furthermore, the results serve as a guide for environmental impact analyses of other aquaculture tools utilizing life cycle assessment (LCA) (Ayuso-Virgili et al., 2022; Turolla et al., 2020; Roy et al., 2020). Keeping in view the above points, the present investigation was focused on the LCA of different types of aerators. The specific objectives of the study are as follows:

- To evaluate the performances of different types of aerators on the basis of standard aeration efficiency.
- To assess the environmental impact of different types of aeration systems using life cycle assessment (LCA).

8.2 MATERIALS AND METHOD

8.2.1 LIFE CYCLE ASSESSMENT (LCA) FRAMEWORK

According to the International Organization for Standardization (ISO) 14040 series, LCA consists of four steps (Figure 8.1): (i) goal and scope definition; (ii) inventory; (iii) impact assessment; and (iv) Interpretation.

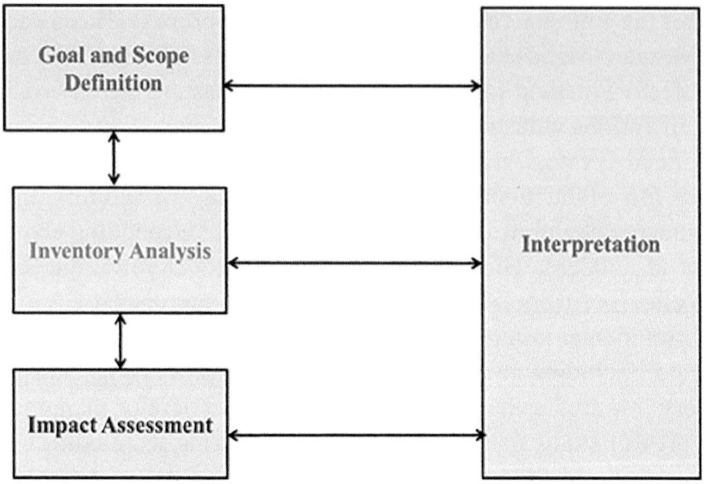

FIGURE 8.1 Phases of a life cycle analysis of aerators.

1. **Goal and Scope Definition:** First, specify the object or product, and then select the boundary condition. The functional unit should be chosen at this stage.
2. **Inventory Analysis:** Collect existing data—inputs and outputs—for various stages of the process. The inventory includes all inputs and outputs across the entire life span of the product.
3. **Impact Assessment:** Impact analysis is the essential step in LCA. It includes evaluation methods, characterization, normalization, and weighting.
4. **Interpretation:** This step involves measuring the emission rate throughout the entire process.

8.2.2 DESCRIPTION OF THE AERATORS

In the past, various kinds of aerators have been used in aquaculture applications. Aerators, as a broader classification, generally fall into three distinct categories: firstly, the splash aerator; secondly, the diffused-air aerator; and finally, the gravity aerator (Roy et al., 2021). Among splash aerators, the most commonly used are the PWA and SA. The PWA involves a revolving wheel splattering water on a vertical surface into the air, allowing the water body to absorb oxygen (Roy et al., 2021). Similarly, the SA is a modified version of the PWA that uses the spiral wheel's tangential spin to shoot water into the air (Roy et al., 2017). A frame, floats, motor, connection, wheel, bearing, and speed reduction mechanism are included in both the PWA and SA.

The propeller aerator (PAA) is a simple type of bubbling aerator (Kumar et al., 2010). The PAA consists of a propeller, hollow pipe, and electric motor. The motor is joined to a revolving hollow pipe, and the propeller is connected to the motor. There is a pressure difference inside the hollow pipe as a result of the propeller's high-velocity rotation into the water body.

The CSCA and PCSCA are common types of gravity aerators. The CSCA is made up of circular steps where water is evenly dispersed and allowed to fall from the top step to the next and lowest step. Beneath the circular stepped cascade, a submersible pump fitted with a control valve regulates the water flow. The pooled circular stepped cascade (PCSCA) is a modified version of the CSCA (Kumar et al., 2013). The zigzag pattern of water flow over the PCSCA aerator increases the water body's dissolved oxygen concentration. The schematic diagrams of aerator manufacturing processes are shown in Figure 8.2.

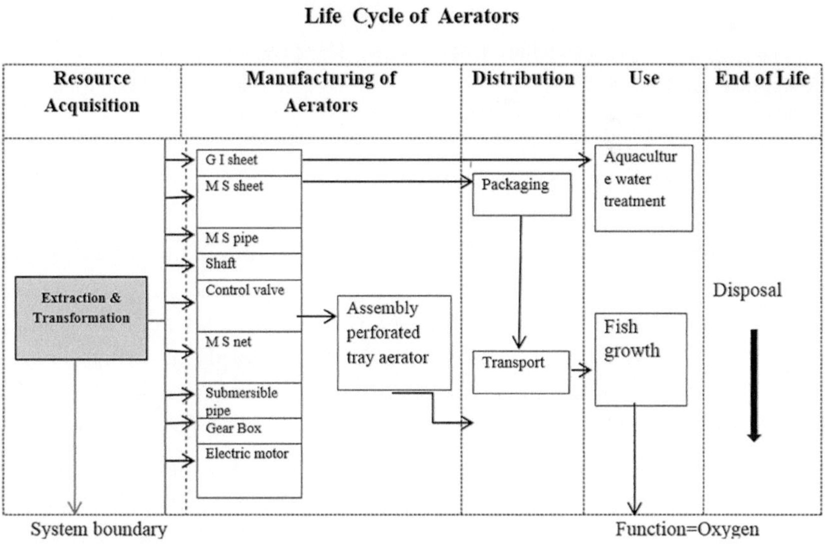

FIGURE 8.2 Schematic diagrams of aerators manufacturing process.

8.2.3 DEFINING THE FUNCTIONAL UNIT OF AERATORS

The functional units of various aerators are shown in Table 8.1. The assessment of impact in LCA is an important step. This step comprises characterization, normalization, and weighting. The characterization (Equation 1) and normalization (Equation 2) steps of LCA are as follows:

$$\text{Characterization Results} = \sum m_i \times \text{Characterization Factor} \quad (1)$$

where; m_i expresses the results of the input or output of the ith substance within the system

$$\text{Normalization Results} = \frac{\textit{Characterization results}}{\textit{Normalization reference value}} \quad (2)$$

TABLE 8.1 Functional Units of Various Aerators

Products	Primary Function	Secondary Function
PWA	Increase efficiency	Remove volatile matter in pond; removes dissolved iron and manganese, carbon dioxide, hydrogen sulfide, etc.
SA	–	
PAA	–	
CSCA	–	
PCSCA		
PTA		
Product	Function unit = service provided	Key parameters
Material of aerator		Oxygen transfer, aeration efficiency

8.3 RESULTS AND DISCUSSION

The performance evaluations of six different aerators were carried out in a laboratory. At first, the DO was dropping to zero, and then operating the aerators, as shown in Figure 8.3, the DO level was increasing linearly with time. In the beginning, the DO level increased sharply and then slowly until it reached the saturation point at around 7.9 mg/L (Beg et al., 2024). It was found that the PWA and SA could increase the DO level rapidly, while the other types of aerators increased the DO level more slowly. This is because the PWA and SA run the paddle vertically at the water surface, which creates more water bubbles containing more DO in the water body. These aerators cover more surface area for aeration. Therefore, paddle wheel and spiral aerators are the first choice of the farmers. In the other four types, the aerator also increases the DO level within a specific time interval and is applicable in intensive aquaculture systems. Generally, CSCA and PCSCA are very useful in small-scale aquaculture operations (Roy et al., 2020).

It is shown that paddle wheel aerators have higher oxygen transfer rate and aeration efficiency. In PCSCA and CCAS aerators, oxygen transfer rate and efficiency are comparatively lower than paddle wheel aerators. These phenomena occur because the paddle wheel impeller diameter is larger, causing an increased water circulation rate. Additionally, in spiral aerators, the cup angle throws water a longer distance, which increases the oxygen transfer rate and covers a larger area. Propeller aspirator aerators are generally used in small pond sizes and also remove volatile organic matter, carbon dioxide, hydrogen sulfide, etc. PTA aerators, attached with submersible pumps, also increase SOTR and remove carbon dioxide, volatile organic matter, H_2S, etc.

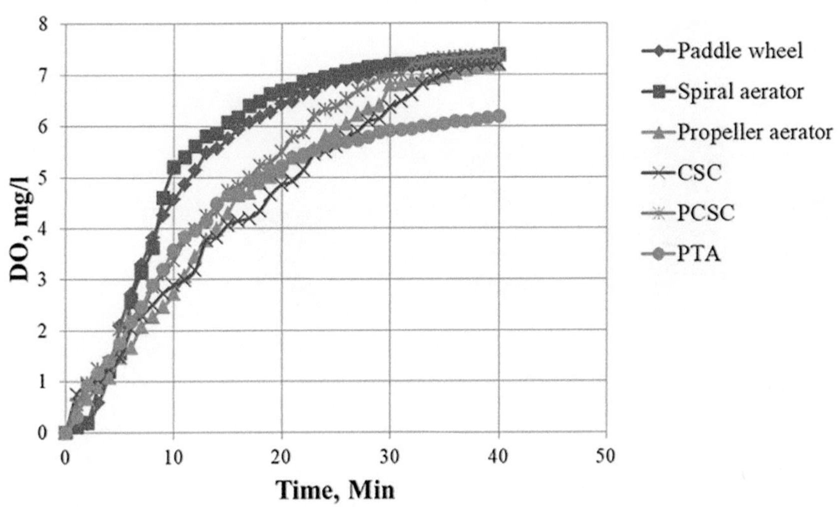

FIGURE 8.3 Relationship between DO and time.

8.3.1 *IMPACT ASSESSMENT OF DIFFERENT AERATORS*

A quantifiable analysis known as life cycle assessment (LCA) is used to examine how products and services affect the environment over the course of their entire life cycles.

8.3.2 PADDLE WHEEL AERATOR

PWA consists of several materials that have caused substantial changes in environmental impacts. It can be seen from Figure 8.4 that CO_2 emission from the PWA aerator is higher than acidification potential (AP) and abiotic depletion potential (ADP). This is due to the greater use of stainless steel, HDPE, oil cleaning, and iron, which contribute to CO_2 emissions. Therefore, it is suggested to use environmentally friendly materials to reduce CO_2 emissions. It is observed that the ADP is 19.87%, which has a harmful effect on the environment.

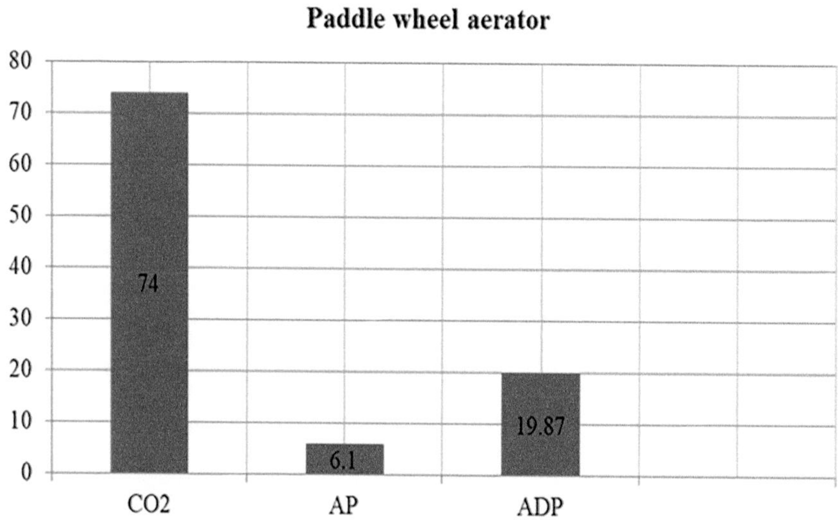

FIGURE 8.4 Emission of paddle wheel.

8.3.3 SPIRAL AERATOR

SA is the most widely used aerator in fish farming. It can be seen from Figure 8.5 that CO_2 and ADP emissions from the spiral aerator are higher than AP. This is due to the increased use of stainless steel, HDPE, oil cleaning, and iron, which contribute to CO_2 and ADP emissions. Therefore, it is essential to replace the materials in spiral aerators to reduce CO_2 emissions.

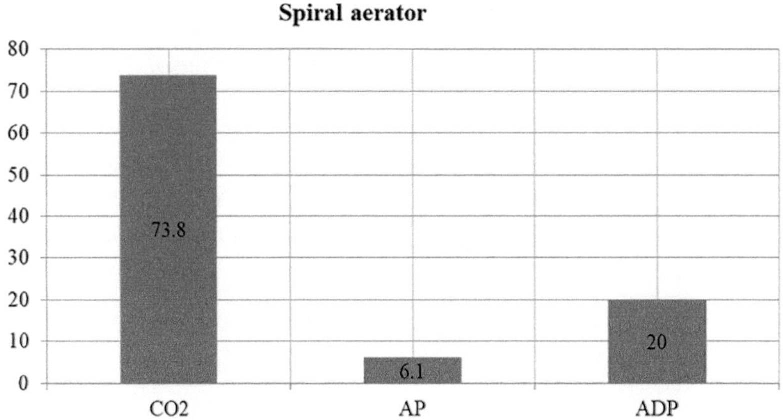

FIGURE 8.5 Emission of spiral aerator.

8.3.4 PROPELLER ASPIRATOR AERATOR

PAA is a bubbling-type aeration system, primarily used in indoor aquaculture systems. It can be seen from Figure 8.6 that 73.8%, 6.15%, and 20% of CO_2, AP, and ADP emissions, respectively, are emitted from propeller aerators, with CO_2 emissions being the highest. This is due to the greater use of stainless steel, oil cleaning, and iron, which contribute to CO_2 and ADP emissions.

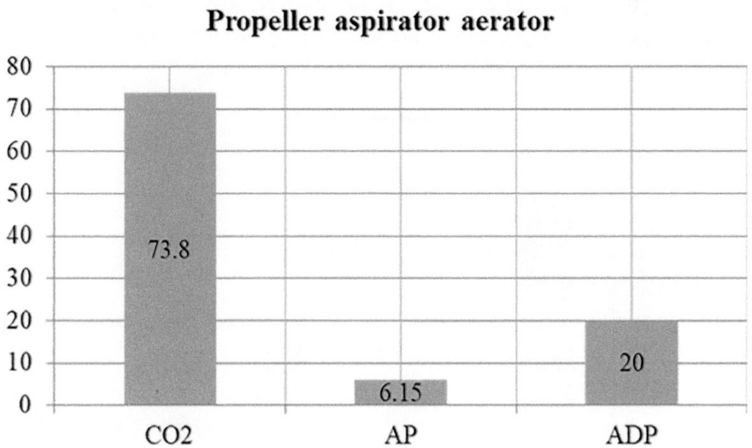

FIGURE 8.6 Emission propeller aspirator aerator.

8.3.5 PERFORATED TRAY AERATOR

PTA is an example of a gravity aeration system, primarily used in wastewater treatment. It can be seen from Figure 8.7 that 71.4%, 6.14%, and 22.44% of CO_2, AP, and ADP emissions, respectively, are observed in the perforated tray aerator. The PTA aerator consists of MS sheet and GI sheet, both of which contribute significantly to CO_2 and ADP emissions. Therefore, it is essential to replace these materials to reduce CO_2 emissions.

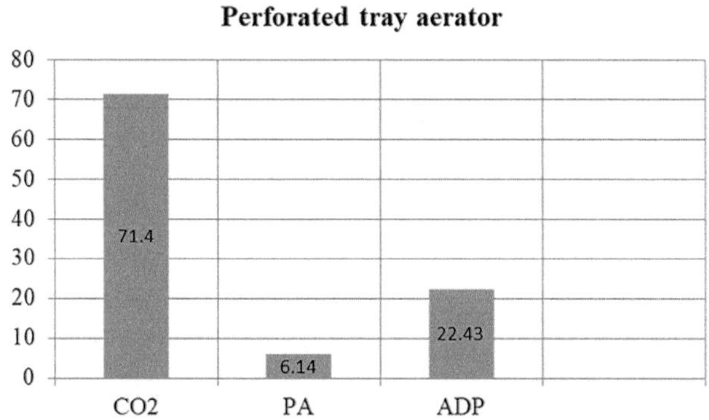

FIGURE 8.7 Emission of perforated tray aerator.

8.3.6 CIRCULAR CASCADE AERATOR

CCA is a gravity aeration system where water falls step by step through gravity. It can be seen from Figure 8.8 that 68.58%, 7.8%, and 23.6% of CO_2, AP, and ADP emissions, respectively, are emitted from circular cascade aerators. The system consists entirely of GI sheet, including the pipe, which is a major contributor to CO_2 emissions.

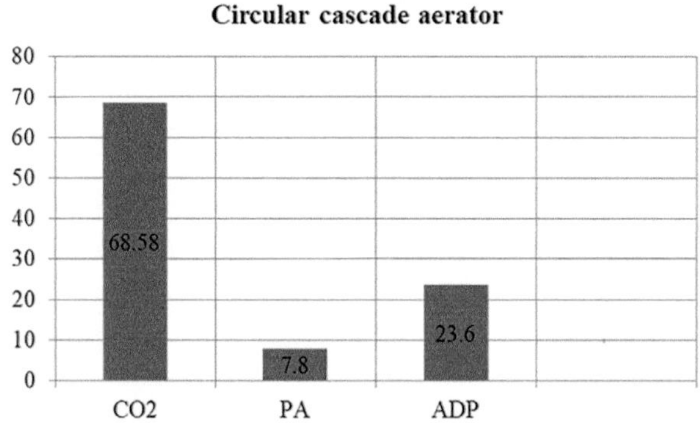

FIGURE 8.8 Emission of circular cascade aerator.

8.3.7 PERFORATED CIRCULAR CASCADE AERATOR

The PCCA is a modified version of the CCA. It can be seen from Figure 8.9 that 68.59%, 8.37%, and 23.03% of CO_2, AP, and ADP emissions, respectively, are generated by perforated circular cascade aerators. It consists of MS sheet in each step and GI sheet in the pipe and barrier. In this case, the MS sheet contributes more to CO_2 emissions, while AP and ADP emissions are primarily from the GI sheet.

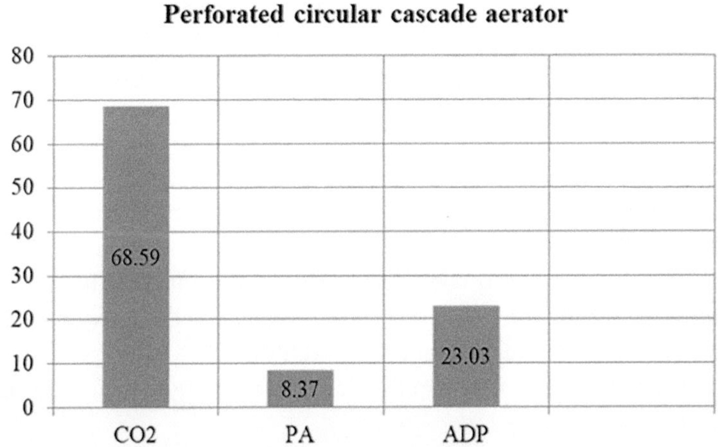

FIGURE 8.9 Emission of perforated circular cascade aerator.

8.3.8 COMPARATIVE LCIA OF DIFFERENT SCENARIOS

Table 8.2 presents the inventory results of six aerators. The materials used for making aerators were traced back and their potential impacts analyzed.

TABLE 8.2 Comparative LCIA of Different Aerators

Impact Category	Global Warming Potential (GWP) (kg CO_2)	Acidification Potential (AP) (kg AP)	Abiotic Depletion Potential (ADP) (kg ADP)	Total Impact
PWA	8793.6	726.8	2362.1	11882.5
SA	9912.4	826.7	2684.5	13423.6
PPA	6748.8	562.4	1827.8	9139
PTA	5710.7	491.4	1794	7996.1
CSCA	5458.12	621.42	1878.16	7957.7
PCSCA	5912.7	721.48	1985.17	8619.35

It can be seen (Table 8.2) that maximum GWP, AP, and ADP emissions are produced by the spiral aerator. This is due to the use of different types of materials such as steel, rubber, plastic PP, and UPVC, which are used in spiral aerators. These types of materials contribute significantly to environmental pollution. On the other hand, PTA, PCCA, and CCA use GI and MS sheets and also employ submersible pumps, which help to reduce environmental pollution.

The LCA results show that the use of aerators in fish farming significantly affects total life cycle impacts (98% to 100%), which is consistent across all six aerators. This is mainly due to the operation of the aerator for an average of 10 hours per day during the fish culture period. The extraction of materials required for aerator assembly contributes very little to the overall impacts, as the lifespan of an aerator is relatively long—approximately 10 years with proper maintenance under operating conditions.

When comparing the impacts of the six aerators (Figure 8.10), the magnitude of impacts caused by the spiral aerator is higher than that of the others in all categories considered. This is a result of its higher energy consumption during operation on the farm, which is the maximum among all aerators. The potential impacts on ADP, GWP, and AP are shown to be significantly higher due to electricity consumption across all aerator categories.

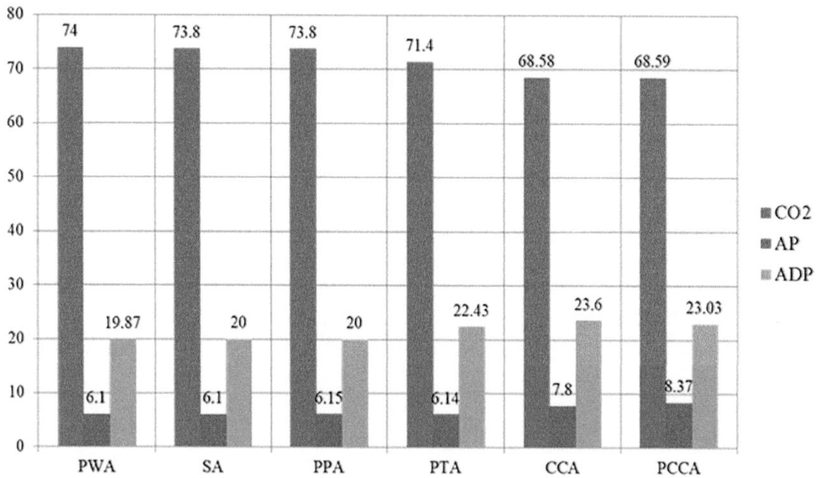

FIGURE 8.10 Comparative LCIA of different aerators.

To avoid emissions during aerator operation, the use of solar energy is the most viable solution. Solar energy is widely adopted in various fields due to its lack of greenhouse gas emissions. Remote locations that are disconnected from the electrical grid often benefit from solar photovoltaic installations. The greatest energy source for fishpond aerators is solar photovoltaic technology because of its adaptability, which directly improves environmental performance and fishery productivity. As a result, the development of novel aerators should focus on enhancing energy sources. It is crucial to produce aerators that utilize solar energy instead of grid electricity or diesel power.

8.4 CONCLUSIONS

In the present study, the environmental effects during fish culture were evaluated based on various artificial aerators, i.e., PWA, SA, PA, CSCA, PCSCA, and PTA. The results showed that paddle wheel aerators provide greater efficiency but also emit more kilograms of CO_2 equivalent due to higher energy use. Spiral aerators offer lower efficiency than paddle wheel aerators, and their environmental impact is also lower. In small ponds, PCSCA, CSCA, and PA are generally used, which provide higher efficiency while also removing ammonia, organic volatile matter, etc., and result in lower CO_2 emissions into the environment. The conclusions drawn from the present investigation are as follows:

- PCSCA provides the maximum oxygen transfer rate and aeration efficiency.
- PCSCA and CCA materials result in low emissions of CO_2, ADP, and AP.
- PWA and SA aerators produce the highest emissions of CO_2, ADP, and AP. Most parts of PPA aerators can be recycled and reused.

ACKNOWLEDGMENT

The research was supported by the Korea Institute of Marine Science and Technology Promotion (KIMST), funded by the Ministry of Oceans and Fisheries (RS-2022-KS221676).

KEYWORDS

- **abiotic depletion**
- **aerators**
- **electrical grid**
- **emissions**
- **life cycle assessment**
- **organic volatile matter**
- **solar photovoltaic installations**
- **spiral aerators**

REFERENCES

Applebaum, J., Mozes, D., Steiner, A., Segal, I., Barak, M., Reuss, M., & Roth, P. (2001). Aeration of fishponds by photovoltaic power. *Progress in Photovoltaics: Research and Applications*, *9*(4), 295–301.

Ayer, N. W., & Tyedmers, P. H. (2009). Assessing alternative aquaculture technologies: Life cycle assessment of salmonid culture systems in Canada. *Journal of Cleaner Production*, *17*(3), 362–373.

Ayuso-Virgili, G., Jafari, L., Lande-Sudall, D., & Lümmen, N. (2023). Linear modelling of the mass balance and energy demand for a recirculating aquaculture system. *Aquacultural Engineering*, *101*, 102330.

Beg, M. M., Roy, S. M., Moulick, S., & Mandal, B. (2023). Quality evaluation of organically farmed fish fillet of Indian major carps. *Aquaculture International*, *31*(5), 2743–2761.

Beg, M. M., Roy, S. M., Moulick, S., Mandal, B., Kim, T., & Mal, B. C. (2024). Economic feasibility study of organic and conventional fish farming systems of Indian major carps. *Scientific Reports*, *14*(1), 7001.

Bohnes, F. A., & Laurent, A. (2021). Environmental impacts of existing and future aquaculture production: Comparison of technologies and feed options in Singapore. *Aquaculture*, *532*, 736001.

Boyd, C. E., & McNevin, A. A. (2021). Aerator energy use in shrimp farming and means for improvement. *Journal of the World Aquaculture Society*, *52*(1), 6–29.

Breitenstein, M., & Hicks, A. (2022). Review and harmonization of the life cycle global warming impact of five United States aquaponics systems. *Aquacultural Engineering*, *96*, 102224.

Cheatham, M., Kumar, G., Tucker, C., & Rutland, B. (2023). Research verification of four commercial scale split-pond designs. *Aquacultural Engineering*, *103*, 102349.

d'Orbcastel, E. R., Blancheton, J. P., & Aubin, J. (2009). Towards environmentally sustainable aquaculture: Comparison between two trout farming systems using life cycle assessment. *Aquacultural Engineering*, *40*(3), 113–119.

Ghamkhar, R., Boxman, S. E., Main, K. L., Zhang, Q., Trotz, M. A., & Hicks, A. (2021). Life cycle assessment of aquaculture systems: Does burden shifting occur with an increase in production intensity? *Aquacultural Engineering*, *92*, 102130.

Haslawati, B., Saadiah, I., Siti-Dina, R. P., Othman, M., & Latif, M. T. (2022). Environmental assessment of giant freshwater prawn, *Macrobrachium rosenbergii* farming through life cycle assessment. *Sustainability*, *14*(22), 14776.

Henriksson, P. J., Guinée, J. B., Kleijn, R., & de Snoo, G. R. (2012). Life cycle assessment of aquaculture systems—A review of methodologies. *The International Journal of Life Cycle Assessment*, *17*, 304–313.

Hou, C., Zhu, L., Zheng, Y., Shi, L., Tan, B., & Zhang, S. (2023). Effects of dietary peptidoglycan on *Litopenaeus vannamei*: Growth performance, disease resistance, non-specific immunity and transcriptome analysis of immune response. *Aquaculture Reports*, *31*, 101676.

ISO. (2006b). *Environmental Management–Life Cycle Assessment–Requirement and Guidelines* (ISO 14044).

Kumar, A., Moulick, S., & Mal, B. C. (2010). Performance evaluation of propeller-aspirator-pump aerator. *Aquacultural Engineering*, *42*(2), 70–74.

Kumar, A., Moulick, S., & Mal, B. C. (2013). Selection of aerators for intensive aquacultural pond. *Aquacultural Engineering*, *56*, 71–78.

Liu, Y., Rosten, T. W., Henriksen, K., Hognes, E. S., Summerfelt, S., & Vinci, B. (2016). Comparative economic performance and carbon footprint of two farming models for producing Atlantic salmon (*Salmo salar*): Land-based closed containment system in freshwater and open net pen in seawater. *Aquacultural Engineering*, *71*, 1–12.

Mungkung, R., Aubin, J., Prihadi, T. H., Slembrouck, J., van der Werf, H. M., & Legendre, M. (2013). Life cycle assessment for environmentally sustainable aquaculture management: A case study of combined aquaculture systems for carp and tilapia. *Journal of Cleaner Production*, *57*, 249–256.

Roy, S. M., Machavaram, R., Moulick, S., & Mukherjee, C. K. (2022). Economic feasibility study of aerators in aquaculture using life cycle costing (LCC) approach. *Journal of Environmental Management*, *302*, 114037.

Roy, S. M., Machavaram, R., Pareek, C. M., & Kim, T. (2024). Investigating the performance of a perforated pooled circular stepped cascade aeration system for intensive aquaculture. *Helion*, 1–12.

Roy, S. M., Moulick, S., & Mal, B. C. (2017). Design characteristics of spiral aerator. *Journal of the World Aquaculture Society*, *48*(6), 898–908.

Roy, S. M., Moulick, S., & Mukherjee, C. K. (2020). Design characteristics of perforated pooled circular stepped cascade (PPCSC) aeration system. *Water Supply*, *20*(5), 1692–1705.

Roy, S. M., P, J., Machavaram, R., Pareek, C. M., & Mal, B. C. (2021). Diversified aeration facilities for effective aquaculture systems—A comprehensive review. *Aquaculture International*, *29*, 1181–1217.

Roy, S. M., Pareek, C. M., Kim, T., & Mal, B. C. (2023). Artificial intelligence-based approach for improving the aeration efficiency of a perforated plate aeration system. *Multiscale and Multidisciplinary Modeling, Experiments and Design*, 1–14.

Roy, S. M., Pareek, C. M., Machavaram, R., & Mukherjee, C. K. (2022a). Optimizing the aeration performance of a perforated pooled circular stepped cascade aerator using hybrid ANN-PSO technique. *Information Processing in Agriculture*, *9*(4), 533–546.

Roy, S. M., Tanveer, M., & Machavaram, R. (2022b). Applications of gravity aeration system in aquaculture—A systematic review. *Aquaculture International*, *30*(3), 1593–1621.

Turolla, E., Castaldelli, G., Fano, E. A., & Tamburini, E. (2020). Life cycle assessment (LCA) proves that Manila clam farming (*Ruditapes philippinarum*) is a fully sustainable aquaculture practice and a carbon sink. *Sustainability*, *12*(13), 5252.

Vasquez-Mejia, C. M., Shrivastava, S., Gudjónsdóttir, M., Manzardo, A., & Ögmundarson, Ó. (2023). Current status and future research needs on the quantitative water use of finfish aquaculture using life cycle assessment: A systematic literature review. *Journal of Cleaner Production*, 139009.

Wilfart, A., Prudhomme, J., Blancheton, J. P., & Aubin, J. (2013). LCA and emergy accounting of aquaculture systems: Towards ecological intensification. *Journal of Environmental Management*, *121*, 96–109.

CHAPTER 9

Water Quality Parameters and Their Management in Aquaculture

SHAFIYA MUSHTAQ,[1] GOWHAR IQBAL,[2,*] SAIMA JAN,[1] and ARYA SINGH[2]

[1]*Faculty of Fisheries, Sher-e-Kashmir University of Agricultural Sciences and Technology of Kashmir (SKUAST-K), Rangil, Ganderbal, Jammu and Kashmir, India*

[2]*ICAR–Central Institute of Fisheries Education, Panch Marg, Mumbai, Maharashtra, India*

*Corresponding author

ABSTRACT

Water quality directly impacts the survival, growth, and overall health of aquatic species raised for food. One of the most important factors influencing the success of aquaculture is water quality. The physical, chemical, and biological characteristics of water are the three primary factors that determine its quality. However, even slight variations in parameters such as pH, temperature, or DO can stress the organism and result in behavioral or physiological effects. Growth and reproductive potential are impacted by alterations in, or deterioration of, water quality. In such conditions, the risk of disease also increases. Solutions for managing water quality are designed to improve the standard of the water. Business owners and fish farmers in the aquaculture sector should be knowledgeable about the fundamentals of water quality management strategies in order to minimize problems and effectively utilize water resources for both economic and environmental sustainability.

Sustainable Aquaculture: Recent Advances and Techniques. Dhanaji W. Patil, Sandip S. Markad (Eds.)

DOI: 10.1201/9781779641113-9

9.1 INTRODUCTION

"Water quality" for aquaculturists refers to the properties of water that facilitate the successful reproduction of the target organisms. The specific organisms to be cultured determine the required water quality, which comprises many interdependent elements. The complex relationships between components necessitate addressing the composition of the full array, even though a component may occasionally be handled independently. Growth and survival are influenced by numerous ecological conditions and management strategies, which ultimately determine the output. High stocking densities of fish or crustaceans usually exacerbate problems with pond water quality and sediment deterioration.

Fish and other aquatic species carry out regular activities such as feeding, swimming, spawning, metabolism, body development, and excretion in the physical, biological, and chemical environments defined by water quality parameters. In many parts of the world, fish and other aquatic species are recognized as significant cash crops and are distributed through aquaculture systems. Aquaculture requires proper water quality because fish life is governed by the aquatic environment. Fish culture may suffer if environmental conditions are not within the required range for healthy growth. Water quality issues that can lead to poor development or even fish mortality should be a primary concern for fish farmers (Boyd, 1978). The goal of water quality management is to control environmental factors to create the ideal habitat for fish survival and growth. Water availability plays a major role in determining the success or failure of an aquaculture enterprise. The aquatic environment consists of numerous interacting variables. Aquaculture practitioners must be aware of potential stressors for fish. According to Das et al. (2015), these variables may also help identify the root causes of problems in fish rearing.

9.2 WATER QUALITY MANAGEMENT

A dynamic network of physical, biological, and chemical elements that make up water quality shapes the aquatic environment and affects fish and other aquatic organisms. Pond fish culture involves many aspects of water quality. In comparison to ponds with low water quality, fish in high-quality ponds are typically larger and healthier. According to Piper et al.

(1982), a fish culture operation's success or failure is largely dependent on the quality of the water. The following are important variables that impact water quality: temperature, turbidity, oxygen, CO_2, nitrogen, ammonia, pH, alkalinity, hardness, and others.

9.2.1 TEMPERATURE

Water temperature is a physical attribute that indicates how hot or cold it is. Solar radiation directly affects both water and air temperatures. The chemical and physical characteristics of water can vary with temperature. Most of the heat is absorbed by dissolved organic matter and particulate matter in the upper layers of the water, as light energy is absorbed exponentially with depth. Water density is influenced by temperature. The development, metabolism, and reproduction of aquatic species are all affected by water temperature. Temperature also serves as a cue for many animals to time their migration and breeding seasons. All living organisms have a maximum temperature threshold, although plants and animals generally grow more rapidly in warmer environments. The ideal growth temperature of target species should be considered when selecting them for aquaculture. The quickest way to measure water temperature is to submerge a calibrated mercury or digital thermometer in the water.

9.2.2 TURBIDITY

The decomposition of organic waste produces humic compounds, suspended particulates, and planktonic organisms, collectively known as "turbidity." Unlike turbidity from suspended particles, planktonic turbidity in aquaculture ponds is typically beneficial (McCombie, 1953). However, excessive blooms reduce the effective area of the productive zone by obstructing light and heat penetration. According to Table 9.1, the optimal Secchi disk visibility in fish ponds is between 30 and 40 cm. Fish may become stressed or even die at night when dissolved oxygen concentrations drop in ponds with Secchi disk visibility of 10 to 20 cm (Romaire and Boyd, 1978). Applications of 250–500 kg/ha of gypsum, 25–50 kg/ha of alum, or 500–1000 kg/ha of organic manure can help reduce turbidity caused by suspended solids.

TABLE 9.1 Guidelines for Assessing the Visibility of Secchi Disks in Pond Aquaculture

Secchi Disk Reading	Remarks
< 20 cm	Pond too turbid. If turbidity is due to phytoplankton, there may be problems with low dissolved oxygen concentrations. If turbidity is from suspended soil particles, productivity will be low.
20–30 cm	Turbidity becoming excessive
30–45 cm	If turbidity is from phytoplankton, the pond is in good condition
45–60 cm	Phytoplankton becoming scarce
> 60 cm	Water is too clear. Inadequate productivity and risk of aquatic weed problems

9.2.3 DISSOLVED OXYGEN (DO)

In water, oxygen is primarily obtained from two sources: photosynthetic plankton and the atmosphere. Adequate oxygenation of aquatic organisms is more challenging than that of terrestrial organisms due to the limited solubility of oxygen in water, which is further reduced by higher temperature, salinity, low air pressure, high humidity, and plankton blooms. According to Bhattagar and Garg (2000), fish that are starved or whose growth is stunted are more likely to die from oxygen deficiency in the water. Dissolved oxygen affects the physiology, behavior, distribution, growth, and survival of aquatic organisms (Solis, 1988). Fish swimming slowly in low DO conditions risk injury if they surface in ponds with Secchi disk readings of less than 20 cm. Fish gulping at the surface is a sign of low DO levels in the pond (Figure 9.1).

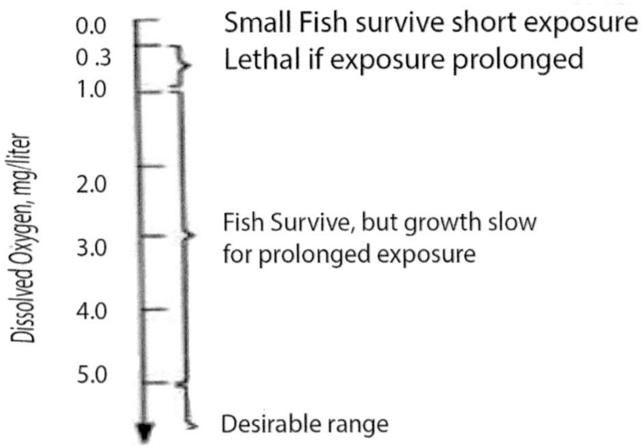

FIGURE 9.1 Relationship between DO and survival rate of fish.

9.2.4 BIOCHEMICAL OXYGEN DEMAND (BOD)

Biochemical oxygen demand (BOD) is the total amount of dissolved oxygen required by bacteria to break down organic materials such as food particles and sewage. Due to excessive industrial and domestic sewage from non-point sources, along with surplus phosphate, rural ponds may carry a high organic load, which increases BOD levels.

9.2.5 CARBON DIOXIDE (CO_2)

Carbon dioxide in fish ponds is mainly produced by minute organisms that form part of the biota and through fish respiration. Ponds also accumulate CO_2 from the decomposition of organic waste. Naturally, fish farmers are concerned about maintaining appropriate dissolved oxygen levels. Due to daily fluctuations in dissolved oxygen and carbon dioxide levels, there is concern about the potential toxicity of CO_2. When dissolved oxygen is low, CO_2 concentrations are usually at their peak. The highest carbon dioxide levels occur in winter, while the lowest are observed in summer. During winter, the concentration of dissolved oxygen often exceeds saturation, so carbon dioxide is rarely a concern. According to Boyd (1978), although freshwater fish can tolerate high CO_2 levels, the ideal concentration of free CO_2 for them is 3 mg/L. To reduce excess CO_2 concentrations, aeration and pH elevation can be applied. As per Adhikari's (2006) experiment, hydrated lime at a concentration of 1.0 mg/L has the potential to eliminate 1.68 mg/L of free CO_2.

9.2.6 pH

pH is calculated as the negative logarithm of the hydrogen ion concentration. The concentration of the acidic gas carbon dioxide has a significant impact on the pH of natural streams (Boyd, 1979). Fish blood pH normally ranges from 7.4 to 8.5, although a slight deviation within the range of 7.0 to 8.5 is generally suitable and beneficial for fish growth and reproduction. In water with a pH range of 4.0 to 6.5 and 9.0 to 11.0, fish can become stressed. At pH values below 4.0 or above 11.0, fish mortality is almost certain (Figure 9.2).

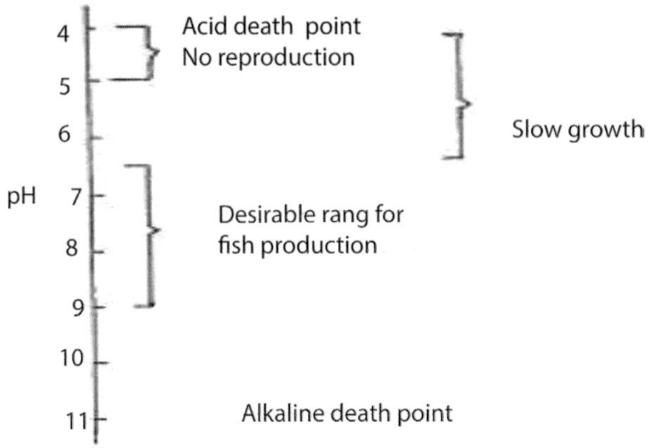

FIGURE 9.2 Relationship between pH and survival rate of fish.

Several factors influence the pH in a pond environment, including:

- Precipitation rate near pond areas;
- Water with inadequate buffering capacity;
- Stocking density;
- Rate of sludge buildup and pond bottom feeding;
- Presence of macro/microorganisms in pond water, including phytoplankton;
- Rate of carbon dioxide production in pond water.

9.2.7 ALKALINITY

Alkalinity is the ability of water to neutralize acids without significantly raising pH. Total alkalinity refers to the sum of bicarbonate and carbonate alkalinities. Some types of water may contain only bicarbonate alkalinity and no carbonate alkalinity. Regardless of the method of production, fish growth depends on the carbonate buffering mechanism. The components of this system—carbonate and bicarbonate—store excess carbon dioxide. They help maintain a stable environment for photosynthesis, which is the main natural source of oxygen in ponds, and prevent daily fluctuations in carbon dioxide levels.

In the absence of a buffering system, pH may drop to around 4.5 at night due to the accumulation of carbonic acid, a weak acid formed from free carbon dioxide. During the peak of photosynthesis, carbon dioxide is mostly consumed by phytoplankton, causing the pH to rise above 10. Pond water containing less than 300 mg/L and less than 20 mg/L of $CaCO_3$ is ineffective. The ideal range for total alkalinity in freshwater fish ponds is between 50 and 300 mg/L of $CaCO_3$.

9.2.8 HARDNESS

Although water hardness and alkalinity are similar, hardness is more complex. It is a crucial factor in fish farming and a commonly discussed issue related to water quality. In addition to measuring calcium and magnesium hardness, other ions such as hydrogen, iron, manganese, strontium, zinc, and aluminum are also considered. Fish require calcium and magnesium for various biological functions. These minerals can be absorbed either from the water or through the food they consume. For optimal growth of aquatic organisms, a water hardness of at least 30 mg/L should be maintained. Low hardness levels can be increased by applying agricultural lime.

9.2.9 AMMONIA

The biological converter's starting state is determined by ammonia readings. The ideal range of unionized ammonia in pond water is 0.02–0.05 mg/liter, as fish are quite sensitive to this nutrient. Fish are unable to properly absorb energy from their food when ammonia concentrations reach dangerous levels. The fish will grow lethargic, eventually enter a state of unconsciousness, and die if the ammonia level rises too high. Ammonia rarely approaches fatal levels in properly managed fish ponds. Still, at concentrations below the lethal threshold, ammonia can have "sub-lethal" effects such as slowed development, poor feed conversion, and decreased disease resistance. Fish feces are the primary source of ammonia in fish ponds. Fish-feeding ponds get their ammonia primarily from the protein in the meal. Diffusion from the sediment is a major source of ammonia in fish ponds. Ammonia seeps into the water through the sediment as a result of this organic material decomposing. Ammonia loss or modification can occur via two primary pathways. Plants, particularly algae, are the primary absorbers of ammonia. For plants to flourish, nitrogen is one nutrient they need. In fish pond ammonia transformation, "nitrification" is one of the most significant processes. Nitrate (NO_3) and nitrite (NO_2) are the two distinct types of ammonia that bacteria can produce. Ammonia's toxicity can be decreased by aerating the air. Well-functioning phytoplankton eliminates ammonia from water. Formalin has the potential to lower ammonia levels (Figure 9.3).

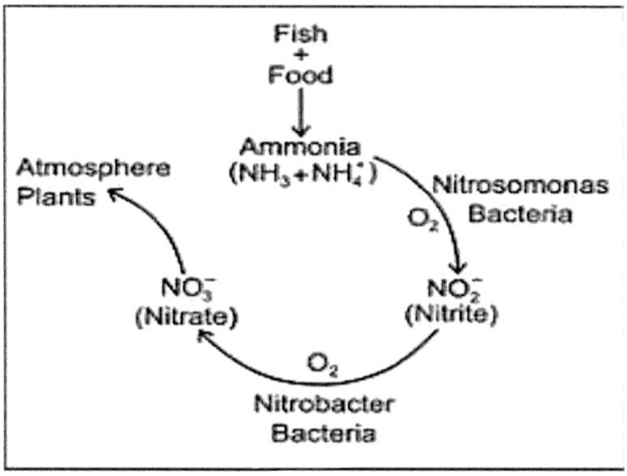

FIGURE 9.3 Relation between ammonia and survival rate of fish.

9.2.10 *NITRITE*

The second chemical indicator used to determine how well the biological converter functions is nitrite. A functional bio-converter should eliminate the presence of nitrite in pond water. *Nitrosomonas*, an autotrophic bacterium, uses ammonia and oxygen in the bioconverter and, to a lesser extent, on the pond walls to produce nitrite. Nitrite is referred to as the "invisible killer." It can be lethal in quantities as low as 0.25 ppm, especially for smaller fish. The most effective methods of preventing hazardous nitrite accumulation in pond culture are effective organic waste collection, suitable aeration, and optimum fertilizer delivery.

9.2.11 *PHOSPHOROUS*

The most crucial component for preserving pond fertility is often known to be phosphorus. It may resemble one of three forms:

- The inorganic phosphate soluble in water, phosphorus (PO_4);

- The particulate organic phosphorus found in plankton, detritus, and sedimentation; and
- Soluble organic phosphorus.

The PO_4 form, sometimes called dissolved phosphorus or soluble inorganic phosphate phosphorus, is the form utilized in production out of the three. The body needs it to divide cells and synthesize fat, protein, and high-energy molecules like AMP, ADP, and ATP. Acidic soil requires the use of phosphate fertilizers in addition to lime because phosphorus becomes inaccessible in acidic pond soil and remains in compound form with Fe, Al, Mn, and Zn.

9.2.12 HYDROGEN SULFIDE

Fish ponds that are freshwater should not contain hydrogen sulfide. Fish undergo sub-lethal stress and lose balance at concentrations of 0.01 mg/liter hydrogen sulfide. Frequent exchange of water can help avoid the accumulation of hydrogen sulfide. Hydrogen sulfide's toxicity can also be reduced by increasing the water's pH through liming. In addition, 6.2 mg/liter of potassium permanganate is added to the water to remove hydrogen sulfide.

9.2.13 REDOX POTENTIAL

Redox potential is useful for determining how much oxidation or reduction has occurred. The anaerobic layer contains chemicals like H_2S, CO_2, NH_3, H_2SO_4, and other substances, while the aerobic layer contains chemical components like oxygen, carbon dioxide, and minerals. The amount of oxidation or reduction is correlated with the existence of microorganisms. The degree of Eh is one measure that shows how effectively soil and water can support prawn biomass. Photosynthetic bacteria (PSB) are important in semi-intensive culture as opposed to the algal population's primary output by absorbing and converting organic matter into minerals and nutrients. PSB can significantly improve the cultural atmosphere and are especially common due to the high illumination and low oxygen levels.

9.3 NUTRIENTS

Nutrients, an essential part of protein, regulate the aquatic environment. Phosphorus is sometimes considered the most crucial element alone in sustaining aquatic productivity while being a very small component. Dissolved inorganic nitrogen in the range of 0.2 to 0.5 mg/liter and phosphorus fertility for aquatic production in the range of 0.05 to 2.0 mg/liter are considered appropriate ranges for fish productivity. Silicate is still found in natural water in its silicate form, which is important for diatoms' structure (Moore, 1946). The nutritional state of the soil and water in fish ponds has the greatest impact on the development of plankton organisms (Banerjea, 1967). The ponds' nutrient levels can be increased by adding organic and inorganic fertilizers in carefully measured doses. However, as they can cause excessive plankton formation, algae blooms, and oxygen deprivation, increased nutrient levels have the potential to be harmful.

9.4 MANAGEMENT

Water quality indicators should be examined to ensure proper pond management and to avert circumstances hindering prawn growth. These elements can help identify problems and help develop solutions by aiding in diagnosis. Even though single statistics are usually not very informative, a combination of factors can indicate dynamic activity in the pond. Microbe and phytoplankton populations control the number of metabolites and oxygen in the pond. CO_2, pH, and DO variation—especially in its vertical profile—are indicators of their population. Because the primary factor influencing pH dial variation is CO_2, monitoring pH fluctuation may be adequate. In addition, quantifying CO_2 is more complicated.

Every day, measures are taken in the afternoon, between 2 and 3 p.m., and in the morning, between 5 and 6 a.m. This is equivalent to the times immediately preceding and following the photosynthetic peak. Thus, these parameters' greatest and minimum values occur throughout this period. Since the other metrics don't have a distinctive dial pattern, they can only be checked once a day, ideally at a consistent time. Alongside indicators of water quality, feeding and growth data must be displayed. This is because feeds provide the nutrients that cause algal blooms, and too much feed can quickly cause the quality of the water to decrease. Collecting data and conducting meticulous observation are useless if they don't influence judgments about water management. This becomes increasingly important as the cost of implementing different management techniques (aeration, water exchange, inputs) increases. Most problems with water quality can be rectified with enough water exchange. Therefore, if vast water supplies suitable for aquaculture were available, monitoring would not be as crucial and high production levels might be pursued. Pursuing a more intense culture increases the likelihood of encountering problems with water quality and sickness when water is scarce.

9.5 WATER EXCHANGE

When stocking density increases, keeping the water supply steady and maintaining high water quality is critical. Water exchange is currently the most efficient and widely used method to maintain good water quality, in addition to water quality boosters like sanitizers, zeolite, etc. Salinity control, excess metabolite removal, oxygen delivery to algae, and pond temperature regulation are all commonly achieved through water exchange. Natural productivity levels, turbidity, stocking density, total biomass, production duration, and water supply and volume all impact the exchange rate. Changing the water in a way that results in a gradual change in quality rather than an abrupt one is the rationale behind water exchange. In semi-intensive systems, water is exchanged at a modest flow rate, often—and sometimes even continuously. Small ponds may experience sudden additions of large volumes of water, which could stress the cultivated organisms by disrupting their environment. Therefore, massive water replacement is not recommended unless there is a rapid die-off of plankton, an extremely low oxygen level, or after chemical treatment. Continuous water exchange and paddlewheel activity are required to mix the pond water thoroughly. There will be wide regional differences in the pond's water quality, and the cultivated organisms' distribution on the pond bottom will be irregular. Reintroducing fresh water after the water level has been lowered is not recommended, especially in the summer. It is possible for the water's oxygen content to be reduced and the pond bottom deterioration process to accelerate when the temperature is raised while the water level is lowered. Before adding fresh water to the pond at the prearranged exchange rate, the paddlewheel should be turned on to homogenize the water throughout.

Surface and bottom water outflow are both worthy of consideration. The water quality at the lowest layer is usually worse than that of the upper layer. It is best to release surface water when there are scums, prawns or prawn droppings, or floating dead plankton. Freshwater with a lower density at the surface layer should be released to avoid salinity fluctuations during and after rainstorms. Reducing salinity by more than 5 ppt at a time with water exchange is not suggested. Sharp variations in salinity may cause the phytoplankton fauna and their population densities to alter, which could lead to ecosystem instability.

9.6 CONCLUSION

Water quality varies greatly between different geographic locations. Aquaculture can be practiced in some salty waters, and fish can even survive in water considered unsafe for human consumption. The quality of the water affects the growth and health of fish. Water quality should, therefore, be a top priority for aquaculture. Since water quality might fluctuate over time, it needs to be periodically checked. Knowing how to evaluate the observed water quality is also essential to preserving the well-being and health of the fish population. Aquaculture practices must maintain critical water quality indicators within optimal ranges to guarantee a sustainable fish yield. In recent years, nanotechnology has profoundly changed every field. Due to its antibacterial properties, nanotechnology has the potential to eradicate illnesses from aquaculture hatcheries and ponds. Nanosensors may provide an additional means of early pathogen and metabolite identification to preserve a healthy pond environment.

KEYWORDS

- **antibacterial properties**
- **aquaculture**
- **fish population**
- **geographic locations**
- **nanosensors**
- **nanotechnology**
- **phytoplankton**
- **water quality management**

REFERENCES

Adhikari, S. (2006). Soil and water quality management in aquaculture (pp. 1–30). In *Hand Book of Fisheries and Aquaculture*. Indian Council of Agricultural Research, New Delhi.

Banerjea, S. M. (1967). Water quality and soil condition of fish ponds in some states of India in relation to fish production. *Indian Journal of Fisheries*, *14*, 115–144.

Bhatnagar, A., & Garg, S. K. (2000). Causative factors of fish mortality in still water fish ponds under sub-tropical conditions. *Aquaculture*, *1*(2), 91–96.

Boyd, C. E. (1978). *Water Quality in Warm Water Fish Ponds* (Technical Bulletin No. 47). Alabama Agricultural Experiment Station, Auburn, Alabama, p. 132.

Boyd, C. E. (1979). *Water Quality in Warmwater Fish Ponds*. Agriculture Experiment Station, Auburn, Alabama, pp. 359.

Cole, B. A., & Boyd, C. E. (1986). Feeding rate, water quality, and channel catfish production in ponds. *Progressive Fish-Culturist*, *81*, 25–29.

Das, P., Khogen, S. S., Mandal, S. C., & Bhagabati, S. K. (2015). Management of water quality in fish ponds for maximizing fish production.

McCombie, A. M. (1953). Factors influencing the growth of phytoplankton. *Journal of the Fisheries Research Board of Canada*, *10*, 253–282.

Moyle, J. B. (1946). Some indices of lake productivity. *Transactions of the American Fisheries Society*, *76*, 322–334.

Piper, R. G., McElwain, I. B., Orme, L. E., McCraren, J. P., Flower, L. G., & Leonard, J. R. (1982). *Fish Hatchery Management*. U.S. Fish and Wildlife Service, Washington, D.C.

Romaire, R. P., & Boyd, C. E. (1978). Predicting night time oxygen depletion in catfish ponds. *Alabama Agricultural Experiment Station Bulletin*, *505*, Auburn University, Auburn.

Solis, N. B. (1988). *The Biology and Culture of Penaeus monodon* (Department Papers, pp. 3–36). SEAFDEC Aquaculture Department, Tigbauan, Iloilo, The Philippines.

CHAPTER 10

The Role of Novel Probiotics in Aquaculture

RAHUL GOGOI,[1] SAMIM DULLAH,[2] MADHURJYA RANJAN SHARMA,[1] MADHURJYA PROTIM BORAH,[3] PRIYADARSHINI DEKA,[4] JUGABRATA DAS,[5] and SUDIPTA SANKAR BORA[6,*]

[1]*Department of Agricultural Biotechnology, Assam Agricultural University, Jorhat, Assam, India*

[2]*Nanda Nath Saikia College, Titabar, Assam, India*

[3]*Department of Biosciences and Bioengineering, Indian Institute of Technology Jammu, Jammu and Kashmir, India*

[4]*Hereditary Biosciences, Guwahati, Assam, India*

[5]*College of Horticulture and Farming System Research, Assam Agricultural University, Nalbari, Assam, India*

[6]*DBT-North East Centre for Agricultural Biotechnology (DBT-NECAB), Assam Agricultural University, Jorhat, Assam, India*

*Corresponding author

ABSTRACT

Aquaculture/aquafarming is emerging as one of the fastest-growing food-producing industries. Aquaculture involves cultivating various water resources, including fish and other aquatic mollusks, insects, and seaweed. The demand for aquaculture products is rising to meet the world's food and nutritional needs. The aquaculture industry faces many challenges, such as increasing feed efficiency, improving disease resistance, and

Sustainable Aquaculture: Recent Advances and Techniques. Dhanaji W. Patil, Sandip S. Markad (Eds.)

DOI: 10.1201/9781779641113-10

supporting aquatic organism development. Probiotics and microbiological strategies can be used to address these obstacles. Probiotics can accelerate the growth of aquatic life. They can offer disease resistance in addition to raising water quality. Probiotics are highly effective in enhancing nutrient digestion. Microorganisms and probiotics are considered a promising alternative for sustainable aquaculture. This book chapter covers various microbial strategies, probiotics, and their mode of action, along with their application to aquaculture. It also emphasizes the precautions to take and the potential benefits of employing microorganisms as a productive tactic in aquaculture.

10.1 INTRODUCTION

Presently, the aquaculture industry is facing the huge challenge of meeting the daily requirement for diet, particularly protein, essential vitamins, and minerals. A fraction of people on the planet depend on aquatic products to meet their daily nutritional requirements (Naylor et al., 2021). Moreover, due to the discharge of pollutants into the aquatic ecosystem, the water quality of various water bodies is changing drastically. Temperature, dissolved oxygen, and salinity fluctuations alter the environments that are beneficial to certain organisms, changing the dominant species and potentially reducing the efficacy of aquaculture production (van Vliet et al., 2023). This change has resulted in deteriorating health conditions of aquatic species, which is another reason why the aquaculture industry is facing serious challenges in meeting the vast need for aquatic food globally, as the majority of the world population is dependent on aquaculture products, particularly for protein intake.

The use of microbe-mediated enhancement for aquatic life has grown in popularity recently. According to the World Health Organization (WHO) and Food and Agriculture Organization (FAO) definitions, such potential microorganisms with host health-promoting traits and activities have been termed probiotics or beneficial microbes (Patil and Markad, 2015; Sehnal et al., 2021). The interaction of microorganisms with aquatic organisms occurs either through the gills or through the food supplied, which ultimately reaches the gut. Many pathogenic microbes, e.g., *Vibrio* spp., *Plesiomonas shigelloides*, *Aeromonas* spp., cause diseases in aquatic organisms. Aquaculture commonly uses antibiotics to prevent infections caused by these types of bacteria. However, the excessive use of antibiotics has led to the emergence of multiple antibiotic-resistant harmful microbes. Therefore, it has become important to find alternative solutions for disease prevention. Numerous nations, including the USA and Canada, have banned specific antibiotics in the aquaculture sector. In certain situations, antibiotics may be substituted with probiotics, as they offer resistance against various pathogens and thus help fight infections (Larsson and Flach, 2022).

Novel vaccinations, immunostimulants, and disinfection techniques have been developed to combat the disease problem in the aquaculture industry. In recent years, microorganisms and probiotics have offered enormous potential for disease management in aquatic organisms (Reverter et al., 2020). Probiotic strains that are administered can prevent pathogen colonization through competitive exclusion. The most commonly used probiotics in aquaculture usually belong to the genera *Bacillus* spp., *Bifidobacterium* spp., *Lactobacillus* spp., and yeasts (Iqbal et al., 2021). The benefits of utilizing probiotics in various aquaculture products, including frogs, shrimp, mollusks, and fish, are currently the subject of research. There are several advantages of using microorganisms and probiotics in aquaculture, such as enhancement of food's nutritional value, microbial enzymatic digestion, infection prevention, growth promotion, immune system stimulation, and water quality improvement for farming (Arun et al., 2023).

This chapter discusses the microbiological communities in aquaculture systems, the relationship between microbial equilibrium and sustainable aquaculture, interactions between probiotics and aquatic organisms, and the role of probiotics in promoting gut health and immune function in aquatic species. It describes the mechanisms of action and application modes of probiotics in aquaculture systems and the potential of indigenous microbial communities for sustainable aquaculture practices. It also outlines the application of microbial antagonists for disease prevention in aquaculture, the influence of probiotics and microbial interventions on water quality, nutritional enhancement of aquatic organisms, and waste management of aquatic ecosystems, as well as sustainability considerations and ecological impacts of microbial aquaculture solutions. It mentions the challenges and regulatory considerations of microbial interventions in aquaculture, along with the technological and research challenges in probiotic development and application. Furthermore, it discusses the prospects and implications for the sustainable development of aquaculture.

10.2 MICROBIAL COMMUNITIES IN AQUACULTURE SYSTEMS

Health management in aquaculture is a challenging aspect since it relies on several factors such as antibiotics, chemotherapeutics, targeted vaccines, and selective breeding to produce disease-resistant lines. Controlling cultured animals and their surrounding environment is a holistic approach to aquaculture health management. Research has demonstrated that microbial populations within and surrounding fishes directly contribute to their productivity and disease resistance (Lorgen-Ritchie et al., 2023). The associated microbial communities within and around the fishes comprise fungi, bacteria, viruses, archaea, and protozoa. Microbiomes can sometimes also contain severe opportunistic pathogens. The surrounding environment shapes the makeup of this type of host microbiome. A metatranscriptomic-based investigation discovered fungal transcripts to be more prevalent than bacterial transcripts in the gut of gilthead sea breams (*Sparus aurata*). These fungi were also shown to be involved in endocrine, digestive, and host immune functions in the fish (Naya-Català et al., 2022). Another investigation into the wood-eating (*Panaque nigrolineatus*) catfish of the Amazon reported the associated mycobiome (fungal diversity) to vary in different gut regions (Marden et al., 2017).

10.2.1 RELATIONSHIP BETWEEN MICROBIAL BALANCE AND SUSTAINABLE AQUACULTURE

The nutritional and physiological benefits of microbial community management for aquaculture's cultivated species are substantial. While using helpful bacteria in aquaculture has great potential for sustainability, other microorganisms pose major risks to both human health and production. Aquatic animals can be easily infected by an assortment of protist, bacterial, and viral diseases that can aggregate and cause mass mortality or accumulate in mollusk tissue, posing a risk for human consumption. These microbial threats also result in significant economic losses (Kunselman et al., 2024).

10.2.1.1 INTERACTIONS BETWEEN PROBIOTICS AND AQUATIC ORGANISMS

The application of probiotics in fish growth is an environment-friendly method. Various commercial probiotic preparations containing one or more live microorganisms have the potential to greatly enhance aquatic organism cultivation. When added to food, probiotics can be mixed with it or added directly to the culture tank. *Biostart* is one of the first commercial probiotics derived from *Bacillus* isolates (Martínez et al., 2012). With the addition of certain microorganisms, the digestive systems of aquatic animals can be improved, as they generate extracellular enzymes that aid in digestion processes—such as lipases and proteases—which help transport essential growth ingredients, including vitamins and fatty acids. Microbiota, in addition to the activity of bacteria in the digestive tract, may function as an additional food source. Vitamins and vital amino acids may also be present in good amounts (Xianghong et al., 2000).

10.2.2 ROLE OF PROBIOTICS IN GUT HEALTH AND IMMUNE FUNCTION

A balanced probiotic microbial community is important for maintaining gut health. The natural microbial populations in the gut are disturbed during disease conditions, leading to several health issues. Fish health can be effectively improved by restoring and maintaining the gut microbial community, supported by appropriate dietary probiotic selection and supplementation (Asaduzzaman et al., 2018). Ingestion of three probiotics—*Shewanella* sp. AFG21, *Bacillus* sp. AHG22, and *Alcaligenes* sp. AFG22—changed the microbial makeup to include more beneficial bacterial populations in the gut of *Tor tambroides*. Probiotics also improved immunostimulation in cultured aquatic species and decreased the impact of diseases. A mixed culture of *Saccharomyces cerevisiae*, *Clostridium butyricum*, *Lactobacillus acidophilus*, and *Bacillus subtilis* improved nonspecific immune parameters, including neutrophil migration, lysozyme activity, and plasma bactericidal activity, among other immunological markers, in Nile tilapia (*Oreochromis niloticus*). Moreover, it strengthened defenses against *Edwardsiella tarda* infections (Taoka et al., 2006). Fish given a diet rich in probiotics (*Flavobacterium sasangense*, *Vibrio lentus*, and *Aeromonas veronii*) exhibited enhanced immune systems, possibly due to humoral and cellular immune responses, as well as the expression of IL-1β, TNF-α, and lysozyme-C. In addition, improvements were observed in blood leukocyte phagocytic activity, complement component C3, albumin and immunoglobulin levels, myeloperoxidase, lysozyme, respiratory burst activity, and several other immune parameters in various fish species (Dawood and Koshio, 2016).

10.2.3 MICROBIAL CONTROL OF AQUATIC PATHOGENS

Aquaculture confronts numerous challenges from pathogens that threaten economic viability and environmental sustainability. In recent years, interest has increased in adopting microbial control strategies to manage these aquatic pathogens effectively (Saxena et al., 2015). This section delves into the mechanisms of biocontrol, the application of microbial antagonists, and the integration of microbial control with probiotic approaches within aquaculture.

10.2.3.1 PROBIOTIC-MEDIATED BIOCONTROL ACTIVITIES

Bacterial response to an environment depends on many factors, including nutrient availability, biotic and abiotic stresses, and the presence of pathogens. Antagonism is the most common phenomenon, as it helps maintain the equilibrium between beneficial and potentially harmful microbes. Thus, antagonism serves as a feasible tool to minimize or eliminate the presence of opportunistic pathogens by (i) competition between colonies and adhesion sites; (ii) competition for nutrient and energy sources; and (iii) competition for iron.

10.2.3.1.1 Competition between colonies and adhesion sites

The alimentary canal can become colonized by microorganisms and remain inhabited for a prolonged period (Gatesoupe, 2008). Probiotic bacteria, particularly *Bacillus coagulans*, *B. mesentericus*, and *Bifidobacterium infantis*, have successfully colonized and persisted within the digestive system of the freshwater ornamental fish *Puntius conchonius*.

These bacteria also demonstrated significant competitive inhibition of harmful gut microbes (Divya et al., 2012). According to a study on fish intestinal mucus comparing five probiotics to two pathogens, the presence of probiotics on the mucus prevented one of the pathogens from adhering. The study concluded that successful probiotic bacteria generally attach to and colonize the gut mucosa, preventing infection spread. Additionally, this colonization facilitates the removal of pathogens from the contaminated gastrointestinal tract (Vine et al., 2004).

10.2.3.1.2 Competition for nutrient and energy sources

The composition of the intestinal tract microbiota or the surrounding environment of cultivated aquatic species can be significantly influenced by competition for vital nutrients. The availability of substrates for other bacterial populations may be reduced by increasing the abundance of certain bacterial strains, such as *Lactobacillus* and *Bacillus*, through probiotic use (Fooks and Gibson, 2002).

10.2.3.1.3 Competition for iron

During intense iron competition, bacteria can intensify siderophore biosynthesis and utilize mechanisms to dissolve precipitated iron, promoting microbial growth. Successful bacterial pathogens are capable of competing for iron from the host's tissues and bodily fluids in extremely iron-limited environments (Verschuere et al., 2000). In one experiment, *Pseudomonas fluorescens* grown in an iron-free culture inhibited the growth of *Vibrio anguillarum*, whereas the supernatant from iron-enriched cultures had no effect—indicating that iron limitation was the inhibitory factor. Additionally, a siderophore-producing probiotic may deprive potential pathogens of iron (Gram et al., 1999).

10.2.3.2 BIOCONTROL AND COMPETITIVE EXCLUSION MECHANISMS IN PATHOGEN MANAGEMENT

In aquaculture, where pathogens pose significant threats to the health and productivity of aquatic organisms, biocontrol strategies leveraging competitive exclusion mechanisms have gained prominence as practical tools for pathogen management. Biocontrol strategies rely on the use of beneficial microbes to reduce the proliferation and activity of pathogenic organisms, thereby enhancing the health and resilience of aquatic populations. Central to biocontrol is competitive exclusion, in which beneficial microbes outcompete pathogenic counterparts for resources and ecological niches within the aquatic environment. This competition for space, nutrients, and attachment sites impedes the establishment and growth of pathogenic organisms, reducing the frequency and severity of illnesses in aquaculture systems (Rahman et al., 2023).

A notable example of competitive exclusion involves certain strains of *Bacillus* spp. and *Pseudomonas* spp., which produce antimicrobial compounds capable of preventing harmful bacteria from growing and becoming more virulent (Lalloo et al., 2010). These beneficial bacteria colonize the surfaces of fish and shellfish, effectively outcompeting pathogens for attachment sites and strengthening the host's defenses against microbial invaders.

Moreover, probiotic microorganisms employed in biocontrol initiatives play a multifaceted role in enhancing host immunity and resilience against pathogenic challenges. Probiotics stimulate the production of antimicrobial peptides, lysozyme, and immunoglobulins within the host, fortifying its innate defense mechanisms. Additionally, probiotics modulate the composition and activity of the host's microbiota, fostering a microbial community conducive to health and disease resistance.

Furthermore, biocontrol strategies extend beyond direct antagonism toward pathogenic organisms to encompass broader ecological interactions within aquaculture ecosystems. By promoting the establishment and persistence of beneficial microbial communities, biocontrol measures bolster ecosystem resilience and stability, reducing the vulnerability of aquaculture systems to disease outbreaks and environmental fluctuations.

Integrating biocontrol and competitive exclusion mechanisms represents a holistic approach to pathogen management in aquaculture, offering sustainable solutions that mitigate reliance on conventional antimicrobial agents and chemical interventions. By harnessing the natural antagonistic interactions between beneficial and pathogenic microorganisms, aquaculturists can foster healthy and productive aquatic environments, safeguarding the sustainability and profitability of aquaculture operations for generations to come (Perez et al., 2016).

10.2.4 APPLICATION OF MICROBIAL ANTAGONISTS FOR DISEASE PREVENTION IN AQUACULTURE

Disease outbreaks in aquaculture systems can lead to significant economic losses and environmental degradation. Applying microbial antagonists stands out as a promising strategy for disease prevention and control. This section explores the diverse applications of microbial antagonists and their role in safeguarding the health and productivity of aquatic organisms in aquaculture systems.

Microbial antagonists, including bacteriophages and predatory bacteria, offer targeted solutions for controlling specific pathogens that threaten aquaculture species. Bacteriophages—often called phages—are viruses that infect and lyse bacterial cells, making them potent candidates for combating bacterial diseases in aquaculture (Oliveira et al., 2012). These bacteriophages can be isolated from the aquatic environment or engineered to target specific pathogenic strains, providing a precise and environmentally friendly approach to disease management.

Predatory bacteria, such as *Bdellovibrio and Like Organisms* (BALOs), exhibit a predatory lifestyle by actively preying on other bacteria through invasion and consumption (George et al., 2023). These predatory bacteria have demonstrated efficacy in controlling bacterial pathogens, including notorious disease-causing *Vibrio* spp. in aquaculture species like shrimp and fish. By selectively targeting pathogenic bacteria, predatory bacteria contribute to suppressing disease outbreaks and maintaining aquaculture health. Furthermore, microbial antagonists offer several advantages over traditional chemical treatments and antibiotics, including greater specificity, reduced environmental impact, and minimal risk of inducing antimicrobial resistance in target pathogens. By harnessing natural antagonistic interactions between microorganisms, aquaculturists can mitigate disease risks while minimizing adverse effects on non-target organisms and ecosystems.

Applying microbial antagonists in aquaculture extends beyond disease prevention to include broader ecological considerations, such as ecosystem resilience and sustainability. By selectively targeting pathogenic organisms, microbial antagonists help maintain a balanced microbial community structure, reducing the risk of dysbiosis and promoting the establishment of a healthy and resilient aquatic ecosystem (Li et al., 2022). Moreover, microbial antagonists contribute to reducing chemical inputs and antibiotic usage in aquaculture, aligning with the principles of sustainable aquaculture and environmental stewardship. By integrating microbial antagonists into disease management strategies, aquaculturists can enhance the sustainability and profitability of aquaculture operations while safeguarding the health and well-being of aquatic organisms and ecosystems.

10.3 MICROBIAL BIOPROSPECTING FOR AQUACULTURE

10.3.1 EXPLORATION OF BENEFICIAL MICROBIAL STRAINS FOR AQUACULTURE

Probiotics are being proposed as an efficient and environmentally responsible substitute for antibiotics. Probiotics are classified as either living or dead microorganisms—or even parts of them—that function in various ways to benefit the host or its surroundings. Probiotics offer numerous benefits for host fish, transforming the study of aquatic animals into an active research field due to their diverse applications. These bacteria generate a variety of compounds, including proteases, siderophores, hydrogen peroxide, bacteriocins, and lysozymes, which exert bacteriostatic or bactericidal effects on other pathogenic microbial populations (El-Saadony et al., 2021). Furthermore, certain bacteria produce volatile fatty acids and organic acids—such as lactic, acetic, butyric, and propionic acids—that can reduce the pH in the gastrointestinal lumen and inhibit opportunistic pathogenic microbes (Zorriehzahra et al., 2016).

It has been shown that several probiotics used in aquaculture exhibit antibacterial properties against specific pathogens. For example, probiotic *Lactococcus lactis* RQ516 fed to tilapia (*Oreochromis niloticus*) showed inhibitory action against *Aeromonas hydrophila* (Simón et al., 2021). The growth of *Listeria innocua* was shown to be inhibited by lactic acid bacteria (*Lactobacillus buchneri*, *L. fermentum*, *L. acidophilus*, *Lactococcus lactis*, and *Streptococcus salivarius*) extracted from the gut of Spanish mackerel (*Scomberomorus commerson*) (Yoon and Kim, 2022). The prawn *Litopenaeus vannamei*, when fed the probiotic *Bacillus megaterium*, developed greater resistance to white spot syndrome virus (WSSV) (Sahu et al., 2008).

The ferric iron-chelating molecules known as siderophores are small-molecular-weight substances that release iron from iron complexes or dissolve precipitated iron. Because siderophore-producing bacteria store ferric iron in iron-limited environments, they can prevent the growth of harmful bacteria. The growth of *Vibrio anguillarum* was suppressed by culture supernatant of *Pseudomonas fluorescens* grown in iron-limited conditions. It has been shown that *P. fluorescens*, by competing for free iron, can competitively suppress the growth of the fish disease agent *Aeromonas salmonicida* (Gram et al., 1999).

The physiological and behavioral characteristics of aquatic species can be directly affected by changes in water parameters. Probiotic bacteria can be added to water or administered through feed to prevent stress, strengthen the immune system, and reduce the negative impacts of various environmental stressors (Taoka et al., 2006). Rearing water temperature, pH, dissolved oxygen, NH_3, and H_2S were all improved by probiotic administration, thereby preserving a favorable and healthy environment for prawn and prawn larvae in green water systems (Hlordzi et al., 2020).

Quorum sensing is the process of regulating gene expression in response to changes in the number of cells in a given population. The marine red alga *Delisea pulchra* yields halogenated furanones, which act as potential quorum sensing (QS) antagonists. These substances protect rainbow trout, *Artemia*, and *Brachionus* from the harmful effects of pathogenic *Vibrio* spp. when added in sufficient quantities (Tinh et al., 2008). A combination of probiotic *Bacillus* species, such as *B. subtilis*, *B. licheniformis*, and *B. pumilus*, in the diet of gilthead sea bream (*Sparus aurata*) larvae was shown to influence overall health, growth, and survival (Avella et al., 2010).

10.3.2 POTENTIAL OF INDIGENOUS MICROBIAL COMMUNITIES FOR SUSTAINABLE AQUACULTURE

Microorganisms are important in aquaculture because both organic and artificial additions can be used to support their diverse functions. Certain bacteria may protect fish and larvae against disease. Therefore, there is great potential for evaluating and enhancing water quality, as well as halting the spread of infections through the observation and modification of microbial populations in aquaculture settings. Similarly, microbial groups—including bacteria and unicellular fungi—grow in aquaculture production systems. These groups can play a beneficial role in transforming organic matter, eliminating pollutant chemicals, and providing microbial biomass for larger species (del Carmen Monroy Dosta et al., 2015).

Beneficial microbes are essential to both artificial and natural aquatic habitats. Their roles include controlling algae populations to prevent undesirable algal blooms; accelerating the breakdown of organic matter; and lowering COD, NH_3–N, and NO_2–N levels in water and sediments to enhance water quality. They also help suppress waterborne diseases and infections that affect fish and prawns (Zhou et al., 2009). Bacterial infections are usually regarded as the most important pathogenic threat in aquaculture. In addition to decontaminators and biocides, antimicrobials are used to treat diseased fish—and, regrettably, sometimes as a preventative strategy (Acosta et al., 2021).

Certain beneficial bacterial strains secrete extracellular enzymes such as lipases and proteases, along with growth-promoting substances that aid the host's digestive process (Zhang et al., 2020). Evaluation of water quality and identification of undesirable microorganisms—such as toxic algae, fish- and human-pathogenic bacteria, and viruses—can be accomplished by deploying a microbiome-based monitoring system as an effective early warning tool (Bentzon-Tilia et al., 2016). Negligent application of chemicals and feed (aqua inputs) increases pollutants and generates large volumes of waste, including metabolic byproducts. Finding a sustainable treatment method is essential for boosting the productivity of affected land and water. In this regard, bioremediation is an innovative and cost-effective alternative to conventional techniques. Microorganisms convert complex pollutants or hazardous wastes into simple, harmless compounds through metabolic processes. By supporting the health of cultured indigenous organisms and enhancing water quality, these microorganisms (especially bacteria) help preserve the sustainability of aquaculture habitats (Nisar et al., 2022).

10.4 INTEGRATION OF MICROBIAL CONTROL STRATEGIES WITH PROBIOTIC APPROACHES

In the dynamic landscape of aquaculture, integrating microbial control strategies with probiotic approaches represents a holistic and multifaceted method for enhancing disease management and promoting overall aquaculture health. This section explores the synergistic potential of combining microbial control strategies with probiotic approaches, emphasizing their collective impact on disease prevention, water quality management, and ecosystem sustainability.

Probiotics—primarily composed of beneficial microorganisms such as *Lactobacillus* spp. and *Bacillus* spp.—play a pivotal role in maintaining the health and resilience of aquatic organisms in aquaculture systems (Li et al., 2022). These probiotic microorganisms exert their beneficial effects through several mechanisms, such as the competitive exclusion of pathogenic organisms, production of antimicrobial compounds, and modulation of host immunity. By integrating specific probiotic strains with known antagonistic activity against critical pathogens, aquaculturists can establish a balanced microbial community that supports a healthy aquatic ecosystem.

Moreover, probiotics enhance the effectiveness of microbial control strategies by promoting the establishment and persistence of beneficial microorganisms in the aquaculture environment. By colonizing the surfaces of fish and shellfish, probiotics outcompete pathogenic counterparts for attachment sites and nutrient resources, thereby reducing the risk of pathogen colonization and infection (Kaushik et al., 2022). Additionally, probiotics stimulate the production of antimicrobial peptides, lysozyme, and immunoglobulins within the host, bolstering its defense against microbial invaders and enhancing overall disease resistance (Goh et al., 2023).

Furthermore, integrating microbial control strategies with probiotic approaches extends beyond disease prevention to include broader ecological considerations, such as water quality management and ecosystem sustainability. Probiotics contribute to maintaining water quality parameters by promoting the degradation of organic waste materials, enhancing nutrient cycling, and suppressing the proliferation of potentially harmful microbial populations. By fostering a balanced microbial community structure, probiotics help maintain ecosystem resilience and stability, reducing the risk of disease outbreaks and environmental degradation. Integrating microbial control strategies with probiotic approaches aligns with the principles of sustainable aquaculture and environmental stewardship, offering targeted and environmentally friendly solutions for disease management and ecosystem health. By harnessing the synergistic potential of beneficial microorganisms, aquaculturists can optimize disease prevention efforts, enhance water quality, and promote the long-term sustainability of aquaculture operations.

10.5 ENVIRONMENTAL AND NUTRITIONAL IMPLICATIONS

Aquaculture practices not only affect the health and productivity of aquatic organisms but also have profound environmental and nutritional implications. This section examines the influence of probiotics and microbial interventions on water quality, the role of microbial applications in nutritional enhancement and waste management, and the sustainability considerations and ecological impacts associated with microbial aquaculture solutions.

10.5.1 INFLUENCE OF PROBIOTICS AND MICROBIAL INTERVENTIONS ON WATER QUALITY

Probiotics and microbial interventions play a significant role in shaping water quality parameters within aquaculture systems. Beneficial microorganisms, which can actively compete with and suppress the growth of harmful bacteria, are introduced into the aquatic environment. By outcompeting pathogens for nutrients and attachment sites, probiotics help reduce the microbial load in water bodies, thereby minimizing the risk of disease outbreaks. Additionally, some probiotic strains produce antimicrobial compounds, further inhibiting pathogenic bacteria and improving water quality (Mohapatra et al., 2013).

Microbial interventions can also impact the chemical parameters of water quality. For example, certain probiotic bacteria can metabolize organic matter, including excess nutrients such as ammonia and nitrate. By assimilating these nutrients, probiotics help mitigate nutrient buildup in aquaculture systems, reducing the risk of water eutrophication and algal blooms. Moreover, some probiotic strains can facilitate the biodegradation of organic pollutants and toxins, enhancing overall water quality and environmental safety (Robles-Porchas et al., 2020).

Probiotics can influence various physical aspects of water quality, such as turbidity and oxygen levels. Probiotic bacteria contribute to the breakdown of organic debris and sedimentation, leading to more transparent water and improved light penetration, which benefits aquatic plant growth. Furthermore, certain probiotic organisms are involved in biofilm formation, which can enhance water filtration and improve overall water clarity. Additionally, probiotics can enhance oxygenation through their metabolic activities, contributing to better oxygen levels and helping prevent hypoxia in aquaculture systems.

The use of probiotics and microbial interventions in aquaculture has broader ecological implications. By promoting a balanced microbial community and reducing reliance on chemical treatments, probiotics support the overall sustainability of aquaculture operations. Moreover, improved water quality resulting from microbial interventions can enhance biodiversity and ecosystem health in surrounding aquatic environments, benefiting cultured species, wild populations, and natural habitats.

The influence of probiotics and microbial interventions on water quality in aquaculture is multifaceted. These strategies contribute to a more stable and balanced nitrogen cycle within the system, enhancing environmental sustainability and reducing the risk of water quality fluctuations. This ultimately improves aquatic health, reduces stress on aquatic organisms, and leads to increased productivity in aquaculture systems.

10.5.2 NUTRITIONAL ENHANCEMENT AND WASTE MANAGEMENT

Microbial applications in aquaculture offer significant potential for enhancing the nutritional quality of aquafeeds. They can also efficiently manage waste generated within aquaculture systems. This section explores how microbial interventions contribute to nutritional enhancement and waste management, promoting sustainable aquaculture practices.

One of the critical roles of microbial applications in aquaculture is the production of enzymes and metabolites that aid in the breakdown of complex organic compounds present in aquafeeds (Wan-Mohtar et al., 2022). Certain microbial species, such as bacteria and fungi, produce diverse enzymes including proteases, lipases, and carbohydrases. These enzymes facilitate the digestion and assimilation of nutrients by aquatic organisms (Babaniyi et al., 2022). By supplementing aquafeeds with these enzyme-producing microbes, aquaculturists can improve the digestibility of feed ingredients and enhance nutrient utilization efficiency. This leads to improved growth performance and reduced feed conversion ratios. For instance, enzymes produced by probiotic bacteria such as *Lactobacillus* spp. can break down complex carbohydrates and proteins in feed ingredients, making them more readily available for digestion and absorption by fish and other aquatic species.

Furthermore, microbial applications play a vital role in waste management by facilitating the decomposition and mineralization of organic matter accumulated within aquaculture systems. Organic waste materials—such as uneaten feed, fecal matter, and metabolic excretions from aquatic organisms—can harm water quality if not effectively managed. Microbial decomposition processes, including aerobic and anaerobic digestion mediated by bacteria and other microorganisms, aid in the breakdown of organic waste materials, releasing nutrients in forms that are more readily available for uptake by aquatic plants and phytoplankton (Manyi-Loh et al., 2019).

In addition, certain microbial species can bioremediate aquaculture effluents by removing pollutants and mitigating the accumulation of harmful compounds. For example, denitrifying bacteria can convert nitrate into harmless nitrogen gas, reducing the risk of nitrogen pollution in aquaculture ponds and recirculating aquaculture systems (Robles-Porchas et al., 2020). Similarly, sulfate-reducing bacteria play a role in sulfate reduction, preventing the accumulation of hydrogen sulfide and sulfide toxicity in aquaculture environments (Zhang et al., 2020).

In addition to their roles in nutrient enhancement and waste management, microbial applications can produce value-added products, such as microbial biomass and biofertilizers, which may be utilized as supplementary feed ingredients or soil amendments in aquaponics and integrated aquaculture-agriculture systems. By harnessing the metabolic capabilities of beneficial microorganisms, aquaculturists can maximize resource utilization efficiency, minimize environmental impacts, and promote circular economy principles within aquaculture production systems.

Overall, microbial applications represent a versatile and sustainable approach for enhancing the nutritional quality of aquafeeds, managing waste, and improving overall resource efficiency in aquaculture operations. By leveraging the diverse metabolic activities of beneficial microorganisms, aquaculturists can optimize feed utilization, mitigate environmental pollution, and foster the development of environmentally friendly and economically viable aquaculture practices. This contributes to improved water quality and reduces the environmental impact of aquaculture operations by lowering the risk of eutrophication and algal blooms in adjacent water bodies.

10.5.3 SUSTAINABILITY CONSIDERATIONS AND ECOLOGICAL IMPACTS

Adopting microbial aquaculture solutions raises important sustainability considerations and ecological concerns that warrant careful evaluation. While microbial interventions offer promising benefits for disease management, water quality improvement, and waste treatment, their long-term ecological implications require thorough assessment. Introducing non-native microbial species into aquatic ecosystems may inadvertently disrupt indigenous microbial communities and alter ecosystem dynamics. The widespread use of microbial interventions in aquaculture may also pose risks of microbial resistance, horizontal gene transfer, and potential impacts on human health (Heuer et al., 2009), similar to the concerns associated with antibiotic use (Okocha et al., 2018).

Therefore, sustainable microbial aquaculture practices should prioritize the selection and use of indigenous microbial strains with proven efficacy and established safety profiles (Bondad-Reantaso et al., 2023). To this end, research efforts are required to assess the ecological consequences of microbial interventions on aquatic biodiversity, ecosystem functioning, and the resilience of native microbial communities.

In summary, while microbial interventions offer promising solutions for enhancing aquaculture productivity and sustainability, careful consideration of their environmental and nutritional implications is essential to ensure the long-term viability and ecological integrity of aquaculture systems. Integrating microbial interventions with holistic management approaches that prioritize ecosystem health and resilience is crucial for advancing sustainable aquaculture development.

10.6 CONCLUSION

Aquaculture is regarded as a viable solution to meet the growing global food demand. However, aquatic organisms are frequently exposed to various pathogenic microbes and pollutants that negatively impact their health and reduce the production of economically important species. In this context, microorganisms and probiotics have emerged as promising tools for addressing these challenges. Probiotics can contribute to ecological system improvement and support technical advancements in sustainable food production.

Further research is urgently needed to evaluate the resilience of probiotics to antibiotics and the potential for genetic factors to be transferred to other bacteria in the gastrointestinal tract (GIT) of fish—and eventually to humans who consume aquaculture products. Additionally, more studies are required to understand how microbial communities present in the GIT of aquatic species and in the surrounding water influence growth and how these interactions relate to overall water quality in aquaculture systems.

KEYWORDS

- **antibiotics**
- **aquaculture**
- **gastrointestinal tract**
- **global food demand**
- **microbial strains**
- **microorganisms**
- **non-native microbial species**
- **probiotics**
- **sustainable food production**

REFERENCES

Acosta, F., Montero, D., Izquierdo, M., & Galindo-Villegas, J. (2021). High-level biocidal products effectively eradicate pathogenic γ-Proteobacteria biofilms from aquaculture facilities. *Aquaculture*, *532*, 736004. https://doi.org/10.1016/j.aquaculture.2020.736004 (accessed on 10th July 2025).

Arun, D., Midhun, S. J., Sheeja, C. C., Maurya, A. K., & Divya, L. (2023). Probiotics and prebiotics in aquaculture. In *Recent Advances in Aquaculture Microbial Technology* (pp. 209–226). Elsevier. https://doi.org/10.1016/B978-0-323-90261-8.00005-5 (accessed on 10th July 2025).

Asaduzzaman, M., Iehata, S., Akter, S., Kader, M. A., Ghosh, S. K., Khan, M. N. A., & Abol-Munafi, A. B. (2018). Effects of host gut-derived probiotic bacteria on gut morphology, microbiota composition and volatile short chain fatty acids production of Malaysian mahseer *Tor tambroides*. *Aquaculture Reports*, *9*, 53–61. https://doi.org/10.1016/j.aqrep.2017.12.003 (accessed on 10th July 2025).

Avella, M. A., Gioacchini, G., Decamp, O., Makridis, P., Bracciatelli, C., & Carnevali, O. (2010). Application of multi-species of *Bacillus* in sea bream larviculture. *Aquaculture*, *305*(1–4), 12–19. https://doi.org/10.1016/j.aquaculture.2010.03.029 (accessed on 10th July 2025).

Babaniyi, G. G., Oyemade, B., & Orija, D. (2022). Effects of aquatic (freshwater and marine) pollution on microbial enzyme activities (pp. 379–403). https://doi.org/10.1007/978-981-19-0155-3_17 (accessed on 10th July 2025).

Bentzon-Tilia, M., Sonnenschein, E. C., & Gram, L. (2016). Monitoring and managing microbes in aquaculture–Towards a sustainable industry. *Microbial Biotechnology*, *9*(5), 576–584. https://doi.org/10.1111/1751-7915.12392 (accessed on 10th July 2025).

Bondad-Reantaso, M. G., MacKinnon, B., Karunasagar, I., Fridman, S., Alday-Sanz, V., Brun, E., Le Groumellec, M., Li, A., Surachetpong, W., Karunasagar, I., Hao, B., Dall'Occo, A., Urbani, R., & Caputo, A. (2023). Review of alternatives to antibiotic use in aquaculture. *Reviews in Aquaculture*, *15*(4), 1421–1451. https://doi.org/10.1111/raq.12786 (accessed on 10th July 2025).

Dawood, M. A. O., & Koshio, S. (2016). Recent advances in the role of probiotics and prebiotics in carp aquaculture: A review. *Aquaculture*, *454*, 243–251. https://doi.org/10.1016/j.aquaculture.2015.12.033 (accessed on 10th July 2025).

del Carmen Monroy Dosta, M., de Oca, G. A. R. M., Mejía, J. C., y, G. C. M., & Cortés, D. B. (2015). Importance and function of microbial communities in aquaculture systems with no water exchange. *Scientific Journal of Animal Science*, *4*, 103–110.

Divya, K. R., Isamma, A., Ramasubramanian, V., Sureshkumar, S., & Arunjith, T. S. (2012). Colonization of probiotic bacteria and its impact on ornamental fish *Puntius conchonius*. *Journal of Environmental Biology*, *33*(3), 551–555.

El-Saadony, M. T., Alagawany, M., Patra, A. K., Kar, I., Tiwari, R., Dawood, M. A. O., Dhama, K., & Abdel-Latif, H. M. R. (2021). The functionality of probiotics in aquaculture: An overview. *Fish & Shellfish Immunology*, *117*, 36–52.

Fooks, L. J., & Gibson, G. R. (2002). Probiotics as modulators of the gut flora. *British Journal of Nutrition*, *88*(S1), s39–s49. https://doi.org/10.1079/BJN2002628 (accessed on 10th July 2025).

Gatesoupe, F. J. (2008). Updating the importance of lactic acid bacteria in fish farming: Natural occurrence and probiotic treatments. *Microbial Physiology*, *14*(1–3), 107–114. https://doi.org/10.1159/000106089 (accessed on 10th July 2025).

George, S., Sharma, R. K., Hazarika, R., Hazarika, S., Hazarika, S., Ahmed, R., & Phangchopi, D. (2023). Predatory bacteria: A novel approach to antibiotic substitution. *International Journal of Bio-resource and Stress Management*, *14*(Jan, 1), 045–053. https://doi.org/10.23910/1.2023.3168 (accessed on 10th July 2025).

Goh, J. X. H., Tan, L. T. H., Law, J. W. F., Khaw, K. Y., Zengin, G., Chan, K. G., Letchumanan, V., Lee, L. H., & Goh, B. H. (2023). Probiotics: Comprehensive exploration of the growth promotion mechanisms in shrimps. *Progress in Microbes and Molecular Biology*, *6*(1). https://doi.org/10.36877/pmmb.a0000324 (accessed on 10th July 2025).

Gram, L., Melchiorsen, J., Spanggaard, B., Huber, I., & Nielsen, T. F. (1999). Inhibition of *Vibrio anguillarum* by *Pseudomonas fluorescens* AH2, a possible probiotic treatment of fish. *Applied and Environmental Microbiology*, *65*(3), 969–973. https://doi.org/10.1128/AEM.65.3.969-973.1999 (accessed on 10th July 2025).

Heuer, O. E., Kruse, H., Grave, K., Collignon, P., Karunasagar, I., & Angulo, F. J. (2009). Human health consequences of use of antimicrobial agents in aquaculture. *Clinical Infectious Diseases*, *49*(8), 1248–1253. https://doi.org/10.1086/605667 (accessed on 10th July 2025).

Hlordzi, V., Kuebutornye, F. K. A., Afriyie, G., Abarike, E. D., Lu, Y., Chi, S., & Anokyewaa, M. A. (2020). The use of *Bacillus* species in maintenance of water quality in aquaculture: A review. *Aquaculture Reports*, *18*, 100503. https://doi.org/10.1016/j.aqrep.2020.100503 (accessed on 10th July 2025).

Iqbal, Z., Ahmed, S., Tabassum, N., Bhattacharya, R., & Bose, D. (2021). Role of probiotics in prevention and treatment of enteric infections: A comprehensive review. *3 Biotech*, *11*(5), 242. https://doi.org/10.1007/s13205-021-02796-7 (accessed on 10th July 2025).

Kaushik, P., Khandelwal, R., Jain, N., Keelka, S., & Jain, P. (2022). Prospectives of prebiotics, probiotics, and symbiotic for sustainable development of aquaculture. In *Prebiotics, Probiotics and Nutraceuticals* (pp. 301–340). Springer Nature Singapore. https://doi.org/10.1007/978-981-16-8990-1_15 (accessed on 10th July 2025).

Kunselman, E., Wiggin, K., Diner, R. E., Gilbert, J. A., & Allard, S. M. (2024). Microbial threats and sustainable solutions for molluscan aquaculture. *Sustainable Microbiology*, *1*(1). https://doi.org/10.1093/sumbio/qvae002 (accessed on 10th July 2025).

Lalloo, R., Moonsamy, G., Ramchuran, S., Görgens, J., & Gardiner, N. (2010). Competitive exclusion as a mode of action of a novel *Bacillus cereus* aquaculture biological agent. *Letters in Applied Microbiology*, *50*(6), 563–570. https://doi.org/10.1111/j.1472-765X.2010.02829.x (accessed on 10th July 2025).

Larsson, D. G. J., & Flach, C. F. (2022). Antibiotic resistance in the environment. *Nature Reviews Microbiology*, *20*(5), 257–269. https://doi.org/10.1038/s41579-021-00649-x (accessed on 10th July 2025).

Li, J., Majzoub, M. E., Marzinelli, E. M., Dai, Z., Thomas, T., & Egan, S. (2022). Bacterial controlled mitigation of dysbiosis in a seaweed disease. *ISME Journal*, *16*(2), 378–387. https://doi.org/10.1038/s41396-021-01070-1 (accessed on 10th July 2025).

Lorgen-Ritchie, M., Uren Webster, T., McMurtrie, J., Bass, D., Tyler, C. R., Rowley, A., & Martin, S. A. M. (2023). Microbiomes in the context of developing sustainable intensified aquaculture. *Frontiers in Microbiology*, *14*. https://doi.org/10.3389/fmicb.2023.1200997 (accessed on 10th July 2025).

Manyi-Loh, C. E., Mamphweli, S. N., Meyer, E. L., & Okoh, A. I. (2019). Microbial anaerobic digestion: Process dynamics and implications from the renewable energy, environmental and agronomy perspectives. *International Journal of Environmental Science and Technology*, *16*(7), 3913–3934. https://doi.org/10.1007/s13762-019-02380-w (accessed on 10th July 2025).

Marden, L., McDonald, R., Schreier, H. J., & Watts, J. E. M. (2017). Investigation into the fungal diversity within different regions of the gastrointestinal tract of *Panaque nigrolineatus*, a wood-eating fish. *AIMS Microbiology*, *3*(4), 749–761. https://doi.org/10.3934/microbiol.2017.4.749 (accessed on 10th July 2025).

Martínez Cruz, P., Ibáñez, A. L., Monroy Hermosillo, O. A., & Ramírez Saad, H. C. (2012). Use of probiotics in aquaculture. *ISRN Microbiology*, *2012*, 1–13. https://doi.org/10.5402/2012/916845 (accessed on 10th July 2025).

Mohapatra, S., Chakraborty, T., Kumar, V., DeBoeck, G., & Mohanta, K. N. (2013). Aquaculture and stress management: A review of probiotic intervention. *Journal of Animal Physiology and Animal Nutrition (Berl.)*, *97*(3), 405–430. https://doi.org/10.1111/j.1439-0396.2012.01301.x (accessed on 10th July 2025).

Naya-Català, F., Piazzon, M. C., Calduch-Giner, J. A., Sitjà-Bobadilla, A., & Pérez-Sánchez, J. (2022). Diet and host genetics drive the bacterial and fungal intestinal meta transcriptome of gilthead sea bream. *Frontiers in Microbiology*, *13*. https://doi.org/10.3389/fmicb.2022.883738 (accessed on 10th July 2025).

Naylor, R. L., Hardy, R. W., Buschmann, A. H., Bush, S. R., Cao, L., Klinger, D. H., Little, D. C., Lubchenco, J., Shumway, S. E., & Troell, M. (2021). A 20-year retrospective review of global aquaculture. *Nature*, *591*(7851), 551–563. https://doi.org/10.1038/s41586-021-03308-6 (accessed on 10th July 2025).

Nisar, U., Peng, D., Mu, Y., & Sun, Y. (2022). A solution for sustainable utilization of aquaculture waste: A comprehensive review of biofloc technology and aquamimicry. *Frontiers in Nutrition*, *8*. https://doi.org/10.3389/fnut.2021.791738 (accessed on 10th July 2025).

Okocha, R. C., Olatoye, I. O., & Adedeji, O. B. (2018). Food safety impacts of antimicrobial use and their residues in aquaculture. *Public Health Reviews*, *39*(1), 21. https://doi.org/10.1186/s40985-018-0099-2 (accessed on 10th July 2025).

Oliveira, J., Castilho, F., Cunha, A., & Pereira, M. J. (2012). Bacteriophage therapy as a bacterial control strategy in aquaculture. *Aquaculture International*, *20*(5), 879–910. https://doi.org/10.1007/s10499-012-9515-7 (accessed on 10th July 2025).

Patil, D. W., & Markad, S. S. (2015). Probiotics as an alternative source for antibiotics and chemotherapeutics in aquaculture. In *Proceedings of National Conference on "Current Trends in Aquaculture (CTIA-2015)"* Organized by Department of Zoology, Indira Gandhi Senior College, CIDCO, Nanded. *International Journal of Advance Research in Basic & Applied Science (IJARBAS), Special Issue*, 16–25.

Pérez, J., Moraleda-Muñoz, A., Marcos-Torres, F. J., & Muñoz-Dorado, J. (2016). Bacterial predation: 75 years and counting! *Environmental Microbiology*, *18*(3), 766–779. https://doi.org/10.1111/1462-2920.13171 (accessed on 10th July 2025).

Rahman, M. A., Ashrafudoulla, M., Akter, S., Park, S. H., & Ha, S. D. (2023). Probiotics and biofilm interaction in aquaculture for sustainable food security: A review and bibliometric analysis. *Critical Reviews in Food Science and Nutrition*, 1–17. https://doi.org/10.1080/10408398.2023.2249114 (accessed on 10th July 2025).

Reverter, M., Sarter, S., Caruso, D., Avarre, J. C., Combe, M., Pepey, E., Pouyaud, L., Vega-Heredía, S., de Verdal, H., & Gozlan, R. E. (2020). Aquaculture at the crossroads of global warming and antimicrobial resistance. *Nature Communications*, *11*(1), 1870. https://doi.org/10.1038/s41467-020-15735-6 (accessed on 10th July 2025).

Robles-Porchas, G. R., Gollas-Galván, T., Martínez-Porchas, M., Martínez-Cordova, L. R., Miranda-Baeza, A., & Vargas-Albores, F. (2020). The nitrification process for nitrogen removal in biofloc system aquaculture. *Reviews in Aquaculture*, *12*(4), 2228–2249. https://doi.org/10.1111/raq.12431 (accessed on 10th July 2025).

Sahu, M. K., Swarnakumar, N. S., Sivakumar, K., Thangaradjou, T., & Kannan, L. (2008). Probiotics in aquaculture: Importance and future perspectives. *Indian Journal of Microbiology*, *48*(3), 299–308. https://doi.org/10.1007/s12088-008-0024-3 (accessed on 10th July 2025).

Saxena, G., Bharagava, R. N., Kaithwas, G., & Raj, A. (2015). Microbial indicators, pathogens and methods for their monitoring in water environment. *Journal of Water and*

Health, *13*(2), 319–339. https://doi.org/10.2166/wh.2014.275 (accessed on 10th July 2025).

Sehnal, L., Brammer-Robbins, E., Wormington, A. M., Blaha, L., Bisesi, J., Larkin, I., Martyniuk, C. J., Simonin, M., & Adamovsky, O. (2021). Microbiome composition and function in aquatic vertebrates: Small organisms making big impacts on aquatic animal health. *Frontiers in Microbiology*, *12*. https://doi.org/10.3389/fmicb.2021.567408 (accessed on 10th July 2025).

Simón, R., Docando, F., Nuñez-Ortiz, N., Tafalla, C., & Díaz-Rosales, P. (2021). Mechanisms used by probiotics to confer pathogen resistance to teleost fish. *Frontiers in Immunology*, *12*. https://doi.org/10.3389/fimmu.2021.653025 (accessed on 10th July 2025).

Taoka, Y., Maeda, H., Jo, J. Y., Kim, S. M., Park, S. I., Yoshikawa, T., & Sakata, T. (2006). Use of live and dead probiotic cells in tilapia *Oreochromis niloticus*. *Fisheries Science*, *72*(4), 755–766. https://doi.org/10.1111/j.1444-2906.2006.01215.x (accessed on 10th July 2025).

Tinh, N. T. N., Dierckens, K., Sorgeloos, P., & Bossier, P. (2008). A review of the functionality of probiotics in the larviculture food chain. *Marine Biotechnology*, *10*(1), 1–12. https://doi.org/10.1007/s10126-007-9054-9 (accessed on 10th July 2025).

van Vliet, M. T. H., Thorslund, J., Strokal, M., Hofstra, N., Flörke, M., Ehalt Macedo, H., Nkwasa, A., Tang, T., Kaushal, S. S., Kumar, R., van Griensven, A., Bouwman, L., & Mosley, L. M. (2023). Global river water quality under climate change and hydroclimatic extremes. *Nature Reviews Earth & Environment*, *4*(10), 687–702. https://doi.org/10.1038/s43017-023-00472-3 (accessed on 10th July 2025).

Verschuere, L., Rombaut, G., Sorgeloos, P., & Verstraete, W. (2000). Probiotic bacteria as biological control agents in aquaculture. *Microbiology and Molecular Biology Reviews*, *64*(4), 655–671. https://doi.org/10.1128/MMBR.64.4.655-671.2000 (accessed on 10th July 2025).

Vine, N. G., Leukes, W. D., & Kaiser, H. (2004). *In vitro* growth characteristics of five candidate aquaculture probiotics and two fish pathogens grown in fish intestinal mucus. *FEMS Microbiology Letters*, *231*(1), 145–152. https://doi.org/10.1016/S0378-1097(03)00954-6 (accessed on 10th July 2025).

Wan-Mohtar, W. A. A. Q. I., Ibrahim, M. F., Rasdi, N. W., Zainorahim, N., & Taufek, N. M. (2022). Microorganisms as a sustainable aquafeed ingredient: A review. *Aquaculture Research*, *53*(3), 746–766. https://doi.org/10.1111/are.15627 (accessed on 10th July 2025).

Xianghong, W., Huirong, L., Xiaohua, Z., Yun, L., Weishang, J., & Huaishu, X. (2000). Microbial flora in the digestive tract of adult penaeid shrimp (*Penaeus chinensis*). *Journal of Ocean University of Qingdao*, *30*(3), 493–498.

Yoon, S. H., & Kim, G. B. (2022). Inhibition of *Listeria monocytogenes* in fresh cheese using a bacteriocin-producing *Lactococcus lactis* CAU2013 strain. *Food Science of Animal Resources*, *42*(6), 1009–1019. https://doi.org/10.5851/kosfa.2022.e48 (accessed on 10th July 2025).

Zhang, K., Zheng, X., He, Z., Yang, T., Shu, L., Xiao, F., Wu, Y., Wang, B., Li, Z., Chen, P., & Yan, Q. (2020). Fish growth enhances microbial sulfur cycling in aquaculture pond sediments. *Microbial Biotechnology*, *13*(5), 1597–1610. https://doi.org/10.1111/1751-7915.13622 (accessed on 10th July 2025).

Zhou, Q., Li, K., Jun, X., & Bo, L. (2009). Role and functions of beneficial microorganisms in sustainable aquaculture. *Bioresource Technology*, *100*(16), 3780–3786. https://doi.org/10.1016/j.biortech.2008.12.037 (accessed on 10th July 2025).

Zorriehzahra, M. J., Delshad, S. T., Adel, M., Tiwari, R., Karthik, K., Dhama, K., & Lazado, C. C. (2016). Probiotics as beneficial microbes in aquaculture: An update on their multiple modes of action: A review. *Veterinary Quarterly*, *36*(4), 228–241. https://doi.org/10.1080/01652176.2016.1172132 (accessed on 10th July 2025).

CHAPTER 11

Recombinant Proteins and Their Applications in Fisheries

PRACHI ASGOLKAR,[1] KIRAN D. RASAL,[1*] JEBARSON SOLOMON,[2] and ANKITA GODAVARIKAR[1]

[1]*Fish Genetics and Biotechnology Division, ICAR–Central Institute of Fisheries Education, Mumbai, Maharashtra, India*

[2]*Fish Nutrition, Biochemistry and Physiology Division, ICAR–Central Institute of Fisheries Education, Mumbai, Maharashtra, India*

*Corresponding author

ABSTRACT

Advancements in aquaculture biotechnology have significantly enhanced the production of aquatic animals. Recombinant technology is one such technique in this field. Recombinant DNA technology is an important tool used to produce recombinant proteins. Proteins generated using this technique can be applied in induced breeding, disease diagnosis, therapeutics, and as growth enhancers. Recombinant proteins with specific functions are primarily produced in laboratories using various expression systems, including prokaryotic systems such as bacterial platforms, and eukaryotic systems such as yeast, mammalian, and insect expression systems.

Various strategies are available for recombinant protein production, including modified plasmids, different host strains, and optimized culture methods. Advances in recombinant protein production have not only reversed previous limitations but also resulted in significantly improved efficiency and reduced costs. The industrial-scale production of recombinant proteins has become increasingly useful for induced breeding and disease treatment, among other applications. This book chapter focuses

Sustainable Aquaculture: Recent Advances and Techniques. Dhanaji W. Patil, Sandip S. Markad (Eds.)

DOI: 10.1201/9781779641113-11

on the recent progress in recombinant protein production in fish and its potential applications and prospects in the aquaculture sector.

11.1 INTRODUCTION

Addressing the critical challenge of ensuring an adequate food supply is a top global priority. Numerous countries have turned to aquaculture to help solve this issue and boost their food production. The cultivation of aquatic animals not only expands food resources but also plays a crucial role in meeting the rising demand for high-quality proteins. Aquaculture holds significant socio-economic importance, contributing to nutrition and economic sustainability (FAO, 2022).

A wide range of substances and hormones has been proposed to enhance the productivity and welfare of aquatic animals and to treat various disorders. Studies have shown that aquaculture production has been improved through selective breeding, transgenesis, ploidy manipulation, and the use of recombinant proteins (Alimuddin et al., 2010). Over the past two decades, the application of recombinant hormones has produced significant outcomes in both the animal and aquaculture sectors (Funkenstein, 2006).

Advancements in recombinant DNA technology (rDNA), along with modified expression systems, have facilitated the large-scale production of various proteins or hormones (e.g., growth factors) in fish species. This progress has eliminated the need to sacrifice large numbers of fish for small quantities of native protein (Kim et al., 2015). Because of the small amount of peptides in target tissues—such as luteinizing hormone (LH) and follicle-stimulating hormone (FSH) in the pituitary—isolating and purifying these peptides becomes challenging. This complicates effective separation, and the need for substantial hormone doses for growth enhancement, particularly with growth hormone (GH), necessitates alternative approaches. To date, recombinant products from fish have been successfully generated across various systems, and their biological significance has been assessed (Mohammadzadeh et al., 2022; Yeganeh et al., 2022).

These recombinant proteins have diverse applications in the aquaculture sector, including growth improvement through growth factors, disease treatment or infection prevention, induced breeding or spawning, and stress mitigation. Furthermore, these recombinant proteins can be useful in developing enzyme-linked immunosorbent assays (ELISA) for assessing the maturity stages of fish during the breeding season or evaluating their stress levels. Overall, this book chapter discusses the recent progress of recombinant proteins and their applications in aquaculture (Figure 11.2).

11.2 RECOMBINANT PROTEINS

The process of recombinant protein/hormone production includes several steps: target gene synthesis and cloning, suitable vector selection and construction, transformation of the cloned vector, expression of the desired protein, and purification of the recombinant protein (Sambrook et al., 1989). Here is a brief overview of each step (Figures 11.1 and 11.2):

1. **Gene Cloning:** The first step involves isolating the gene encoding the hormone of interest. This can be achieved either by isolating the gene from the organism that naturally produces the hormone or by synthesizing it artificially based on a known sequence. The gene is then inserted into a suitable vector for further manipulation.
2. **Vector Construction:** A vector—often a plasmid or viral vector—is selected for cloning the target gene. The gene is typically inserted into the vector using restriction enzymes and DNA ligase. Additionally, regulatory sequences such as promoters and terminators may be included to control gene expression.
3. **Transformation:** The desired gene of interest is cloned into the vector and transformed into a host organism such as bacteria, yeast, or mammalian cells. This can be achieved through various methods, including heat shock for bacterial cells or transfection for mammalian cells.
4. **Expression:** Once inside the host cell, the recombinant vector replicates and expresses the gene of interest. The host cell's machinery transcribes and translates the gene, producing the hormone as a protein product.

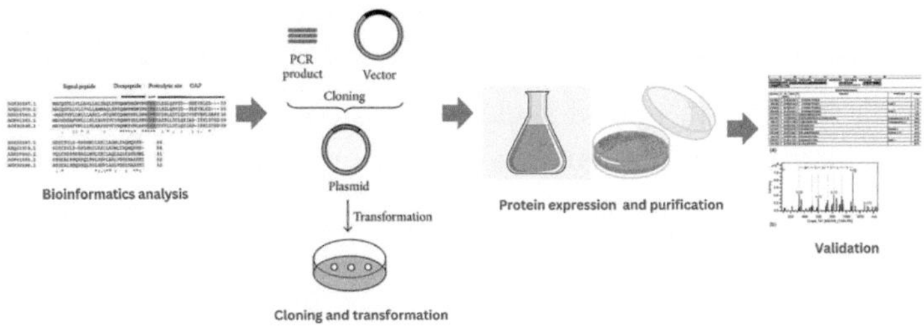

FIGURE 11.1 Process of production for recombinant proteins.

5. **Purification:** The expressed protein is purified from the host organism using standard procedures, including chromatography, filtration, and precipitation techniques.
6. **Characterization:** The purified hormone or protein is then characterized to ensure it meets the required quality standards. This may involve biochemical assays, mass spectrometry, and other analytical techniques to confirm its identity, purity, and biological activity.

Recombinant DNA technology has emerged as a cost-effective and transformative tool for synthesizing desired proteins. The use of microorganisms as host systems for recombinant protein production has indisputably revolutionized the field of biochemistry. This innovative approach has not only streamlined the production process but has also significantly contributed to the accessibility and affordability of producing a wide range of proteins with diverse applications (Rosano and Ceccarelli, 2014).

Commercial protein production in the industry relies significantly on the combined techniques of protein engineering and genetic engineering. This capability not only facilitates detailed studies of the biochemical properties of proteins but also opens avenues for their use across various industries, including large-scale production. Molecular tools and protocols are now available for the high-level heterologous synthesis of proteins.

The challenges posed by the short half-life of synthetic hormones—along with the difficulty, time constraints, and high costs associated with obtaining significant quantities of these proteins from natural sources—underscore the importance of focusing on recombinant protein production in alternative host systems. Using genetic engineering techniques, researchers can overcome these limitations and achieve efficient and sustainable production of recombinant proteins with enhanced stability and longevity. This advancement expands possibilities for both scientific research and industrial applications (Mohammadzadeh et al., 2020).

11.2.1 RECOMBINANT PROTEINS FOR ENHANCING FISH GROWTH

Enhancing the growth rate of skeletal muscle is paramount for the aquaculture industry, with particular emphasis on fostering hypertrophy and hyperplasia—two key stages of fish skeletal muscle development (Daczewska and Saczko, 2005). In the embryonic stage, hypertrophy involves the elongation of myofiber formation, while hyperplasia entails increasing the number of myofibers within the skeletal muscle layer. Unlike in mammals, where myofiber proliferation ceases at birth, this process continues in fish (Soto et al., 2017).

Growth regulation is significantly influenced by the growth axis involving growth hormone (GH) and insulin-like growth factor (IGF) (Yakar et al., 2005). In response to environmental cues, the fish's hypothalamus governs the secretion of GH by the pituitary gland. GH then stimulates the synthesis and release of IGF in respective tissues via circulation, thereby promoting cell growth, differentiation, and ultimately organ development (Yang et al., 2011).

Growth, characterized as a polygenic trait and somatic event, is shaped by various physiological and environmental factors (Canosa and Bertucci, 2020). The rate of growth depends on a diverse array of influences, including genetics, diet, and environmental conditions. While selective breeding is a widely applied method for improving growth performance, it has limitations, including the long duration required to achieve measurable growth improvements (Bolivar and Newkirk, 2002). In contrast, transgenesis offers faster results where fertilized eggs are available.

Studies have documented that recombinant growth hormones (rGH) from various fish species can be synthesized using *Escherichia coli* or yeast systems. Supplementation with rGH has been shown to enhance growth performance across different species. For instance, in *Barbodes binotatus*, dietary supplementation with 2 mg/kg of rGH led to significant improvements in growth rate, specific growth rate, feed conversion ratio, and survival rate compared to the control group. A similar study in *Puntius* spp. (Tawes's fish) reported increased growth and survival rates following administration of the same rGH dosage (Fauzannadi et al., 2020).

Positive outcomes from recombinant growth hormone treatments have also been observed in multiple species, including *Oncorhynchus kisutch* (Kawauchi et al., 1991), *Oreochromis niloticus* (Chen et al., 2016), *Cyprinus carpio* (Alimuddin et al., 2010), *Epinephelus coioides* (Shi et al., 2002), and *Cynoglossus semilaevis* (Xu et al., 2018).

11.2.2 RECOMBINANT PROTEINS FOR AQUATIC ANIMAL HEALTH MANAGEMENT

The conventional approach to disease control in aquaculture involves the use of antibiotics and chemical drugs. However, this method often leads to the continual emergence of "superbugs" and the accumulation of chemicals within the food chain (Williams et al., 2005). For this reason, the use of vaccines is consistently recommended for the prevention of bacterial diseases in fish (Pridgeon and Klesius, 2011). Recently developed vaccine production using recombinant DNA technology offers a promising and safer alternative to conventional antibiotics.

Thirumalaikumar et al. (2022) developed a vaccine candidate targeting *Aeromonas hydrophila* infection in rohu fish (*Labeo rohita*), using genes encoding outer membrane protein (OMP) and haemolysin (hly). These genes were cloned into a vector, expressed in *Escherichia coli*, and purified. Fish were vaccinated with the recombinant proteins and challenged with *A. hydrophila*. The vaccinated fish showed improved immune responses and higher survival rates compared to non-vaccinated controls. The OMP and hly proteins demonstrated potential as vaccine candidates for controlling *A. hydrophila* in aquaculture.

Similar research has explored other recombinant proteins for aquatic animal health. For example, a recombinant lectin from *Agrocybe aegerita*, expressed in *E. coli*, was shown to protect shrimp against white spot syndrome virus (WSSV). Recombinant interferon-gamma (IFN-γ) has been reported to enhance the immune response in carp against spring viremia virus. Recombinant capsid proteins from *Macrobrachium rosenbergii* nodavirus were also effective in protecting freshwater prawns against white tail disease (WTD).

A recent study investigated the biochemical properties of marine virophage endolysins. Specifically, the VpKK5 Vibrio phage, isolated from Sabah, Malaysia, demonstrated bactericidal activity against *Vibrio parahaemolyticus* strains. Genome analysis identified a putative endolysin, LysVpKK5, homologous to *N*-acetylmuramoyl-*L*-alanine amidase, suggesting its potential applications. The study biochemically characterized the muralytic activity of recombinant LysVpKK5 against permeabilized *V. parahaemolyticus* strains, demonstrating its promise as a therapeutic agent for combating *Vibrio*-related diseases in aquatic species (Melo-López et al., 2021).

Likewise, multiple recombinant proteins have been produced for therapeutic use in aquaculture, including:

- Hepcidin for treating *Flavobacterium columnare* (Chen et al., 2020);
- IFN-γ from common carp against spring viremia of carp virus (Fu et al., 2019);

- Modified LBD domain of anti-lipopolysaccharide factor FcALF2 from *Fenneropenaeus chinensis* to treat WSSV and both Gram-positive and Gram-negative bacteria (Yang et al., 2016);
- Aerolysin and haemolysin to treat *A. hydrophila* in catfish (Zhang et al., 2020);
- VP28 protein of WSSV as a vaccine candidate (Le Linh et al., 2021).

11.2.3 RECOMBINANT PROTEINS FOR ENHANCING REPRODUCTION IN FISH

In aquatic invertebrates, reproduction is orchestrated by the BPG axis—a complex interplay of neuroendocrine and endocrine systems (Biran and Levavi-Sivan, 2018). Unlike in vertebrates, where direct communication between the brain and gonads is absent, the pituitary gland serves as the primary endocrine organ, influenced by neurohormones, particularly gonadotropin-releasing hormone (GnRH) from the hypothalamus. This stimulates the production of gonadotropins (GtHs) such as luteinizing hormone (LH) and follicle-stimulating hormone (FSH), which are crucial for reproductive regulation. These hormones are transported through the bloodstream to gonadal target cells, where they control maturation and spawning (Planas et al., 2008).

In aquaculture, induced breeding plays a pivotal role in enhancing production. Traditional methods rely on hormones extracted from animal sources or synthetic alternatives, which pose challenges related to purity and specificity. However, advances in recombinant DNA technology offer a promising solution for hormone production. This approach minimizes the need to sacrifice numerous fish to obtain limited quantities of native proteins. Recent studies have focused on producing recombinant gonadotropins, including recombinant FSH for Japanese eel (*Anguilla japonica*) (Kamei et al., 2006), common carp (*Cyprinus carpio*) LH (Aizen et al., 2017), and beluga (*Huso huso*) GnRH (Mohammadzadeh et al., 2021). This shift toward recombinant hormones holds great potential for revolutionizing induced breeding practices in aquaculture by streamlining gamete collection, reducing costs, and enhancing overall efficiency.

GnRH, a pivotal neurohormone in reproduction, stimulates gonadotropin secretion. Its therapeutic potential in infertility treatment has led to the development of recombinant GnRH proteins, particularly in animal reproduction (Gifre et al., 2017). However, due to the peptide's short half-life, high doses are required for efficacy, making natural sources costly. To address this, researchers are focusing on producing recombinant GnRH proteins with extended half-lives. The conserved decapeptide, along with the GnRH-associated peptide (GAP), shows potential for producing rGnRH/GAP with suitable biological activity, offering a promising approach to effective hormone therapy in aquaculture (Mohammadzadeh et al., 2019). The successful induction of maturation in fish during the breeding season further underscores its potential. The biological activity of rGnRH has been tested for inducing ovulation and spawning in various species, including sterlet sturgeon (*Acipenser ruthenus*) (Mohammadzadeh et al., 2021) and female koi carp (*Cyprinus carpio*) (Yeganeh et al., 2022).

Under the control of the hypothalamus and gonadotropin-releasing hormone (GnRH), gonadotrophs of the pituitary synthesize and release gonadotropins (GTHs), including luteinizing hormone (LH) and follicle-stimulating hormone (FSH), which play important roles in fish reproduction (Swanson et al., 1991). GTH-I is associated with vitellogenesis, while GTH-II is linked to oocyte maturation and ovulation, highlighting the complexity of teleost pituitary hormone production (Kawauchi et al., 1989; Suzuki et al., 1988b; Yaron et al., 2003). The biosynthesis of GTH occurs in the rough endoplasmic reticulum, followed by glycosylation and assembly in the Golgi apparatus, ultimately resulting in the release of functional dimers (Kah and Dufour, 2011). Protein glycosylation patterns affect heterodimerization, secretion, half-life, and receptor binding. GTHs are pivotal in treating reproductive disorders, and recombinant protein production has revolutionized animal reproduction. Before recombinant technology, crude pituitary extracts or heterologous hormones were used, presenting challenges such as batch variation, purity concerns, and immunogenicity (Cahoreau et al., 2015). Advances in molecular biology enabled the isolation of complementary DNA (cDNA) for GTH expression studies, though functional analyses using recombinant GTHs are essential for understanding their biological activity. Recombinant GTHs offer several advantages over native hormones, including uniformity, ease of purification, and the absence of pathogens. Their coding sequences can also be modified to enhance potency or prolong half-life, supporting biotechnological applications in reproductive medicine and animal breeding (Adams and Boime, 2008).

Understanding the distinct roles of LH and FSH during critical reproductive phases remains challenging. Earlier research methods included enzyme-linked immunosorbent assays (ELISA) and radioimmunoassays (RIA). However, the laborious and costly process of purifying native GTHs from large numbers of pituitary glands limited research advancements. Although GTH isolation and characterization were achieved in various fish species by the late 1980s, homologous immunoassays for FSH were developed only for select species such as chum salmon

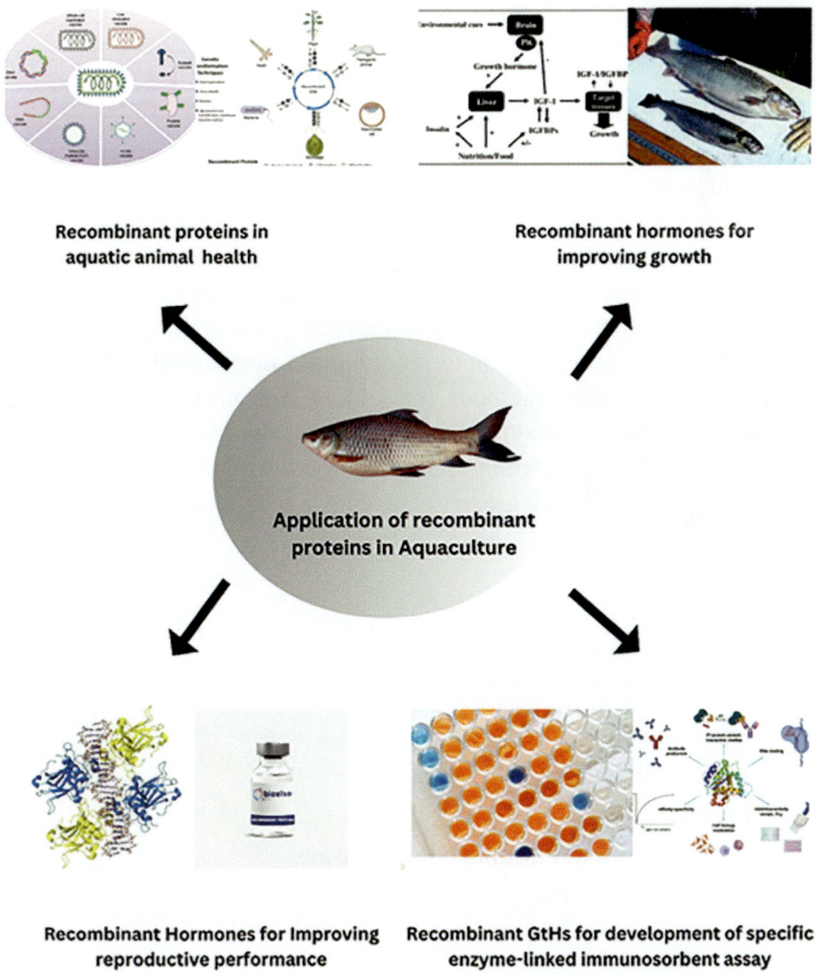

FIGURE 11.2 Summary of recombinant protein applications in aquaculture.

(*Oncorhynchus keta*) (Suzuki et al., 1988a), coho salmon (*Oncorhynchus kisutch*) (Swanson et al., 1991), and rainbow trout (*Oncorhynchus mykiss*) (Saligaut et al., 1998). Pituitary mRNA levels of LH and FSH may not accurately reflect plasma hormone levels, necessitating more precise methods to measure GTH levels in both the pituitary and plasma for assessing biological activity. Recombinant GTHs offer valuable applications in fish reproductive biology with minimal cross-contamination (Chauvigné et al., 2015). The use of recombinant gonadotropins in fish research addresses challenges such as asynchronous ovarian development and advances our understanding of these crucial reproductive hormones.

11.3 CONCLUSION AND FUTURE PERSPECTIVES

The production of recombinant proteins and their use as advanced technology in aquaculture present significant potential. This technology will be beneficial for aquaculturists as well as researchers by enabling improved growth performance, enhanced breeding outcomes, and effective disease treatment—all aimed at increasing aquaculture productivity. Key factors to consider when producing new recombinant proteins include the cost of production, effective methods for folding and purifying recombinant proteins or peptides, and appropriate assays for detecting their biological activity. Recombinant hormones may outperform traditional hormonal treatments in fish, offering greater efficiency and helping to stimulate fish production.

KEYWORDS

- **aquaculture**
- **biological activity**
- **breeding**
- **enzyme-linked immunosorbent assays**
- **expression system**
- **hormones**
- **recombinant protein**

REFERENCES

Adams, T. E., & Boime, I. (2008). The expanding role of recombinant gonadotropins in assisted reproduction. *Reproduction in Domestic Animals*, *43*, 186–192.

Aizen, J., Hollander-Cohen, L., Shpilman, M., & Levavi-Sivan, B. (2017). Biologically active recombinant carp LH as a spawning-inducing agent for carp. *Journal of Endocrinology*, *232*, 391–402.

Alimuddin, A., Lesmana, I., Sudrajat, A. O., Carman, O., & Faizal, I. (2010). Production and bioactivity potential of three recombinant growth hormones of farmed fish. *Indonesian Aquaculture Journal*, *5*, 11–17.

Biran, J., & Levavi-Sivan, B. (2018). Endocrine control of reproduction, fish. In *Encyclopedia of Reproduction* (Second Edition), 362–368.

Bolivar, R. B., & Newkirk, G. F. (2002). Response to within family selection for body weight in Nile tilapia (*Oreochromis niloticus*) using a single-trait animal model. *Aquaculture*, *204*, 371–381.

Cahoreau, C., Klett, D., & Combarnous, Y. (2015). Structure–function relationships of glycoprotein hormones and their subunits' ancestors. *Frontiers in Endocrinology*, *6*, 26.

Canosa, L. F., & Bertucci, J. I. (2020). Nutrient regulation of somatic growth in teleost fish. The interaction between somatic growth, feeding and metabolism. *Molecular and Cellular Endocrinology*, *518*, 111029.

Chauvigné, F., Verdura, S., Mazón, M. J., Boj, M., Zanuy, S., Gómez, A., & Cerdà, J. (2015). Development of a flatfish-specific enzyme-linked immunosorbent assay for Fsh using a recombinant chimeric gonadotropin. *General and Comparative Endocrinology*, *221*, 75–85.

Chen, D., Wang, W., & Ru, S. (2016). Effects of dietary genistein on GH/IGF-I axis of Nile tilapia *Oreochromis niloticus*. *Chinese Journal of Oceanology and Limnology*, *34*, 1004–1012.

Chen, T., Zhou, J., Qu, Z., Zou, Q., Liu, X., Su, J., Fu, X., & Yuan, G. (2020). Administration of dietary recombinant hepcidin on grass carp (*Ctenopharyngodon idella*) against *Flavobacterium columnare* infection under cage aquaculture conditions. *Fish & Shellfish Immunology*, *99*, 27–34.

Daczewska, M., & Saczko, J. (2005). Myotomal myogenesis of axial muscle in the sturgeon *Acipenser baeri* (Chondrostei, Acipenseriformes). *Folia Biologica (Kraków)*, *53*, 29–38.

Fauzannadi, F., Irawan, H., & Yulianto, T. (2020). Pengaruh pemberian hormon rGH (recombinant Growth Hormone) melalui metode perendaman dengan dosis yang berbeda terhadap pertumbuhan ikan kerapu cantang *Epinephelus fuscoguttatus* × *Epinephelus lanceolatus*. *Intek Akuakultur*, *4*, 24–36.

Fu, J., Yi, Z., Cui, H., Song, C., Yu, M., & Liu, Y. (2019). Intein-mediated expression and purification of common carp IFN-γ and its protective effect against spring viremia of carp virus. *Fish & Shellfish Immunology*, *88*, 403–406.

Funkenstein, B. (2006). Production of recombinant hormones and growth factors for use in aquaculture. *Israeli Journal of Aquaculture-Bamidgeh*, *58*.

Gifre, L., Arís, A., Bach, À., & Garcia-Fruitós, E. (2017). Trends in recombinant protein use in animal production. *Microbial Cell Factories*, *16*, 1–17.

Kah, O., & Dufour, S. (2011). Conserved and divergent features of reproductive neuroendocrinology in teleost fishes. In *Hormones and Reproduction of Vertebrates*. Elsevier.

Kamei, H., Kaneko, T., & Aida, K. (2006). *In vivo* gonadotropic effects of recombinant Japanese eel follicle-stimulating hormone. *Aquaculture*, *261*, 771–775.

Kawauchi, H., Suzuki, K., Itoh, H., Swanson, P., Naito, N., Nagahama, Y., Nozaki, M., Nakai, Y., & Itoh, S. (1989). The duality of teleost gonadotropins. *Fish Physiology and Biochemistry*, *7*, 29–38.

Kawauchi, K., Mori, H., Sugiyama, H., Oshimi, K., & Hirayama, A. (1991). Multiple myeloma with coexistent myelofibrosis: Improvement of myelofibrosis following recovery from multiple myeloma after treatment with melphalan and prednisolone. *Japanese Journal of Medicine*, *30*, 483–486.

Kim, H., Yoo, S. J., & Kang, H. A. (2015). Yeast synthetic biology for the production of recombinant therapeutic proteins. *FEMS Yeast Research*, *15*.

Le Linh, H., Thu, N. P. A., Dung, T. T. X., Van Hau, N., Nghia, N. H., & Thao, D. T. P. (2021). Yeast cell surface displaying VP28 antigen and its potential application for shrimp farming. *Applied Microbiology and Biotechnology*, *105*, 6345–6354.

Melo-López, F. N., Zermeño-Cervantes, L. A., Barraza, A., Loera-Muro, A., & Cardona-Félix, C. S. (2021). Biochemical characterization of LysVpKK5 endolysin from a marine vibriophage. *Protein Expression and Purification*, *188*, 105971.

Mohammadzadeh, S., Ahmadifar, E., Masoudi, E., Milla, S., Nahed, A., Alagawany, M., Emran, T. B., Michalak, I., & Dhama, K. (2022). Applications of recombinant proteins in aquaculture. *Aquaculture*, *561*, 738701.

Mohammadzadeh, S., Moradian, F., Yeganeh, S., Falahatkar, B., & Milla, S. (2020). Design, production and purification of a novel recombinant gonadotropin-releasing hormone associated peptide as a spawning inducing agent for fish. *Protein Expression and Purification*, *166*, 105510.

Mohammadzadeh, S., Yeganeh, S., Moradian, F., Falahatkar, B., & Milla, S. (2019). Transformation and gene expression study of recombinant fish GnRH in *E. coli* BL21 in order to produce recombinant hormone. *Iranian Scientific Fisheries Journal*, *28*, 137–147.

CHAPTER 12

Biotechnological Advancements in Aquaculture: Promoting Sustainable Aquaculture

MAYUR BHADARKA,[1,*] D. T. VAGHELA,[1] PINAK BAMANIYA,[1] HARDIK SIKOTARIYA,[2] SAYAN BISWAS,[3] JORA KIRANKUMAR,[3] MOHIT BAMANIYA,[4] and NEHA KHARADI[4]

[1]*Department of Aquatic Environment Management, College of Fisheries Science, Kamdhenu University, Veraval, Gujarat, India*

[2]*Department of Fisheries Resource Management, College of Fisheries Science, Kamdhenu University, Veraval, Gujarat, India*

[3]*Department of Aquaculture, College of Fisheries Science, Kamdhenu University, Veraval, Gujarat, India*

[4]*Department of Fish Processing Technology, College of Fisheries Science, Kamdhenu University, Veraval, Gujarat, India*

*Corresponding author

ABSTRACT

The aquaculture industry has experienced remarkable progress thanks to rapid breakthroughs in biotechnology. This progress spans genetic and genomic improvements, innovative disease control methods, advanced reproductive techniques, and sustainable approaches to nutrition and environmental management. Modern genetic and genomic tools have strengthened selective breeding, enabling the development of fish varieties with higher growth performance, stronger disease resistance, and improved

Sustainable Aquaculture: Recent Advances and Techniques. Dhanaji W. Patil, Sandip S. Markad (Eds.)

DOI: 10.1201/9781779641113-12

feed conversion. Molecular diagnostic tools and marker-assisted breeding now play a key role in safeguarding the genetic quality and health of farmed stocks. Biotechnological advances have also transformed disease prevention and treatment, providing rapid disease detection, cutting-edge vaccines, and health-promoting measures like probiotics and prebiotics to boost immunity. Reproductive biotechnologies, such as hormone treatments, in-vitro fertilization, sex manipulation, and gamete cryopreservation, help maintain genetic diversity and support sustainable stock enhancement. In nutrition, biotechnology promotes alternative proteins like microbial and insect-based meals, enzyme additives, and functional feeds enriched with health-boosting compounds to maximize feed efficiency and fish wellbeing. Biofloc systems, which rely on managing beneficial microbes, help maintain water quality and create more sustainable production environments. Nanotechnology further contributes with innovations in targeted drug delivery, water monitoring, and durable farming materials. Lastly, molecular tools for product traceability ensure transparent and responsible aquaculture.

12.1 INTRODUCTION

Aquaculture involves the raising of aquatic animals or the cultivation of aquatic plants primarily for food production. Global aquaculture production has been steadily increasing over time. Aquaculture has emerged as the fastest-growing and most efficient agricultural sector (Gui, 2015). Sustainable growth in aquaculture is essential for alleviating pressure on wild fish stocks, which are globally strained due to many factors. Thus, the rapid expansion of aquaculture is evidently attributed to advancements in fish biology and biotechnology. Biotechnology involves leveraging biological principles to create new products, techniques, and organisms aimed at enhancing human health and societal well-being. Biotechnology, employing living organisms and their components, is widely utilized across various sectors such as health, agriculture, food, and numerous industrial fields. With its substantial potential to enhance agricultural production, biotechnology has gained increasing significance in the realm of aquaculture in recent years. Biotechnology offers effective instruments for the sustainable advancement of aquaculture, fisheries, and the food sector (Mayekar et al., 2013). The swift growth of aquaculture has benefited significantly from both conventional and biotechnological approaches. It is expected that further advancements in biotechnology will continue to support the industry in meeting the rising global demand for aquatic food in the coming decades. Biotechnological applications have bolstered the efficiency and cost-effectiveness of aquaculture by boosting its productivity. Potential applications of biotechnology in aquaculture encompass utilizing synthetic hormones for induced breeding, generating monosex and polyploid populations, employing molecular biology techniques, creating transgenic fish, establishing gene banks, enhancing feed quality, managing health, and exploring natural products derived from marine organisms (Lakra and Ayyappan, 2003). Biotechnology in aquaculture enables the manipulation of fish sex and breeding cycles, the development of stocks with elevated nutrient content, increased disease resistance, enhanced growth rates, and greater tolerance to environmental fluctuations. Utilizing contemporary biotechnological techniques to enhance fish production holds substantial promise, not only for fulfilling demand but also for improving the quality and quantity of fish within aquaculture setups (Goswami et al., 2023). In recent times, genetics has made notable contributions to fish farming by applying the latest techniques developed in biotechnology and genetic engineering (Foresti, 2000). Currently, the primary concern in aquaculture is disease management. Diseases can be effectively controlled through the development of biotechnologically produced vaccines. There is a growing demand to utilize biotechnology for creating vaccines against bacterial, viral, and other diseases that impact various commercially significant cultured fish species.

12.2 BIOTECHNOLOGICAL APPLICATIONS IN AQUACULTURE

12.2.1 GENETIC AND GENOMIC INTERVENTIONS

The application of genetic and genomic biotechnology is progressively expanding in aquaculture to enhance various facets of production. This encompasses the utilization of DNA markers like single nucleotide polymorphisms (SNPs) for genetic enhancement via marker-assisted selection (MAS) and genomic selection (Su et al., 2023). The discovery of particular genomic regions correlated with economically significant traits has facilitated the integration of markers associated with quantitative trait loci (QTL) into breeding programs (Yanez et al., 2023).

Implementing these technologies carries the potential to expedite genetic advancements, fortify disease resistance, and elevate the sustainability of aquaculture.

12.2.1.1 SELECTIVE BREEDING PROGRAMS AND THEIR IMPACT ON DISEASE RESISTANCE, GROWTH RATES, AND FEED EFFICIENCY

Biotechnology has found extensive application in selective breeding programs aimed at enhancing disease resistance, growth rates, and feed efficiency. Various biotechnological methods, including artificial insemination, embryo transfer, and marker-assisted selection, have been utilized to improve reproductive efficiency and propagate superior germplasm (Das et al., 2022). These technologies have had a substantial influence on genetic advancement, precision in selection, and the preservation of genetic diversity. In aquaculture, genetic diversity in host resistance has been harnessed through genetic technologies to address disease and parasitism challenges. This has resulted in the creation of methods for enhancing disease resistance via selective breeding, gene editing, and targeted treatments (Poland and Rutkoski, 2016). Reproductive and genetic technologies applied to aquaculture species have demonstrated promise in enhancing the effectiveness of selective breeding programs. This is notably achieved through marker-assisted selection and *in vitro* fertilization techniques (Bigini et al., 2021). Genetic technologies, including selective breeding, gene editing, and targeted feed treatments, are being investigated to enhance disease resistance in aquaculture stocks. These approaches aid in identifying the genetic mechanisms associated with resistance and developing tools to mitigate the disease burden. Biotechnological tools are used to control fish sex, manipulate breeding cycles, and cultivate stocks with superior nutrient content and accelerated growth rates (Su et al., 2023). Enhancements in feed efficiency in *Nile tilapia* have been accomplished through genomic selection, resulting in notable progress within the aquaculture sector (Labh, 2023). In general, biotechnology in aquaculture holds promise for enhancing feed efficiency and growth via selective breeding initiatives, thereby driving improvements in productivity and quality across the industry.

12.2.1.2 MOLECULAR DIAGNOSTICS AND MARKER-ASSISTED SELECTION METHODS FOR PRESERVING HEALTH AND GENETIC INTEGRITY

Biotechnology is employed within aquaculture to apply molecular diagnostics and marker-assisted selection techniques, aimed at safeguarding the health and genetic consistency of fish populations. Molecular markers, including DNA-centric methods like PCR, qPCR, and microarrays, play a pivotal role in swiftly and precisely identifying fish diseases. These methods offer swifter and more sensitive capabilities compared to conventional techniques, enabling early detection and pre-emptive measures against disease outbreaks (Rao, 2023). Biotechnology is utilized to identify and merge desirable traits in fish, aiming to enhance productivity and elevate quality standards (Su et al., 2023). Genetic markers are employed to target favorable characteristics like disease resistance and growth rate in breeding programs (Mishra et al., 2023). By leveraging these biotechnological tools, aquaculture can generate disease-free or high-health broodstock for seed production, ultimately enhancing aquaculture yields. The application of biotechnology in aquaculture is pivotal for safeguarding the health and genetic consistency of fish populations, thus promoting sustainable aquaculture practices.

12.2.2 DISEASE MANAGEMENT STRATEGIES

Biotechnology is actively applied in aquaculture for disease management strategies. Nanotechnology and biotechnology present innovative and effective solutions to address infectious diseases in shrimp aquaculture. The adoption of nanotechnology exhibits potential in mitigating viral and bacterial infections in fish and other crustaceans, providing precise drug and gene delivery, vaccination, and diagnostics (Raj et al., 2023). Genetic diversity in host resistance is being harnessed to address disease challenges in aquaculture. Swift and dependable diagnostic methods, including serological tests and DNA-based diagnostic tests, have been developed to rapidly and precisely detect pathogens (Mishra et al., 2023). Microalgal technology presents various potential applications in disease control within aquaculture.

12.2.2.1 INTEGRATION OF BIOTECHNOLOGICAL SOLUTIONS FOR DISEASE DETECTION AND VACCINE DEVELOPMENT

Biotechnological advancements are being incorporated into aquaculture to detect diseases and develop vaccines. Utilizing vaccines is regarded as the safest and most efficient method for preventing aquatic animal illnesses, thereby securing the sustainability of aquaculture (Du et al., 2022). Conventional approaches to pathogen detection and diagnosis in aquaculture, like culturing and serological methods, face limitations regarding accuracy and efficiency. Molecular techniques such as metabarcoding and droplet digital PCR provide swift and sensitive pathogen detection alternatives (Mondal and Thomas, 2022). In aquaculture, high-throughput genotyping methods like SNP arrays and genotyping by sequencing are employed to pinpoint genetic variations linked to disease resistance in aquatic species (Mičúchová et al., 2022). Gene editing technologies present new opportunities for improving disease resistance, but their utilization demands careful consideration of ethical concerns and challenges related to dissemination (Li and Jiang, 2023). Incorporating these biotechnological solutions holds promise for enhancing our comprehension of genetic mechanisms related to disease resistance and facilitating the creation of more resilient aquaculture stocks.

12.2.2.2 ROLE OF PROBIOTICS, PREBIOTICS, AND IMMUNE SYSTEM MODULATION IN DISEASE PREVENTION

In aquaculture biotechnology, probiotics, prebiotics, and immune system modulation play pivotal roles in disease prevention. Probiotics are living microorganisms ingested orally that confer health benefits. These microorganisms, present in sufficient quantities, can modify the microflora within a specific host's compartment, leading to beneficial health effects for the host through implantation or colonization (Akhter et al., 2015). Prebiotics are food ingredients that cannot be digested by the host but have a beneficial effect by stimulating the growth and activity of intestinal bacteria (Okey et al., 2018). Several methods of immune system modulation have been investigated to bolster the immune response of fishes and prevent disease outbreaks. Immunostimulants, including vitamins E and C, have been identified to activate the innate immune system and enhance the nonspecific immunity of fishes (Singh and Jain, 2022). The utilization of probiotics, prebiotics, and synbiotics in aquaculture has demonstrated advantageous outcomes, such as enhanced growth, bolstered immunity, preservation of intestinal health, and heightened disease resistance. These interventions, when incorporated into feeds, can promote holistic well-being, shield against infectious agents, and bolster the sustainability of aquaculture operations. Moreover, incorporating prebiotics into fish diets and rearing water has been found to boost fish health, immunity, and resistance to diseases. Therefore, probiotics, prebiotics, and immune system modulation offer promising alternatives to decrease reliance on antibiotics and improve disease prevention in aquaculture.

12.2.3 REPRODUCTIVE BIOTECHNOLOGIES

Reproductive biotechnologies are increasingly employed in aquaculture to elevate fish production and enhance quality. Among these biotechnologies are the utilization of molecular markers for assessing genetic diversity and genetic tagging in breeding programs. A variety of biotechnologies have been developed for the production of commercially significant fish species. These include artificial manipulation of photoperiod, hormone treatments to induce oocyte maturation, techniques for semen storage and cryopreservation, as well as strategies for managing monosex populations.

12.2.3.1 HORMONE-INDUCED SPAWNING, IN VITRO FERTILIZATION, AND SEX CONTROL TECHNIQUES

Biotechnology is applied in aquaculture to regulate hormone-induced spawning and enhance reproductive processes in various fish species. Hormone therapies, including recombinant follicle-stimulating hormone (rFSH) and luteinizing hormone (rLH), have proven effective in stimulating and completing vitellogenesis in females, as well as in facilitating the production of viable sperm in males (Ramos-Júdez et al., 2022). Hormone administration has been instrumental in the development of fish farming, utilizing pituitary extracts and gonadotropin-releasing hormone analogs (GnRHa) as standard methods to trigger spawning in both freshwater and marine species throughout history (Mechaly et al., 2024). Biotechnology methods, including chromosome manipulation and hormonal treatments, have been employed to generate sterile, monosex, and highly inbred fish populations for commercial applications as well (Foresti, 2000). Biotechnology is applied in the production of embryos through *in vitro* methods, encompassing their maturation, fertilization, and subsequent culture outside the organism. This approach enables the replication of genetically superior animal lines and provides alternative options such as transfer, cryopreservation, vitrification, cloning, and genetic engineering for the embryos (Oliveira-Araújo et al., 2016). Sex control in aquaculture is attained through a variety of mechanisms, including genetic, epigenetic, endocrine, and environmental factors, along with techniques like chromosome set manipulation and gene knockout approaches. Additionally, biotechnology contributes to nutrition, disease and health management, and water quality control in aquaculture. Hormonal manipulations, chromosome set manipulation, and gene transfers are among the techniques applied in fish culture to achieve sex control and enhance breeding practices.

12.2.3.2 CRYOPRESERVATION OF GAMETES FOR CONSERVING GENETIC RESOURCES

The cryopreservation of gametes stands as a crucial biotechnological method utilized to conserve genetic resources within the realm of aquaculture. The cryopreservation of fish gametes stands as a crucial technique for conserving genetic resources in aquaculture. It offers a solution to challenges such as inbreeding depression, genetic drift, and introgressive hybridization, which can have detrimental effects on the production of high-quality seeds (Ekico et al., 2022). This technique facilitates the enduring preservation of gametes and embryos, thereby safeguarding genetic diversity and securing future food production. Cryopreservation of fish gametes has been accomplished successfully across numerous species, with the sperm of 200 fish species being cryopreserved effectively (Siva, 2021). Preserving fish genetic resources through cryopreservation is critical not only for ensuring the sustainability of freshwater fish species but also for developing and maintaining high-quality strains essential for sustainable aquaculture practices (Li, 2022). Therefore, cryopreservation of fish gametes plays a vital role in both conserving and effectively utilizing genetic resources within the field of aquaculture.

12.2.4 INNOVATIONS IN NUTRITION

Biotechnology has been extensively employed within the aquaculture industry to enhance the nutritional quality of feeds. Utilizing alternative ingredients like microorganisms, yeasts, fungi, and bacteria in aquafeeds has demonstrated potential in offering protein, lipids, and sources of omega-3 fatty acids (Sarker, 2023). The aquaculture industry has been shifting away from relying solely on fish meal towards incorporating plant, terrestrial animal, and microbial proteins as alternative protein sources. This transition aims to mitigate the over-exploitation of marine resources (Thazeem et al., 2022). Biotechnology has significantly contributed to innovating aquaculture nutrition feed, enhancing productivity, sustainability, and disease control measures within the industry.

12.2.4.1 EXPLORATION OF ALTERNATIVE PROTEIN SOURCES AND FUNCTIONAL FEEDS

Biotechnology has been utilized in aquaculture to create new protein options and enhanced feeds. The incorporation of functional feed additives (FuFAs) like nucleotides, organic acids, probiotics, prebiotics, and specific amino acids has demonstrated positive outcomes in enhancing fish growth and overall health (Hossain et al., 2024). Research has been conducted on the utilization of industrial food wastes as a sustainable protein source to substitute fishmeal (Sampathkumar et al., 2023). Researchers have investigated nonmicrobial feed components like plant-based additives, insect-derived additives, and animal byproducts, alongside microbial feed components such as microbial fermentation products and beneficial microbes, as potential substitutes for traditional feed ingredients (Midhun and Arun, 2023).

12.2.4.2 ENZYME SUPPLEMENTATION FOR IMPROVED NUTRIENT UTILIZATION

The enhancement of nutrient utilization in fish feed through enzyme supplementation has been investigated in aquaculture. Exogenous enzymes, such as protease, lipase, carbohydrate enzymes, and phytase, have been commonly employed as supplements in fish feed to regulate growth performance, digestibility, feed utilization, and overall body composition (Zheng et al., 2020). The addition of enzymes can enhance the digestibility and absorption of ingredients sourced from both plants and animals in fish feed, consequently resulting in improved growth parameters for aquatic animals in aquaculture (Liang et al., 2022). The inclusion of enzymes like phytase and xylanase can augment nutrient digestion kinetics, elevate nutrient availability in fish, and concurrently promote gut health, fostering a more stable gut microbiome environment (Maas et al., 2021). The utilization of microbial phytase as a feed supplement has demonstrated improvements in fish growth by boosting nutrient bioavailability and enhancing digestion (Virmani et al., 2023). Enzyme supplementation in aquaculture holds potential for enhancing nutrient utilization and improving both fish growth and health performances.

12.2.5 BIOFLOC TECHNOLOGY

Biofloc technology is employed in aquaculture to bolster fish farming productivity, ameliorate water quality, and bolster the growth and immunity of cultured animals. It represents a sustainable approach that mitigates pollution, reduces additional water treatment expenses, and enhances the growth performance of raised animals. Introducing biofloc systems can elevate water quality and production performance of fish, while also enhancing their physiological activity and resilience to diseases (Ghimire et al., 2023). The technology harnesses microbial interactions to establish an artificial environment that facilitates natural nutrient recycling and mitigates the risk of common diseases (Yu et al., 2023). Biofloc technology holds promise for sustainable aquaculture and has the potential to contribute to the intensification of fish farming in the future.

12.2.5.1 MANIPULATION OF MICROBIAL COMMUNITIES FOR SUSTAINABLE AQUACULTURE ENVIRONMENTS

Biotechnology is employed to manipulate microbial communities within aquaculture environments, aiming to achieve sustainable production practices. The microbiome plays a pivotal role in animal health, and its manipulation can improve health, welfare, and production outcomes in aquaculture systems (Lorgen-Ritchie et al., 2023). Observing and controlling microbial communities in aquaculture systems can aid in evaluating and enhancing water quality, managing microbial infections, and diminishing the necessity for antibiotics (Bentzon-Tilia et al., 2016). High-throughput sequencing technology has been utilized to analyze bacterial communities in biofloc-based aquaculture systems, offering valuable insights into microbiota composition and the potential for antibiotic resistance (Chen et al., 2022). Such studies significantly enhance our comprehension of microbiomes within aquaculture environments, serving as a foundation for devising efficient management strategies to promote sustainable aquaculture practices.

12.2.5.2 OPTIMIZATION OF BIOFLOC SYSTEMS USING PROBIOTICS AND PREBIOTICS

Biofloc technology in aquaculture seeks to maximize fish farming productivity through the utilization of microbial inoculums containing beneficial microbes endowed with probiotic, immunostimulatory, and bioremediation properties. These microbial inoculums, including *Bacillus subtilis* and *Pseudomonas putida*, have demonstrated substantial enhancements in water quality, gut histomorphology, growth performance, and immune responses in fish species like *Heteropneustes fossilis* (Kumar et al., 2023). Utilizing laboratory-developed probiotics within the biofloc system has been noted to improve growth, feed efficiency, intestinal microbiota, and immune responses in stinging catfish (Sohel et al., 2023). Incorporating probiotics such as *Bacillus subtilis* and effective microorganisms has been demonstrated to have a positive effect on water quality, growth performance, and microbial community dynamics in *Macrobrachium rosenbergii* (Qui et al., 2023). Optimizing biofloc systems with probiotics and prebiotics shows potential for sustainable aquaculture by enhancing water quality, fostering growth, boosting immune responses, and mitigating environmental impacts (Hancz, 2022).

12.2.6 NANOTECHNOLOGY APPLICATIONS

Nanotechnology has emerged as a promising solution for sustainable aquaculture, offering a range of biotechnological applications to address field challenges. Nanosensors, nanoencapsulations, and other nanotechnology methods can enhance aquaculture efficiency by detecting pathogens, improving nutrient absorption, and preserving bioactive compounds (Ramesh et al., 2023). Nanotechnology-based alternatives such as algal extracts, phytobiotics, probiotics, prebiotics, and synbiotics can be used to combat infectious diseases in shrimp aquaculture, reducing the reliance on antibiotics (Ogunkalu, 2019). Nanoparticles have demonstrated potential in drug delivery and vaccine development, offering protection to farmed fish against diseases (Dar et al., 2020). The integration of nanotechnology in aquaculture has the potential to foster more sustainable practices, enhance food safety, and bolster production levels.

12.2.6.1 DRUG DELIVERY, WATER QUALITY MONITORING, AND INFRASTRUCTURE DEVELOPMENT

Nanotechnology finds diverse applications in areas such as drug delivery, water quality monitoring, and infrastructure development. In drug delivery, nanobiotechnology merges biotechnology and nanotechnology to produce nanoparticles with enhanced therapeutic efficacy, improved drug stability, targeted cell delivery, and controlled release within the cell (Jackson et al., 2021). Nanomaterials, especially those derived from microbial systems, have been engineered for precise and effective drug delivery, offering distinct advantages over conventional drug delivery systems in treating chronic and acute diseases (Srivastava et al., 2022). In the realm of water quality monitoring, nanotechnology provides efficient techniques for environmental assessment and treatment. Nanoparticles, nanomembranes, nanofilters, and nano-catalysts offer enhanced efficiency and selectivity owing to their smaller size and higher surface area-to-volume ratio. In infrastructure development, nanotechnology holds promise in the realm of nanomedicine, especially within drug delivery systems. Nano systems like nanobots, injectable nanoparticle generators, and nanocrystals present enhanced efficacy and targeted drug delivery capabilities (Rather et al., 2022).

12.2.6.2 CONTRIBUTION OF MOLECULAR MARKERS TO SUPPLY CHAIN TRANSPARENCY

Molecular markers have played a substantial role in enhancing transparency within the aquaculture supply chain. Molecular markers have been utilized for genetic characterization, parentage determination, and population analyses in aquaculture species (Lo Presti et al., 2009). In penaeid shrimp, both mitochondrial and nuclear markers have been employed to evaluate genetic variation, population structure, and the efficacy of genetic tag propagation-assisted rehabilitation initiatives (Baskaralingam et al., 2013). Molecular markers have been utilized in verifying the authenticity of fish and seafood species, enabling the detection of food mislabeling and the confirmation of species identity (Kotsanopoulos et al., 2021). Molecular genetic markers have been employed for the identification of species and populations, thereby enhancing fisheries management and facilitating genetic enhancements in species (Ayyappan et al., 2014). The application of molecular markers in aquaculture has bolstered supply chain transparency, facilitated genetic advancements, and promoted sustainability within the industry.

12.3 CONCLUSION

The integration of biotechnological methods has sparked a significant shift in the aquaculture industry, presenting novel solutions to longstanding issues. Genetic and genomic interventions have strengthened selective breeding programs, improving disease resistance, growth rates, and feed efficiency. Disease management techniques have progressed with the introduction of swift molecular diagnostics, vaccine innovation, and the application of probiotics and prebiotics. Reproductive biotechnologies have enabled precise regulation of reproduction, fostering sustainable aquaculture approaches. Developments in nutrition, biofloc technology, and nanotechnology have additionally enhanced productivity, water quality, and disease resistance. Molecular markers have played a pivotal role in enhancing supply chain transparency and advancing genetic research, promoting sustainability in aquaculture practices. Together, these biotechnological advancements highlight the potential for more effective, resilient, and sustainable aquaculture methods to address the escalating global demand for aquatic food while addressing environmental concerns.

KEYWORDS

- **aquaculture**
- **aquatic food**
- **biotechnology**
- **disease management**
- **growth performance**
- **molecular markers**
- **reproductive biotechnology**
- **supply chain transparency**

REFERENCES

Akhter, N., Wu, B., Memon, A. M., & Mohsin, M. (2015). Probiotics and prebiotics associated with aquaculture: A review. *Fish & Shellfish Immunology*, *45*(2), 733–741.

Ayyappan, S., Jena, J. K., & Gopalakrishnan, A. (2014). Molecular tools for sustainable management of aquatic germplasm resources of India. *Agricultural Research*, *3*, 1–21.

Bentzon-Tilia, M., Sonnenschein, E. C., & Gram, L. (2016). Monitoring and managing microbes in aquaculture–Towards a sustainable industry. *Microbial Biotechnology*, *9*(5), 576–584.

Bigini, V., Camerlengo, F., Botticella, E., Sestili, F., & Savatin, D. V. (2021). Biotechnological resources to increase disease-resistance by improving plant immunity: A sustainable approach to save cereal crop production. *Plants*, *10*(6), 1146.

Chen, X., He, Z., Zhao, J., Liao, M., Xue, Y., Zhou, J., & Sun, C. (2022). Metagenomic analysis of bacterial communities and antibiotic resistance genes in *Penaeus monodon* biofloc-based aquaculture environments. *Frontiers in Marine Science*, *8*, 762345.

Dar, A. H., Rashid, N., Majid, I., Hussain, S., & Dar, M. A. (2020). Nanotechnology interventions in aquaculture and seafood preservation. *Critical Reviews in Food Science and Nutrition*, *60*(11), 1912–1921.

Das, D. N., Paul, D., & Mondal, S. (2022). Role of biotechnology on animal breeding and genetic improvement. In *Emerging Issues in Climate Smart Livestock Production* (pp. 317–337). Academic Press.

Du, Y., Hu, X., Miao, L., & Chen, J. (2022). Current status and development prospects of aquatic vaccines. *Frontiers in Immunology*, *13*, 1040336.

Ekici, A., Yamaner, G., & Demircan, M. D. (2022). Cryopreservation studies in aquaculture from past to present: Scientific techniques and quality controls for commercial applications.

Foresti, F. (2000). Biotechnology and fish culture. *Hydrobiology*, *420*, 45–47.

Ghimire, R. R., Ghimire, A., Karki, D., Basyal, D., & Rai, K. B. (2023). Determination of ammonia level and its protein conversion in the water of biofloc fish farming technology. *Prithvi Academic Journal, 11–20*.

Goswami, M., Trudeau, V. L., & Lakra, W. S. (2023). Biotechnology in modern aquaculture: Innovations, advancements, and challenges. In *Frontiers in Aquaculture Biotechnology* (pp. 1–13). Academic Press.

Gui, J. F. (2015). Fish biology and biotechnology is the source for sustainable aquaculture. *Science China Life Sciences*, *58*, 121–123.

Hancz, C. (2022). Application of probiotics for environmentally friendly and sustainable aquaculture: A review. *Sustainability*, *14*(22), 15479.

Hossain, M. S., Small, B. C., Kumar, V., & Hardy, R. (2024). Utilization of functional feed additives to produce cost-effective, ecofriendly aquafeeds high in plant-based ingredients. *Reviews in Aquaculture*, *16*(1), 121–153.

Jackson, T. C., Obiakor, N. M., Iheanyichukwu, I. N., Ita, O. O., & Ucheokoro, A. S. (2021). Biotechnology and nanotechnology drug delivery: A review. *Journal of Pharmacy and Pharmacology*, *9*, 127–132.

Kaur, J., Pathak, T., Singh, A., & Kumar, K. (2017). Application of nanotechnology in the environment biotechnology. In *Advances in Environmental Biotechnology* (pp. 155–165).

Kotsanopoulos, K. V., Exadactylos, A., Gkafas, G. A., Martsikalis, P. V., Parlapani, F. F., Boziaris, I. S., & Arvanitoyannis, I. S. (2021). The use of molecular markers in the verification of fish and seafood authenticity and the detection of adulteration. *Comprehensive Reviews in Food Science and Food Safety*, *20*(2), 1584–1654.

Kumar, V., Swain, H. S., Vuong, P., Roy, S., Upadhyay, A., Malick, R. C., & Das, B. K. (2023). Microbial inoculums improve growth and health of *Heteropneustes fossilis* via biofloc-driven aquaculture. *Microbial Cell Factories*, *22*(1), 106.

Labh, S. N. (2023). Applications of biotechnology in aquaculture nutrition and fisheries. In *Frontiers in Aquaculture Biotechnology* (pp. 163–170). Academic Press.

Lakra, W. S., & Ayyappan, S. (2003). Recent advances in biotechnology applications to aquaculture. *Asian-Australasian Journal of Animal Sciences*, *16*(3), 455–462.

Li, C. (2022). Conservation of genetic resources for sustainable aquaculture. *Journal of the World Aquaculture Society*, *53*(1), 4–7.

Li, C., & Jiang, J. (2023). High-throughput genotyping in estimating genetic resources and detecting pathogens in aquaculture. *Journal of the World Aquaculture Society*, *54*(3), 558–565.

Liang, Q., Yuan, M., Xu, L., Lio, E., Zhang, F., Mou, H., & Secundo, F. (2022). Application of enzymes as a feed additive in aquaculture. *Marine Life Science & Technology*, *4*(2), 208–221.

Lo Presti, R., Lisa, C., & Di Stasio, L. (2009). Molecular genetics in aquaculture. *Italian Journal of Animal Science*, *8*(3), 299–313.

Lorgen-Ritchie, M., Webster, T. U., McMurtrie, J., Bass, D., Tyler, C. R., Rowley, A., & Martin, S. A. (2023). Microbiomes in the context of developing sustainable intensified aquaculture. *Frontiers in Microbiology*, *14*.

Maas, R. M., Deng, Y., Dersjant-Li, Y., Petit, J., Verdegem, M. C., Schrama, J. W., & Kokou, F. (2021). Exogenous enzymes and probiotics alter digestion kinetics, volatile fatty acid content and microbial interactions in the gut of Nile tilapia. *Scientific Reports*, *11*(1), 8221.

Mayekar, T. S., Salgaonkar, A. A., Koli, J. M., Patil, P. R., Chaudhari, A., Pawar, N., & Kapse, P. (2013). Biotechnology and its applications in aquaculture and fisheries. *Aquafind-Aquat Fish Database*.

Mechaly, A. S., Batlouni, S. R., Elisio, M., Sanches, E. A., Guzmán, J. C., García, M. M., & Duncan, N. (2024). Spawning induction for Latin American fishes. *Reviews in Aquaculture*, *16*(1), 337–356.

Mičúchová, A., Piačková, V., Frébort, I., & Korytář, T. (2022). Molecular farming: Expanding the field of edible vaccines for sustainable fish aquaculture. *Reviews in Aquaculture*, *14*(4), 1978–2001.

Midhun, S. J., & Arun, D. (2023). Alternative feed technology in aquaculture. In *Recent Advances in Aquaculture Microbial Technology* (pp. 291–306). Academic Press.

Mishra, S., Seshagiri, B., Rathod, R., Sahoo, S. N., Choudhary, P., Patel, S., & Swain, P. (2023). Recent advances in fish disease diagnosis, therapeutics, and vaccine development. In *Frontiers in Aquaculture Biotechnology* (pp. 115–145).

Mondal, H., & Thomas, J. (2022). A review on the recent advances and application of vaccines against fish pathogens in aquaculture. *Aquaculture International*, *30*(4), 1971–2000.

Ogunkalu, O. A. (2019). Utilization of nanotechnology in aquaculture and seafood sectors. *Eurasian Journal of Food Science and Technology*, *3*(1), 26–33.

Okey, I. B., Gabriel, U. U., & Deekae, S. N. (2018). The use of symbiotic (prebiotic and probiotic) in aquaculture development. *Sumerianz Journal of Biotechnology*, *1*(2), 51–60.

Oliveira-Araújo, M. S., Salmito-Vanderley, C. S. B., Almeida-Monteiro, P. S. D., Lopes, J. T., & Leite-Castro, L. V. (2016). Dose inseminante e resfriamento de embriões de peixes de água doce. *R. Bras. Reprod. Anim.*, 35–40.

Poland, J., & Rutkoski, J. (2016). Advances and challenges in genomic selection for disease resistance. *Annual Review of Phytopathology*, *54*, 79–98.

Qiu, Z., Xu, Q., Li, S., Zheng, D., Zhang, R., Zhao, J., & Wang, T. (2023). Effects of probiotics on the water quality, growth performance, immunity, digestion, and intestinal flora of giant freshwater prawn (*Macrobrachium rosenbergii*) in the biofloc culture system. *Water*, *15*(6), 1211.

Raj, N. S., Arivarasan, V. K., Hameed, A. S. S., & Swaminathan, T. R. (2023). Nanotechnologies in the health management of aquatic animal diseases. In *Nanotechnological Approaches to the Advancement of Innovations in Aquaculture* (pp. 157–181). Cham: Springer International Publishing.

Ramesh, A. S., Meghana, N. V. P., Tripathy, A., Karthik, L., Kiran, G., & Arivarasan, V. K. (2023). Future prospects of nanotechnology in aquaculture. In *Nanotechnological Approaches to the Advancement of Innovations in Aquaculture* (pp. 89–97). Cham: Springer International Publishing.

Ramos-Júdez, S., Giménez, I., Gumbau-Pous, J., Arnold-Cruañes, L. S., Estévez, A., & Duncan, N. (2022). Recombinant FSH and LH therapy for spawning induction of previtellogenic and early spermatogenic arrested teleost, the flathead grey mullet (*Mugil cephalus*). *Scientific Reports*, *12*(1), 6563.

Rao, P. K. (2023). Biotechnology tools are an important for improving the fish production. *World Journal of Advanced Research and Reviews*, *17*(1), 912–916.

Rather, M. A., Bashir, S. M., Nabi, S. U., Ahmad, S. U., Zhang, J., & Prasad, M. (2022). Application of nanotechnology in drug delivery. In *Diverse Applications of Nanotechnology in the Biological Sciences* (p. 39).

Sampathkumar, K., Yu, H., & Loo, S. C. J. (2023). Valorization of industrial food waste into sustainable aquaculture feeds. *Future Foods*, 100240.

Sarker, P. K. (2023). Microorganisms in fish feeds, technological innovations, and key strategies for sustainable aquaculture. *Microorganisms*, *11*(2), 439.

Seethalakshmi, P. S., Rajeev, R., Kiran, G. S., & Selvin, J. (2021). Shrimp disease management for sustainable aquaculture: Innovations from nanotechnology and biotechnology. *Aquaculture International*, *29*, 1591–1620.

Singh, A., & Jain, S. (2022). Immunomodulatory effect of vitamin C and E on non-specific immune parameters in fishes: A review. *South Asian Journal of Experimental Biology*, *12*(4), 468–475.

Siva, C. (2021). Cryopreservation and its application in aquaculture. In *Animal Reproduction*. IntechOpen.

Sohel, A. M., Shahjahan, M., Hossain, M. K., Sumi, K. R., Hossain, M. S., Abdul Kari, Z., & Téllez-Isaías, G. (2023). Effects of multispecies probiotics on growth, hematology, and gut health of stinging catfish (*Heteropneustes fossilis*) in biofloc system. *Water*, *15*(14), 2519.

Srivastava, N., Singh, M., & Kumar, V. (2022). Biomedical applications of nanotechnology in development of novel drug delivery system to tackle future challenges. In *Microbial Nanotechnology: Advances in Agriculture, Industry and Health Sectors* (p. 95).

Su, B., Wang, X., & Dunham, R. A. (2023). The application of genetic and genomic biotechnology in aquaculture. *Biology*, *12*(1), 127.

Thazeem, B., Umesh, M., Sarojini, S., Allwyn Vyas, G., Adhithya Sankar, S., Sapthami, K., & Stanly, L. M. (2022). Developments in feeds in aquaculture sector: Contemporary aspects. In *Aquaculture Science and Engineering* (pp. 35–78). Singapore: Springer Nature Singapore.

Vaseeharan, B., Rajakamaran, P., Jayaseelan, D., & Vincent, A. Y. (2013). Molecular markers and their application in genetic diversity of penaeid shrimp. *Aquaculture international*, *21*(2), 219–241.

Virmani, I., Goswami, R. K., Singh, B., Sharma, J. G., & Giri, B. (2023). Role of microbial phytases in improving fish health. *Reviews in Aquaculture*, *15*(4), 1480–1500.

Yanez, J. M., Barria, A., Lopez, M. E., Moen, T., Garcia, B. F., Yoshida, G. M., & Xu, P. (2023). Genome-wide association and genomic selection in aquaculture. *Reviews in Aquaculture*, *15*(2), 645–675.

Yu, Y. B., Choi, J. H., Lee, J. H., Jo, A. H., Lee, K. M., & Kim, J. H. (2023). Biofloc technology in fish aquaculture: A review. *Antioxidants*, *12*(2), 398.

Zheng, C. C., Wu, J. W., Jin, Z. H., Ye, Z. F., Yang, S., Sun, Y. Q., & Fei, H. (2020). Exogenous enzymes as functional additives in finfish aquaculture. *Aquaculture Nutrition*, *26*(2), 213–224.

CHAPTER 13

Vaccination for Disease Prevention in Sustainable Aquaculture

R. R. GAMIT* and P. V. PARMAR

Department of Aquaculture, College of Fisheries Science, Kamdhenu University, Veraval, Gujarat, India

*Corresponding author

ABSTRACT

Aquaculture has faced obstacles from numerous infectious viral, bacterial, and parasitic diseases. The successful development and expansion of the aquaculture business is heavily reliant on the control and prevention of emerging and re-emerging pathogens. Infectious diseases can cause economic loss, food safety issues, and environmental issues. Immunoprophylaxis and vaccination strategies are highly effective and economical means of protection from a wide range of infectious agents. This is particularly important during growing concerns about food safety, human health, and environmental sustainability resulting from the rise of antibiotic-resistant microorganisms. Vaccination appears to be the most promising long-term strategy for stopping and managing outbreaks of diseases in aquaculture. Fish vaccines should be potent, safe, and effective. Although there are immersion and oral delivery routes for fish vaccination, injectable delivery is currently the most extensively used method in the business. The emphasis of this chapter is on vaccines and vaccination within the aquaculture sector, covering a spectrum from conventional inactivated and attenuated vaccines to advanced types like recombinant, subunit, vectored, genetically engineered, DNA, and peptide vaccines. It

Sustainable Aquaculture: Recent Advances and Techniques. Dhanaji W. Patil, Sandip S. Markad (Eds.)

DOI: 10.1201/9781779641113-13

also delves into innovative approaches such as reverse vaccinology, plant-based edible vaccines, and nano vaccines.

13.1 INTRODUCTION

Aquaculture stands as the globe's most rapidly expanding food production industry (FAO, 2018). Over the past three decades, it has experienced consistent growth of 8–10%, and aquaculture today provides more than half of the seafood consumed by humans. Growth is likely to continue as the global population increases and more people seek to eat seafood. Aquaculture relies heavily on land, water, feed, and energy inputs, just like other animal production systems. Historically, aquaculture consisted of larger systems that required less maintenance and input (FAO, 2016).

Larger water bodies/ponds were used to produce aquatic species at lower stocking densities with minimal additional feeding. The need to boost production efficiency has led to the development of more intensive aquaculture systems, such as raceways, offshore cages, net pens, and recirculating aquaculture systems (RAS). High-value species are typically raised in intensive aquaculture at high densities and fed commercial feed under closely monitored conditions. This level of intensification may increase the likelihood of disease outbreaks, which can negatively impact output and lead to significant financial losses (Subasinghe et al., 2001).

For aquaculture, preventing and controlling disease outbreaks has emerged as the primary challenge. Infectious diseases, including bacterial, viral, mycotic, and parasitic illnesses, continue to hinder aquaculture growth. In light of this, effective fish health management is crucial for disease prevention, increased yields, and sustainable aquaculture practices.

Antibiotics and chemicals can be used to treat diseases, but fish producers have limited options for these treatments. Antibiotic use to control bacterial infections was common in the early phases of intensive aquaculture development. However, concerns about public health and the environment have arisen due to the potential for bacteria to become resistant to antibiotics and for antibiotic residue to contaminate aquaculture products or the environment (Fernández-Alarcón et al., 2010).

In addition to antibiotics, fish farmers have implemented various methods for disease control and prevention. These approaches include the use of chemicals and disinfectants, the implementation of optimal aquaculture practices, improvements in environmental and feed management, the enforcement of rigorous biosecurity measures, and the use of vaccines, probiotics, and immunostimulants. Among these methods, vaccination stands out as one of the most pragmatic, cost-efficient, and environmentally sustainable strategies for preventing or mitigating losses from infectious diseases in aquaculture, as supported by scientific research (Sommerset et al., 2005).

While fish immune systems are relatively basic compared to mammals, there are more parallels than disparities. The predominant antibody form in the limited fish species studied is an immunoglobulin (Ig) M-like isotype, which typically appears in its secretory form as a tetramer. Fish do not possess isotypes that correspond to mammalian IgG, IgA, or IgE. If fish have a secondary humoral immune response, it is less pronounced than in mammals. Limited understanding of the immune systems of various fish species restricts research opportunities regarding pathogen-induced immunity and vaccine-induced immunity. One major barrier to the development of commercial vaccines is cost-effectiveness in the field. Some species may be too delicate to withstand the stress induced during vaccination, or they may experience significant adverse effects afterward. Fish typically require a higher antigen dose than terrestrial animals, and generating affordable inactivated viral vaccines has been challenging. Additionally, in some species, primary illness issues arise in the larval and fry stages, before the animal reaches a size suitable for vaccination or before its immune system has fully developed. The potential for parental immunization to protect offspring is further limited by fish's apparent lack of maternal immunity. One of the most significant methods of disease control is vaccination.

It is possible to generate active immunity against illnesses through vaccination. The process involves using germs that have been killed, weakened, or attenuated. The immune system produces antibodies in response to recognition. Vaccination offers several advantages over chemotherapy in controlling diseases.

13.1.1 HISTORY OF FISH VACCINATION

Vaccines have been instrumental in preventing infectious diseases for nearly two centuries. In 1796, Edward Jenner, an English physician, observed that dairy farmers who had contracted cowpox appeared to develop immunity to smallpox. Jenner subsequently inoculated a boy with cowpox, leading to immunity against smallpox (Jenner, 1798). The term "vaccine inoculation," coined by Jenner, later evolved into "vaccination," derived from the Latin word *vacca*, meaning cow. Louis Pasteur and his collaborators further advanced vaccination science by demonstrating that administering attenuated and live microorganisms could confer protection against pathogenic bacteria (Fenner et al., 1997). In 1881, Pasteur proposed the term "vaccination" to describe the preventive inoculation of germs, paying homage to Jenner's pioneering work (Fenner et al., 1997).

Snieszko et al. (1938) published a paper discussing protective immunity in *carp* immunized with *Aeromonas punctata*, marking the earliest known account of disease prevention through vaccination. However, due to the publication being in Polish, the information was not widely accessible. Duff is credited with the first English reports on *Aeromonas salmonicida* protection in trout through parenteral injection and oral administration (Duff, 1942).

Vaccination is the technique of exposing fish to a non-pathogenic form or components of microorganisms in order to stimulate protective immune responses against pathogenic germs. Since fish vaccination was created for the first time in the 1940s, it has shown to be the most efficient and economical method of protecting farmed finfish from infectious pathogens (Snieszko, 1949). A vaccination induces immunity to disease by stimulating the development of antibodies and memory cells. The process involves exposing a healthy animal's immune system to an antigen, which creates a "memory" that accelerates the immune system's response to future infections by the specific pathogen (Ellis, 1988).

13.1.2 USE OF VACCINATION IN AQUACULTURE

The year 1938 saw the publication of the first report on fish vaccine research when *carp* that had been vaccinated against *Aeromonas punctata* demonstrated protective immunity. After an oral vaccination against *A. salmonicida* for trout was published in English (Duff, 1942), the field of fish immunization remained largely ignored for another thirty years, mostly because antibacterial chemicals became widely available following World War II. It was demonstrated in a preliminary investigation that immunizing trout by injection would cause a specific antibody response against *A. salmonicida* (Krantz, 1963). Concurrent advances in human vaccination and the successful use of oral vaccination in the poultry industry sparked renewed interest in fish vaccination (Horne, 1982; Snieszko, 1970). Growing public health and environmental concerns about antibiotic usage in aquaculture are driving the search for more efficient antibiotic substitutes, as well as enhanced disease prevention strategies like immunization. The gram-negative bacterial pathogens *Yersinia ruckeri* and *Vibrio anguillarum* were the targets of the first US-licensed vaccinations used in commercial aquaculture (Midtlyng, 1996; Shao, 2001).

13.2 PROPERTIES OF VACCINE

The "ideal fish vaccine" should possess the capability to elicit an immunological response. For commercial and practical viability, vaccines must confer long-term immune response, specificity, protection, and memory. In vaccine development, it is crucial to target multiple disease strains for comprehensive protection. Additionally, the vaccine should be user-friendly and cost-effective. Safety is paramount, ensuring the well-being of fish, those administering the vaccine, and consumers (Bedekar et al., 2022).

13.3 TYPES OF VACCINE

A vaccine's classification is determined by the methodology employed during its development. Every strategy has unique benefits and different modes of action. The characteristics of illnesses and manufacturing feasibility are taken into consideration when designing vaccines. When designing vaccines, fundamental knowledge about the microorganism is crucial, including its mechanisms of cell infection and the immune system's response to it. Pragmatic factors such as the size and significance of the fish species targeted for vaccination are also considered. Generally speaking, antigen delivery techniques can be used to categorize vaccinations.

13.3.1 ATTENUATED VACCINE

Live-attenuated bacteria, which are rendered "weakened" or non-pathogenic yet retain the ability to replicate and express their immunogenic properties within the host, serve as the basis for this type of vaccination. Preparation of these vaccines involves continuous passaging of the pathogen in various heterologous systems (such as different animal hosts, tissue cultures, and embryonated eggs), as well as methods like chemical/

heat attenuation and genetic manipulation (such as mutation by deletion, disruption, or insertion of metabolic pathways or virulence genes) (Desmettre, 1997). Because these vaccines are self-replicating, they do not necessitate booster shots and can effectively stimulate both the host's humoral and cell-mediated immune responses, thereby contributing to the establishment of robust, long-lasting protective immunity.

➢ **Advantages:**

- Booster vaccination is not necessary for self-replicating organisms. It gives the host durable protective immunity that is simple to deliver by oral or immersion.
- Infectious microorganisms have the capability to induce both humoral and cellular immune response memory.
- Fewer are needed to produce protection since they can proliferate within the host.
- Long-lasting immunity is frequently produced by a single vaccination treatment with great effectiveness.
- In their native shape, whole microorganisms respond to antigens by stimulating the immune system. They bolster the body's defenses against all protective antigens.
- Some live vaccines can be given orally, enhancing protection at the typical site of entry by eliciting mucosal immunity and IgA production (Rathore, 2017).

➢ **Disadvantages:**

- Carries the potential for strains to recombine, giving rise to new strains. Possesses the potential to revert to a virulent strain, posing a significant risk to the aquatic environment and unintended wildlife.
- Unsuitable for animals with impaired immune systems because they depend on an active immune system.
- May sporadically convert to its pathogenic form and spread illness.
- Appropriate storage is essential because they are living organisms whose survival depends on their activity.
- Immunocompromised individuals cannot safely receive live vaccinations (Rathore, 2017).

13.3.2 WHOLE-CELL INACTIVATED VACCINE

Louis Pasteur's "isolate, inactivate, and inject" method is the foundation for whole-cell inactivated vaccines (Zhao et al., 2014). These vaccinations contain dead microbes (bacteria, viruses, or parasites) that have been rendered inactive by chemical or physical methods, including heating, formaldehyde, or radiation therapy. While the inactivated pathogens are rendered incapable of causing illness, they retain the ability to provoke an immune response in the host. Upon recognition of the foreign antigens of the deactivated pathogen, the host initiates its immune response, particularly activating the humoral immune system.

However, due to the relatively weaker immune responses elicited by these inactivated vaccines compared to live vaccines, they require administration with an appropriate adjuvant and multiple booster doses to sustain adequate protective immunity over time. Commercially available inactivated vaccines for salmon and carp have been reported worldwide. The first documented fish vaccine trials involved an oral vaccine for cutthroat trout (*Oncorhynchus clarkii*) and an inactivated vaccine against *Aeromonas salmonicida* (Duff, 1942). The first fish vaccine to receive commercial approval was an inactivated vaccine that demonstrated effective immune protection against *Yersinia ruckeri* (Evelyn, 1997).

While this approach was successful in creating vaccines against certain harmful bacteria found in fish, it encountered significant challenges in creating vaccines against the majority of other fish diseases, including viruses. Nevertheless, a Czechoslovakian company called Bioveta developed the first inactivated viral vaccine for fish in 1982. It provided protection against a carp rhabdovirus responsible for spring viremia of carp (SVC).

➢ **Advantages:**

- Inactivated vaccines do not have the same risk of reverting to their disease-causing form as live attenuated vaccines do.
- Do not require a cold chain for storage, and the freeze-dried form is convenient for transportation.
- These vaccines are affordable and simple to produce.

➢ **Disadvantages:**

- Since these vaccines are inactivated, they produce comparatively weaker immune responses, necessitating multiple

booster doses to sustain a sufficient degree of protective immunity over an extended period of time in order to optimize their efficacy with an appropriate adjuvant.
- High expense of production in cell culture, difficult delivery, and a complicated purifying process.
- In general, vaccines composed exclusively of inactivated agents predominantly elicit the humoral immune response rather than the cell-mediated immune response.
- Additionally, it creates protective immunity, which eventually decreases and requires booster doses.
- The injectable mode of delivery is generally efficient.

13.3.3 TOXOID VACCINE

Exotoxins and endotoxins are substances released by bacteria as part of their pathogenic response. Exotoxins, typically utilized in the creation of toxoid vaccines, undergo chemical or heat treatment to preserve immunogenicity while reducing toxicity, resulting in a toxoid. Toxoids are capable of inducing an immune response, generating immunity, and influencing memory. Upon administration of a vaccine containing a harmless toxoid, the humoral immune system is activated, producing antibodies that bind to and neutralize the toxin. This is also known as an anatoxin. There are few reports of poor antibody responses in toxoid vaccine trials in aquaculture (Bedekar et al., 2022).

A toxoid-enriched inactivated vaccine containing *Photobacterium damselae* subspecies *piscicida* was observed to offer 37–41% protection. Additionally, toxoid vaccine trials against *A. salmonicida* are currently under investigation (Austin et al., 1981).

➢ **Advantages:**

- A toxoid can stimulate an immune response, establish immunity, and influence memory.
- These vaccination methods are highly safe and unlikely to result in any adverse side effects.
- They can also function in immunocompromised individuals.

➢ **Disadvantages:**

- Usually requires an adjuvant and may need multiple doses.

- The limited experimental trials of toxoid vaccination in aquaculture have shown relatively low antibody responses, thereby constraining the vaccine's applicability.

13.3.4 VECTOR VACCINE

The vaccine antigen, representing the encoded protein of another harmful bacterium, is expressed within the recipient host through the utilization of live virus vectors for transferring antigenic genes (Adam et al., 2008). These subunit vaccines, founded on virus-like particles (VLPs), were conceived due to the capacity of viral structural proteins to self-assemble into structures resembling authentic viruses (Dhar and Allnutt, 2011). The baculovirus expression system has proven to be a more efficient method for rapidly producing numerous recombinant proteins (VLPs) and is proposed as a cost-effective and efficient approach for generating heterologous proteins (Adam and Thompson, 2006). While the vector can actively replicate within host cells and stimulate the immune system akin to an adjuvant, the vaccination antigens possess the ability to provoke both humoral and cell-mediated immune responses. Competent hosts for VLP creation include bacteria, plants, or fungi. Another method for VLP production involves genetic recombination of a chimera that yields a distinct virus. Recently, several experimental vaccines based on virus-like particles (VLPs) have been developed. These include a vaccine against infectious pancreatic necrosis, derived from the IPNV capsid protein VP2 expressed in yeast, which also triggers an immune response in rainbow trout (Dhar et al., 2010). The vaccine against Atlantic cod NNV (ACNNV) for seabass is grounded on the expression of the coat protein in the plant *Nicotiana benthamiana* (Marsian et al., 2019). Vaccines targeting grouper nervous necrosis (Chien et al., 2018) and viral nervous necrosis (Thiéry et al., 2006) were also developed for orange-spotted grouper and European seabass, respectively.

- **Advantages:**
  - Alongside the antigen, the vector possesses the ability to actively replicate within host cells and can serve as an adjuvant to activate the immune system.

- An advantage of the alphavirus-based replicon is its cessation of spreading and infecting new cells following its initial replication.
- The capability of the alphavirus replicon to bolster mucosal immunity.

➢ **Disadvantages:**

- Pre-existing antibodies against the vector virus can neutralize or inhibit the viral vector, thereby reducing the targeted immune response to the foreign antigen.
- When it comes to fish vaccine research, vector vaccine technology is still very young and has not undergone much testing.

13.3.5 SUB-UNIT VACCINE

The subunit vaccine is formulated using recombinant technology, where the immunogenic target areas of a pathogen are exclusively produced in a heterologous host to purify the protective antigen for use in the vaccine (Adams et al., 2008). Biotechnological tools are employed to identify and create the gene sequence of the pathogen's protective antigen. Antigenic genes are then designed, introduced into prokaryotic (Noonam and Enzmann, 1995) or eukaryotic production hosts, and cultivated on a large scale using fermentation technology under carefully regulated laboratory conditions.

Microalgae, transgenic plants (Muktar et al., 2016), yeast (Vakharia, 2005), insect cells, bacteria, and cell culture (Acosta et al., 2006) are among the production hosts. However, when it comes to fish vaccines, the administration of recombinant antigens made by fermentation has been found to be ineffective in eliciting protective immunity, possibly due to the antigens' low immunogenicity.

One of the most effective instances of subunit vaccination is against infectious pancreatic necrosis (IPN), which involves the fused IPN-VP2 gene. In Chile, the ISAV vaccine, known as Centrovet, is available as an oral vaccination containing recombinant hemagglutinin-esterase protein. Yeast expression and baculovirus systems have been utilized to develop vaccines against IHNV and viral hemorrhagic septicemia (Biering et al., 2005).

➢ **Advantages:**

- Possess no living components and, hence, pose no risk of illness. Simple, reliable, and safe to produce.
- Can be safely administered to immunosuppressed animals.
- Less likely to induce side effects (Rathore, 2017).

➢ **Disadvantages:**

- The delivery of recombinant antigens in the fish vaccine has been found ineffective in eliciting protective immunity, despite being highly successful against infections that affect humans and animals.
- A weaker immune response is induced by poorly immunogenic stimuli.
- Although an immune response can frequently be elicited, immunological memory formation is not always guaranteed (Rathore, 2017).

13.3.6 PEPTIDE VACCINE

Synthetic peptides or short amino acid sequences displayed on the surface of a carrier protein, termed peptide vaccines, possess the capability to activate the immune system of the recipient host. The initial step in producing immunogenicity is to identify the short amino acid domain with bioinformatic tools like SwissProt, Prosite, Predict Protein, and Epitope Mapper.

Subsequently, the peptide is synthesized and employed as a vaccine to elicit an immunological response. Because they can elicit an immunological response and memory, these are known as peptide vaccines. The brief peptides are employed as a vaccination and are attached to a few surface carrier proteins. Despite their simplicity and safety, their limited immunogenicity limits their use in fish (Bedekar et al., 2022).

➢ **Advantages:**

- Very simple and safe.

➢ **Disadvantages:**

- Their poor immunogenicity limits their use in fish immunology.

13.3.7 DNA VACCINE

The DNA vaccine consists of an extra-chromosomal plasmid that replicates itself and contains the pathogen's immunogenic gene. The process of DNA vaccination involves giving the host plasmid DNA, which is produced in microorganisms like bacteria and encodes a vaccine antigen (Heppell, 2000). Experimental development of DNA vaccines has targeted various fish diseases, including infectious hematopoietic necrosis virus (IHNV), hirame rhabdovirus (HIRRV), spring viremia of carp virus (SVCV), and viral hemorrhagic septicemia virus (VHSV). Additionally, other viruses such as infectious salmon anemia virus (ISAV), nervous necrosis virus (NNV), salmonid alphavirus, grass carp reovirus (GCRV), infectious pancreatic necrosis virus (IPNV), Koi herpes virus (KHV), Channel catfish virus (CCV), and Lymphocystis disease virus (LCDV), as well as bacterial pathogens like *E. tarda*, *Aeromonas* sp., *Vibrio* sp., and *Streptococcus* sp., have been targeted for experimental DNA vaccine development (Bedekar et al., 2022). DNA vaccines exploit the recipient's plasmid DNA, which undergoes transcription into mRNA and translation into the encoded protein of the gene, both regulated by eukaryotic promoters.

- **Advantages:**
  - DNA vaccines exhibit stability in dried powder or solution forms and do not necessitate refrigeration, thus eliminating the need for a cold chain (Bedekar et al., 2022).
  - The vector can encode for a multivalent vaccine targeting multiple diseases, enabling single-administration convenience (Bedekar et al., 2022).
  - DNA vaccines are comparatively cost-effective and straightforward to manufacture using standardized production processes. Upon administration, the host develops robust and enduring protective immunity without the risk of accidental infection. Moreover, DNA vaccines activate both humoral and cellular immunity, benefiting from intrinsic immunostimulatory properties due to CpG motifs, and allowing for the possibility of multivalent vaccination by simple mixing (Collins et al., 2019).
  - They are particularly effective when administered during the early life stage, provide effective cross-protection, and offer

temperature-independent protection for poikilothermic organisms. Storage and transport are facilitated by DNA's exceptional stability against extreme temperatures. Furthermore, DNA sequences can be readily modified in the laboratory, and the inserted DNA does not replicate, encoding only the proteins of interest.

- The absence of a protein component precludes any immune response against the vector itself. Due to the manner in which antigens are presented, DNA vaccines elicit a cell-mediated response capable of targeting any antigen within the pathogen (Rathore, 2017).

➢ **Disadvantages:**

- The potential integration of DNA into the host genome may result in insertional mutagenesis.
- The introduction of DNA may induce autoimmune responses, potentially leading to the production of anti-DNA antibodies (Rathore, 2017).
- Expression of the antigen in the host could lead to the development of immunologic tolerance, resulting in specific non-responsiveness to that antigen.
- Licensing and commercialization of DNA vaccines for food fish are impeded by regulations in many countries, primarily concerning genome integration (Bedekar et al., 2022).

13.3.8 EDIBLE VACCINE

Edible vaccines are made from plants via a process called molecular farming, in which entire plants or plant tissues are cultivated in vitro to produce immunogenic proteins (Schillberg et al., 2013). These can be produced at a reasonable cost and provide a strong substitute for conventional production methods. Antigenic proteins produced by edible vaccines are then taken up by specialized M-cells and delivered to dendritic cells, initiating a coordinated immune response that involves B-cells and T-helper cells. Although fish vaccines are currently in their nascent phase of development (Shin et al., 2013), progress is expected to accelerate in the near future.

- **Advantages:**
  - These could be produced at a low cost and provide a competitive option to conventional manufacturing methods like growing microorganisms and mammalian cells in large-scale bioreactors.
  - They don't include any unwanted elements, such as endotoxins found in bacteria or yeast-produced hyperglycosylated proteins, like other recombinant technologies do.
  - Their production scale is infinite, and scaling up comes at a low cost.

- **Disadvantages:**
  - Although fish vaccination technology is still in its infancy, it is expected to advance in the near future.

13.4 VACCINATION STRATEGIES AND PROCEDURES

For vaccination rates to be as low as possible, vaccinations must be developed and administered correctly. The choice of vaccination process and the diseases to be vaccinated against are included in the phrase "vaccination strategy" (Lillehaug, 1997). The vaccination process includes the type and technique of vaccinations used, when to vaccinate in relation to the fish's size and water temperature, and when to employ revaccination. Since these criteria are interdependent, it is impossible to decide on one without considering the others.

When there is a high probability of a disease outbreak, the amount of protective immunity that has been developed should be sufficient to withstand the infectious pressure. The duration of the risk should be covered by protection. Furthermore, the immunization plan should consider the potential for side effects and the associated expenses (Lillehaug, 2014).

13.5 TIMING OF VACCINATION

The vaccination plan that is selected should also align with the fish species' life cycle, ideally being completed well in advance of any possible contact with the pathogenic bacterium. Production methods for intensive aquaculture are frequently divided into two separate stages. The first stage, which includes hatching, beginning feeding, and the initial growth phase, is controlled and occurs in tank systems with minimal water usage, allowing pathogen entry to be managed (i.e., disinfection systems). During this period, vaccinations should be administered prior to the subsequent stage, which involves grow-out in ponds or net enclosures. Vaccination against infections prevalent in the marine environment should be conducted during the freshwater phase for anadromous species, such as salmonids (Press and Lillehaug, 1995). Polyvalent vaccines against many illnesses are the most common choice for vaccination strategies in certain aquaculture productions, with the goal of shielding the fish from multiple diseases with just one immunization procedure. Oil-adjuvanted vaccinations against many infectious agents are the foundation for effective Atlantic salmon production since they only require a single injection during the freshwater phase of the vaccination process. In seawater, this single action alone provides significant levels of protection that persist throughout the entire growth period (Lillehaug, 2014).

13.6 WATER TEMPERATURE

It is commonly recognized that fish's immune response and immune system function are influenced by their surroundings, particularly the temperature of the water (Bowden, 2008). In the past, it was recommended to vaccinate fish, especially salmonids, at relatively high temperatures to prevent potential compromised immune responses at low environmental temperatures (Bly and Clem, 1992). However, ambient water temperature may only increase just before sea transfer in the spring for cold-water anadromous fish species like salmonids, and smoltification may have already commenced during this period of rising temperatures.

Fingerlings were assessed at various intervals after immunization against cold-water vibriosis at water temperatures ranging from 2 to 10°C (Lillehaug et al., 1993). The results indicated that vaccination of Atlantic salmon against cold-water vibriosis can be conducted at low water temperatures. In fact, groups of fish vaccinated at lower temperatures, specifically 2–6°C, exhibited tendencies toward higher levels of protection for both dip- and injection-vaccinated fish.

Martins et al. (2011) observed differing outcomes among groups of fish immunized with live theronts of *Ichthyophthirius multifiliis* and then exposed to virulent challenge, particularly with the warm-water species channel catfish. Fish vaccinated at 15°C showed reduced antibody response and increased mortality compared to groups vaccinated at higher temperatures, up to 30°C.

Overall, it seems that the ultimate level of protection against ambient temperature does not significantly appear to be affected by fish vaccination, at least not when the fish are within the "normal" range for the species in question. The effects of temperature may vary among fish species (warm vs. cold water) and depending on the type of antigen. Generally, vaccine manufacturers do not always specify the water temperature for administering a shot; however, Merck Animal Health recommends a range of 21–29°C for their channel catfish vaccines.

13.7 SIZE OF FISH

Only a few fish species' age or size at which the immune system is fully established has been documented in published data. As per findings by Zapata et al. (2006), economically important species like carp and rainbow trout typically develop a full immune response within approximately 2-3 months. Conversely, salmonids should not undergo vaccination until they reach immunological maturity, estimated to occur at around 5 grams of weight (Hart et al., 1988).

It is advised that salmonids be 20 g before receiving an injectable vaccination for handling considerations, and this recommendation most likely holds true for other fish species as well. According to Schroder et al. (1998), the cod's immune system appears to be completely formed at two months of age; however, Overgard et al. (2011) reports that the Atlantic halibut takes nearly a month to attain immunological maturity. Conversely, in the case of the subtropical freshwater species channel catfish, when vaccinated with a live *Edwardsiella ictaluri* vaccine just seven days after hatching, a full response appears to occur (Shoemaker et al., 1999). This highlights the considerable variation observed in this domain among fish species and possibly between different antigens.

13.8 VACCINE ADMINISTRATION

Effective vaccination relies on the development of protective vaccines and their proper administration. It is best to deliver vaccinations prior to pathogen exposure in order to give the body enough time to acquire immunity. The method of immunization that is employed is also selective; it is determined by the type of vaccine, its cost, its kind, size, number of fish that need to be immunized, and its intended application. Three immunization techniques are frequently employed in aquaculture: injection, immersion, and oral immunization (Vallejos-Vidal et al., 2017).

13.8.1 INJECTION IMMUNIZATION

Injectable vaccination stands as the most effective method of immunization. By ensuring that each immunized fish receives an equal dose of the vaccine, it effectively triggers the body's immune response and establishes long-term, consistent immune protection. This immunization schedule uses fewer shots and works best with mixed, monovalent, multivalent, and inactivated vaccines as well as subunit and nucleic acid vaccines. Fish under 20 g are often the most susceptible to disease; however, there are a number of problems with the injectable immunization procedure, including handling challenges and the inconvenient nature of immunizing small fish. Furthermore, it can lead to adhesions between internal organs and the peritoneal wall, as well as ulceration at the injection site, posing mechanical harm to the fish.

Secondly, when vaccinating a large number of fish, the handling, anesthesia, and injection procedures not only induce stress in the fish but also escalate labor costs. In contrast to hand injection, machine injection can minimize injection times, lessen labor effort, stress fish less, and circumvent some drawbacks of the injection immunization technique itself (Liang, 2015). The commercial availability of various patented vaccination apparatus inventions has not only simplified large-scale vaccination operations for farmers but also enhanced vaccine utilization efficiency and streamlined the vaccine development process.

Intramuscular and intraperitoneal injections are common injectable immunizations. Adjuvants are typically needed for subunit vaccines, inactivated vaccines are often given intraperitoneally, and certain nucleic acid vaccines are specifically suitable for intramuscular immunization due to their high absorption by muscle tissue, which can elicit a systemic immune response. Corbeil et al. (2000) assessed the vaccination effectiveness through various routes, including scarification of the skin, intraperitoneal injection, intrabuccal administration, gene gun, or immersion with pIHNVw-G, and found that only intramuscular injection and the use of a gene gun induced protective immunity in rainbow trout fry.

13.8.2 IMMERSION IMMUNIZATION

In aquatic vaccinations, immersion immunization stands as a unique method. Antigens are absorbed by mucosal tissues such as the skin, gills, lateral line, and gastrointestinal tract, swiftly and efficiently triggering the fish's mucosal immune system. Subsequently, through blood circulation, the antigens disseminate to the kidney, spleen, peripheral blood, and other systemic immune tissues, eliciting a systemic immune response (Liang et al., 2015).

Its advantages include minimizing mechanical damage and stress stimulation to fish bodies, facilitating herd immunity, reducing labor intensity, and requiring less time for immunity development. It proves effective particularly with smaller fish and fry. Several fish species, including *Anguilla japonica*, Atlantic cod, grouper, crucian carp, and rainbow trout, have been reported to undergo immersion vaccination. Direct immersion (DI), hypertonic immersion (HI), flushing, and spraying, among others, represent the primary forms of immersion immunity (Du et al., 2015).

13.8.3 ORAL IMMUNIZATION

In aquatic vaccinations, oral administration is considered the most practical method of immunization. It works well with a variety of vaccine types, such as nucleic acid, live attenuated, and inactivated vaccines. When administering a vaccination orally as opposed to injectably, there are fewer limitations related to the animal's age and size, less chance of mechanical harm and stress, potential labor cost savings, and compliance with animal welfare regulations (Embregts and Forlenza, 2016).

Oral immunization is deemed more suitable for large-scale farm animals compared to other immunization techniques, given that aquaculture represents the most intensive and factory-based form of farming. However, oral vaccinations for fish are still relatively novel and not widely adopted. The primary drawback of these vaccines is their susceptibility to degradation by digestive tract enzymes, their inability to elicit systemic immunological responses and intestinal mucosal immunity, and their potential to produce immune tolerance. Furthermore, it is currently not possible to adequately guarantee the quantity of fish that will receive an oral vaccination or the dosage of the vaccine, which results in a lesser level of immunological protection than that of injectable immunization (Du et al., 2022).

It is vital to minimize exposure to naked antigens in order to enhance the immunological impact of oral vaccinations. Currently, liposomes, chitosan, alginate, and other polymeric microspheres and biofilms are frequently utilized to encapsulate vaccines (Sukkarun, 2022). Studies have demonstrated that DNA vaccines can enhance fish defenses against disease; however, the immunological efficacy of orally administered naked DNA vaccines is suboptimal. Instead, encapsulating the vaccine in sodium alginate can significantly enhance its immunological impact (Angulo et al., 2022; Ballesteros et al., 2015).

Furthermore, recombinant subunit vaccines, when encapsulated, can also be administered orally. Caruffo et al. (2016) encapsulated the Hemagglutinin-esterase virus (HEAV) and the salmon anemia virus (ISAV) F protein in a cationic polysaccharide matrix capsule by simultaneously expressing the two proteins in brewer's yeast.

13.9 EXAMPLES OF VACCINE USED IN FINFISH

Vaccines used to build robust immunity and prevent illness in finfish against major infectious bacterial diseases are shown in Table 13.1, while examples of vaccines against major infectious viral diseases of finfish are listed in Table 13.2.

TABLE 13.1 Vaccine Against Major Infectious Bacterial Diseases of Finfish (Bondad-Reartaso et al., 2022)

Target Disease	Target Pathogen	Target Fish Species	Type of Vaccine	Product Name	Route of Administration
Bacterial kidney disease (BKD)	*Renibacterium salmoninarum*	Salmonids	*Arthrobacter davidanieli*, live culture	Elanco: Renogen	Injection
Edwardsiellosis/ Enteric septicaemia of catfish (ESC)	*Edwardsiella ictalurid*	Catfish spp. (channel catfish, freshwater catfish, striped catfish, brown bullhead, *Danio* spp.)	*Edwardsiella ictaluri*, avirulent live culture	MSD Animal Health: AquaVac-ESC™	Immersion
Edwardsiellosis/ Enteric septicaemia of catfish (ESC)	*Edwardsiella ictalurid*	Pangasius	*Edwardsiella ictaluri*, inactivated live culture	Pharmaq: ALPHAJECT Panga 1 and 2	Injection
Flavobacteriosis/ Rainbow trout fry syndrome/ Columnaris disease	*Flavobacterium columnare*	Cyprinids, salmonids, catfish, carp, trout, perch, tilapia	*Flavobacterium columnare*, attenuated bacterin	FryVacc1 and 2	Immersion
Flavobacteriosis/ Rainbow trout fry syndrome/ Columnaris disease	*Flavobacterium columnare*	Catfish, largemouth bass	*Flavobacterium columnare*, avirulent, live culture	MSD Animal Health: AquaVac-Col™	Immersion
Furunculosis	*Aeromonas salmonicida*	Salmonids, flounder, turbot, carp, tilapia, sole	*Aeromonas salmonicida*, inactivated bacterin	Elanco: Furogen Dip	Injection
Furunculosis	*Aeromonas salmonicida*	Salmonids	Iron-regulated outer membrane protein (IROMP) antigens of 2 strains of *Aeromonas salmonicida*, non-mineral oil based	MSD Animal Health: AquaVac® FNM	Injection

TABLE 13.1 *(Continued)*

Target Disease	Target Pathogen	Target Fish Species	Type of Vaccine	Product Name	Route of Administration
Furunculosis	*Aeromonas salmonicida*	Salmonids	*Aeromonas salmonicida*	Pharmaq: AlphaJect 1200	Injection
Lactococcosis	*Lactococcus garvieae*	Salmonids, European sea bass, gilthead sea bream, Seriola spp., yellowtail (hiramasa), amberjack	*Lactococcus garvieae*, inactivated	MSD Animal Health: Amalin™ Rensa	Oral
Lactococcosis	*Lactococcus garvieae*	Rainbow trout	*Lactococcus trucha*, inactivated	Hipra: ICTHIOVAC® LG	Injection
Pasteurellosis	*Photobacterium damselae* spp. *Piscicida*	European sea bass and gilthead seabream	*Photobacterium piscida*, inactivated	MSD Animal Health: AquaVac Photobac Prime™	Immersion
Pasteurellosis	*Photobacterium damselae* spp. *Piscicida*	Gilthead seabream	*Photobacterium piscida*, inactivated	Hipra: ICTHIOVAC® PD	Immersion
Streptococcosis	*Streptococcus agalactiae*	Grouper, salmonids, turbot, flounder, sturgeon	*Streptococcus agalactiae* biotype 2 bacterin, inactivated, oil adjuvant	MSD Animal Health: AquaVac® Strep Sa; AquaVac® Strep Sa1	Injection
Streptococcosis	*Streptococcus iniae*	Warm-water marine and freshwater finfish	*Streptococcus iniae*, inactivated	MSD Animal Health: AquaVac® Strep Si	Injection
Streptococcosis	*Streptococcus iniae*	Turbot	*Streptococcus iniae*, inactivated	Hipra: ICTHIOVAC® STR	Injection
Vibriosis	*Vibrio anguillarum*, *V. ordalii*	Rainbow trout, European seabass	*Vibrio anguillarum* (biotype I and II), *V. ordalii*, inactivated	MSD Animal Health: AquaVAC® Vibrio	Injection

TABLE 13.1 *(Continued)*

Target Disease	Target Pathogen	Target Fish Species	Type of Vaccine	Product Name	Route of Administration
Vibriosis	*Vibrio anguillarum*, *V. ordalii*	Rainbow trout, European seabass	*Vibrio anguillarum* 01 and 02+a (*V. ordalii*), inactivated	MSD Animal Health: AQUAVAC® Vibrio Ora	Injection
Vibriosis, ISA, Wound disease	*Vibrio anguillarum*, *V. salmonicida*, *Aeromonas salmonicida* subsp. *salmonicida*	Salmonids	*Vibrio anguillarum*, serotypes O1 and O2 α, *V. salmonicida* and *Aeromonas salmonicida* subsp. *salmonicida*, inactivated	Pharmaq: AlphaJect 5200	Injection
Yersiniosis/ERM	*Yersinia ruckeri*	Rainbow trout	Inactivated	AquaVac® ER; AquaVac® ERM	Oral Immersion/ oral
Tenacibaculosis	*Tenacibaculum maritimum*	Turbot	Inactivated	ICTHIOVAC®TM	Injection
Vibriosis and Pasteurellosis	*Photobacterium damselae* subs. *piscicida*, *Listonella anguillarum*	European seabass	Inactivated	ICTHIOVAC VR®/PD	Injection
Lactococcosis, Pseudotuberculosis, and Vibriosis	*Lactococcus garvieae*, *Photobacterium damselae* sp. *Piscicida*, *Vibrio anguillarum*	Yellowtail, amberjack	Inactivated oil adjuvant	NORVAX® PLV 3-way	Injection
Pasteurellosis, Streptococcosis	*Photobacterium damselae*, *Lactococcus garvieae*	Yellowtail, amberjack	Inactivated oil adjuvant	NORVAX® Ruiketsu Rensa Oil	Injection

TABLE 13.1 *(Continued)*

Target Disease	Target Pathogen	Target Fish Species	Type of Vaccine	Product Name	Route of Administration
Furunculosis, Classical Vibriosis, Cold-water *Listonella* (*Vibrio*) anguillarum	*Listonella* (*Vibrio*) anguillarum	Atlantic salmon	Inactivated	Norvax® Minova 6	Injection
Vibriosis, Wound or Winter Ulcer Disease, and IPN	*Aeromonas salmonicida* subsp. *Salmonicida*, *Vibrio salmonicida*, *Moritella viscosa*, *IPNV*	Atlantic salmon	Inactivated	Not specified	Injection
Vibriosis, Pasteurellosis	*Photobacterium damselae*, *Vibrio anguillarum*, *V. ordalii*	European seabass	Inactivated	QUAVAC® Vibrio Pasteurella	Injection

TABLE 13.2 Vaccines Against Major Infectious Viral Diseases of Finfish (Bondad-Reantaso et al., 2022)

Disease	Target Pathogen	Target Fish Species	Type of Vaccine	Product Name	Route of Administration
Infectious haematopoietic necrosis	Infectious haematopoietic necrosis virus	Salmonids	DNA vaccine	Apex-IHN (Canada)	Injection
Infectious pancreatic necrosis (IPN)	Infectious pancreatic necrosis virus	Atlantic salmon	VP2 and VP3 subunit proteins	AQUAVAC® IPN	Oral
Pancreatic disease (PD)	SAV alphaviruses	Salmonids	Inactivated SAV F93-125	Norvax® Compact PD	Injection
Pancreatic disease (PD)	SAV alphaviruses	Salmonids	Inactivated strain AL V405	Alpha Ject Micro 1 Pd	Injection
Koi herpesvirus (KHV) disease	KHV Herpesvirus	Koi carp	Live, attenuated viral vaccine	KV-3	Immersion or injection
Infectious spleen and kidney necrosis	ISKNV Iridovirus	Asian seabass, grouper, pompano, Japanese yellowtail	Inactivated ISKNV	AQUAVAC®IridoV	Injection
Viral Nervous Necrosis (VNN)	Betanodavirus	European sea bass	Inactivated Betanodavirus strain	ICTHIOVAC® VNN	Injection

13.10 CONCLUSION

A fish vaccination should prioritize safety for both the animal and the environment, affordability for mass production, simplicity in administration, and the capacity to build robust immunity during peak susceptibility periods. Alternative and new fish vaccinations are developing, and modern technologies have shown great promise for aquaculture, although they were frequently created with demands in human or animal medicine. The greatest value and commercialization potential will accrue to those that satisfy the requirements for an efficacious aquaculture vaccination. Vaccination stands as the most effective method for preventing illness, and there are many choices available for developing and using vaccines. These approaches include the newest molecular vaccines as well as traditional live vaccinations. Every vaccine type has pros and cons of its own, and the selection of vaccination type is shaped by factors such as the type of target pathogen, recipient safety, immune response, and practicality of application. Given the current landscape of emerging diseases significantly impacting aquaculture productivity, there is a crucial need to prioritize the development of effective vaccines to mitigate infectious diseases and minimize production losses.

KEYWORDS

- **aquaculture**
- **aquaculture vaccination**
- **immunization**
- **pathogen**
- **peptide**
- **toxoid**
- **vaccination**
- **vector**
- **viral nervous necrosis**

REFERENCES

Acosta, F., Collet, B., Lorenzen, N., & Ellis, A. E. (2006). Expression of the glycoprotein of viral haemorrhagic septicaemia virus (VHSV) on the surface of the fish cell line RTG-P1 induces type 1 interferon expression in neighboring cells. *Fish & Shellfish Immunology*, *21*(3), 272–278.

Adams, A., & Thompson, K. D. (2006). Biotechnology offers revolution to fish health management. *Trends in Biotechnology*, *24*(5), 201–205.

Adams, A., Aoki, T., Berthe, C., Grisez, L., & Karunasagar, I. (2008). Recent technological advancements on aquatic animal health and their contributions toward reducing disease risks—a review. In *Diseases in Asian Aquaculture VI* (pp. 71–88). Fish Health Section, Asian Fisheries Society, Colombo.

Angulo, C., Sanchez, V., Delgado, K., Monreal-Escalante, E., Hernandez-Adame, L., Angulo, M., et al. (2022). Oral organic nanovaccines against bacterial and viral diseases. *Microbial Pathogenesis*, *169*, 105648. https://doi.org/10.1016/j.micpath.2022.105648 (accessed on 10th July 2025).

Austin, B., Austin, D. A., & Munn, C. B. (2007). *Bacterial Fish Pathogens: Disease of Farmed and Wild Fish* (Vol. 26, p. 552). Dordrecht, The Netherlands: Springer.

Ballesteros, N. A., Alonso, M., Saint-Jean, S. R., & Perez-Prieto, S. I. (2015). An oral DNA vaccine against infectious hematopoietic necrosis virus (IHNV) encapsulated in alginate microspheres induces dose-dependent immune responses and significant protection in rainbow trout (*Oncorhynchus mykiss*). *Fish & Shellfish Immunology*, *45*, 877–888. https://doi.org/10.1016/j.fsi.2015.05.045 (accessed on 10th July 2025).

Bedekar, M., Kole, S., & Marappan, M. (2022). Types of vaccines used in aquaculture. https://doi.org/10.1007/978-981-19-1268-9_3 (accessed on 10th July 2025).

Biering, E., Villoing, S., Sommerset, I., & Christie, K. E. (2005). Update on viral vaccines for fish. *Developmental Biology*, *121*, 97–113.

Bly, J. E., & Clem, L. W. (1992). Temperature and teleost immune functions. *Fish & Shellfish Immunology*, *2*, 159–171.

Bondad-Reantaso, M. G., MacKinnon, B., Karunasagar, I., Fridman, S., Alday-Sanz, V., Brun, E., & Caputo, A. (2023). Review of alternatives to antibiotic use in aquaculture. *Reviews in Aquaculture*.

Bowden, T. J. (2008). Modulation of the immune system of fish by their environment. *Fish & Shellfish Immunology*, *25*, 373–383.

Caruffo, M., Maturana, C., Kambalapally, S., Larenas, J., & Tobar, J. A. (2016). Protective oral vaccination against infectious salmon anaemia virus in *Salmo salar*. *Fish & Shellfish Immunology*, *54*, 54–59. https://doi.org/10.1016/j.fsi.2016.03.009 (accessed on 10th July 2025).

Chien, M. H., Wu, S. Y., & Lin, C. H. (2018). Oral immunization with cell-free self-assembly virus-like particles against orange-spotted grouper nervous necrosis virus in grouper larvae, *Epinephelus coioides*. *Veterinary Immunology and Immunopathology*, *197*, 69–75.

Collins, C., Lorenzen, N., & Collet, B. (2019). DNA vaccination for finfish aquaculture. *Fish & Shellfish Immunology*, *85*, 106–125.

Corbeil, S., Kurath, G., & LaPatra, S. E. (2000). Fish DNA vaccine against infectious hematopoietic necrosis virus: Efficacy of various routes of immunization. *Fish & Shellfish Immunology*, *10*, 711–723. https://doi.org/10.1006/fsim.2000.0286 (accessed on 10th July 2025).

Desmettre, P., Martinod, S., Pastoret, P. P., Blancou, J., Vannier, P., & Verschueren, C. (1997). Research and development in veterinary vaccinology. In *Veterinary Vaccinology* (pp. 175–194). Amsterdam: Elsevier Press.

Dhar, A. K., Bowers, R. M., Rowe, C. G., & Allnutt, F. T. (2010). Expression of a foreign epitope on infectious pancreatic necrosis virus VP2 capsid protein subviral particle (SVP) and immunogenicity in rainbow trout. *Antiviral Research*, *85*, 525–531.

Dhar, A., & Allnutt, F. (2011). Challenges and opportunities in developing oral vaccines against viral diseases of fish. *Journal of Marine Science: Research & Development*, *1*, 2.

Du, Y., Hu, X., Miao, L., & Chen, J. (2022). Current status and development prospects of aquatic vaccines. *Frontiers in Immunology*, *13*, 1040336.

Du, Y., Tang, X., Sheng, X., Xing, J., & Zhan, W. (2015). Immune response of flounder (*Paralichthys olivaceus*) was associated with the concentration of inactivated *Edwardsiella tarda* and immersion time. *Veterinary Immunology and Immunopathology*, *167*, 44–50. https://doi.org/10.1016/j.vetimm.2015.07.002 (accessed on 10th July 2025).

Duff, D. C. B. (1942). The oral immunization of trout against *Bacterium salmonicida*. *Journal of Immunology*, *44*, 87–94.

Ellis, A. E. (1988). Fish vaccination. *Aquaculture Information Series*, Issue 4. Cambridge: Academic Press.

Embregts, C. W. E., & Forlenza, M. (2016). Oral vaccination of fish: Lessons from humans and veterinary species. *Developmental & Comparative Immunology*, *64*, 118–137. https://doi.org/10.1016/j.dci.2016.03.024 (accessed on 10th July 2025).

Evelyn, T. P. T. (1997). A historical review of fish vaccinology. In R. Gudding, A. Lillehaug, P. J. Midtlyng, & F. Brown (Eds.), *Fish Vaccinology. Developments in Biological Standardization* (Vol. 90, pp. 3–12). Basel: Karger.

FAO. (2016). *The State of World Fisheries and Aquaculture 2016: Contributing to Food Security and Nutrition for All*. Rome: Food and Agriculture Organization of the United Nations.

FAO. (2018). *The State of Food and Agriculture 2018: Migration, Agriculture and Rural Development*. Rome: Food and Agriculture Organization of the United Nations. Licence: CC BY-NC-SA 3.0 IGO.

Fenner, F. (1997). Historical introduction. In P. P. Pastoret, J. Blancou, P. Vannier, & C. Verschueren (Eds.), *Veterinary Vaccinology*.

Fernández-Alarcón, C., Miranda, C. D., Singer, R. S., López, Y., Rojas, R., Bello, H., Domínguez, M., & González-Rocha, G. (2010). Detection of the *floR* gene in a diversity of florfenicol-resistant Gram-negative bacilli from freshwater salmon farms in Chile. *Zoonoses and Public Health*, *57*, 181–188.

Hart, S., Wrathmell, A. B., Harris, J. E., & Grayson, T. H. (1988). Gut immunology in fish: A review. *Developmental & Comparative Immunology*, *12*, 453–480.

Heppell, J., & Davis, H. L. (2000). Application of DNA vaccine technology to aquaculture. *Advanced Drug Delivery Reviews*.

Horne, M. T., Tatner, M., McDerment, S., Agius, C., & Ward, P. (1982). Vaccination of rainbow trout, *Salmo gairdneri* Richardson, at low temperatures and the long persistence of protection. *Journal of Fish Diseases*, *5*, 343–345.

Jenner, E. (1798). *An Inquiry into The Causes and Effects of The Variolate Vaccinee, A Disease Discovered in Some of The Western Counties of England particularly Gloucestershire, and known by the name of cow pox*. London: Sampson Low.

Krantz, G. E., Reddecliff, J. M., & Heist, C. E. (1963). Development of antibodies against *Aeromonas salmonicida* in trout. *Journal of Immunology*, *91*, 757–760.

Liang, R., Li, J., Ruan, Y., & Zhu, S. (2015). The design and parameter optimization of automatic positioning mechanism for automatic injection system of fry vaccine. *American Society of Agricultural and Biological Engineers Annual International Meeting*, *2015*(1), 1–18. https://doi.org/10.13031/aim.20152188971 (accessed on 10th July 2025).

Lillehaug, A. (1997). Vaccination strategies in seawater cage culture of salmonids. *Developments in Biological Standardization*, *90*, 401–408.

Lillehaug, A. (2014). Vaccination strategies and procedures. In *Fish Vaccination* (pp. 140–152).

Lillehaug, A., Ramstad, A., Bækken, K., & Reitan, L. J. (1993). Protective immunity in Atlantic salmon (*Salmo salar* L.) vaccinated at different water temperatures. *Fish & Shellfish Immunology*, *3*, 143–156.

Marsian, J., Hurdiss, D. L., Ranson, N. A., Ritala, A., Paley, R., Cano, I., & Lomonossoff, G. P. (2019). Plant-made nervous necrosis virus-like particles protect fish against disease. *Frontiers in Plant Science*, *10*, 880.

Martins, M. L., Xu, D. H., Shoemaker, C. A., & Klesius, P. H. (2011). Temperature effects on immune response and hematological parameters of channel catfish *Ictalurus punctatus* vaccinated with live theronts of *Ichthyophthirius multifiliis*. *Fish & Shellfish Immunology*, *31*, 774–780.

Midtlyng, P., Reitan, L., & Speilberg, L. (1996). Experimental studies on the efficacy and side-effects of intraperitoneal vaccination of Atlantic salmon (*Salmo salar* L.) against furunculosis. *Fish & Shellfish Immunology*, *6*, 335–350.

Muktar, Y., Tesfaye, S., & Tesfaye, B. (2016). Present status and future prospects of fish vaccination: A review. *Journal of Veterinary Science & Technology*, *7*(2), 299.

Noonam, B., & Enzmann, P. J. (1995). Recombinant infectious hematopoietic necrosis virus and viral hemorrhagic septicemia virus glycoprotein epitopes expressed in *Aeromonas salmonicida* induce protective immunity in rainbow trout (*Oncorhynchus mykiss*). *Applied and Environmental Microbiology*, *61*(10), 3586–3591.

Overgård, A. C., Fiksdal, I. U., Nerland, A. H., & Patel, S. (2011). Expression of T-cell markers during Atlantic halibut (*Hippoglossus hippoglossus* L.) ontogenesis. *Developmental & Comparative Immunology*, *36*, 203–213.

Press, C. McL., & Lillehaug, A. (1995). Vaccination in European salmonid aquaculture: A review of practices and prospects. *British Veterinary Journal*, *151*, 45–69.

Rathore, G. (2017). Bacterial vaccines for fishes: Current status. In *Prophylaxis in Aquaculture* (p. 87).

Schillberg, S., Raven, N., Fischer, R., Twyman, R. M., & Schiermeyer, A. (2013). Molecular farming of pharmaceutical proteins using plant suspension cell and tissue cultures. *Current Pharmaceutical Design*, *19*(31), 5531–5542.

Schrøder, M. B., Villena, A. J., & Jørgensen, T. O. (1998). Ontogeny of lymphoid organs and immunoglobulin producing cells in Atlantic cod (*Gadus morhua* L.). *Developmental & Comparative Immunology*, *22*, 507–517.

Shao, Z. J. (2001). Aquaculture pharmaceuticals and biologicals: Current perspectives and future possibilities. *Advanced Drug Delivery Reviews*, *50*, 229–243.

Shin, Y. J., Kwon, T. H., Seo, J. Y., & Kim, T. J. (2013). Oral immunization of fish against iridovirus infection using recombinant antigen produced from rice callus. *Vaccine*, *31*(45), 5210–5215.

Shoemaker, C. A., Klesius, P. H., & Bricker, J. M. (1999). Efficacy of a modified live *Edwardsiella ictaluri* vaccine in channel catfish as young as seven days post hatch. *Aquaculture*, *176*, 189–193.

Snieszko, S. F. (1970). Immunization of fishes: A review. *Journal of Wildlife Diseases*, *6*, 24–30.

Snieszko, S. F., & Friddle, S. B. (1949). Prophylaxis of furunculosis in brook trout (*Salvelinus fontinalis*) by oral immunization and sulfamerazine. *Progressive Fish Culturist*, *11*, 161–168.

Snieszko, S., Piotrowska, W., Kocylowski, B., & Marek, K. (1938). Badania bakteriologiczne i serogiczne nad bakteriami posocznicy karpi. *Mémoires de l'Institut d'Ichtyobiologie et Pisciculture de la Station de Pisciculture Expérimentale à Mydlniki de l'Université Jagiellonienne à Cracovie*, *38*.

Sommerset, I., Krossoy, B., Biering, E., & Frost, P. (2005). Vaccines for fish in aquaculture. *Expert Review of Vaccines*, *4*(1), 89–101.

Subasinghe, R. P., Bondad-Reantaso, M. G., & McGladdery, S. E. (2001). Aquaculture development, health and wealth. In R. P. Subasinghe, P. Bueno, M. J. Phillips, C. Hough, S. E. McGladdery, & J. R. Arthur (Eds.), *Aquaculture in the Third Millennium: Technical Proceedings of the Conference on Aquaculture in the Third Millennium* (pp. 167–191). Bangkok: NACA and FAO.

Sukkarun, P., Kitiyodom, S., Yostawornkul, J., Chaiin, P., Yata, T., Rodkhum, C., et al. (2022). Chitosan-polymer based nanovaccine as promising immersion vaccine against *Aeromonas veronii* challenge in red tilapia (*Oreochromis* sp.). *Fish & Shellfish Immunology*, *129*, 30–35. https://doi.org/10.1016/j.fsi.2022.08.035 (accessed on 10th July 2025).

Thiéry, R., Cozien, J., Cabon, J., Lamour, F., Baud, F., & Schneemann, F. (2006). Induction of a protective immune response against viral nervous necrosis in the European sea bass *Dicentrarchus labrax* by using Betanodavirus virus-like particles. *Journal of Virology*, *80*(20), 10201–10207.

Vakharia, V. N. (2005). University of Maryland Biotechnology Institute (UMBI). Sub-unit vaccine for infectious pancreatic necrosis virus. *US Patent 6,936,256*.

Vallejos-Vidal, E., Reyes-López, F., & MacKenzie, S. (2017). Immunostimulant diets and oral vaccination in fish. In *Diagnosis and Control of Diseases of Fish and Shellfish* (pp. 147–184). https://doi.org/10.1002/9781119152125.ch6 (accessed on 10th July 2025).

Zapata, A., Diez, B., Cejalvo, T., et al. (2006). Ontogeny of the immune system of fish. *Fish & Shellfish Immunology*, *20*, 126–136.

Zhao, L., Seth, A. A., Wibowo, N., Zhao, C. X., Mitter, N., Yu, C., & Middelberg, A. P. J. (2014). Nanoparticle vaccines. *Vaccine*, *32*(3), 327–337.

CHAPTER 14

Optimizing Sustainability: A Comprehensive Analysis of Nutrient Cycling and Environmental Impact in Aquaponic Systems

APARNA RATHORE,[1,*] DHANAJI PATIL,[2] and SANDIP MARKAD[2]

[1]*Department of Botany, Kanoria PG Mahila Mahavidyalaya, Jaipur, Rajasthan, India*

[2]*Department of Fishery Science, Toshniwal Arts, Commerce and Science College, Sengaon, Hingoli, Maharashtra, India*

*Corresponding author

ABSTRACT

As global food demand continues to rise, sustainable agricultural practices are essential for meeting these challenges. Aquaponics, a synergistic integration of aquaculture and hydroponics, presents a promising solution by maximizing resource efficiency and minimizing environmental impact. This research paper aims to provide a comprehensive analysis of nutrient cycling and environmental sustainability in aquaponic systems. Through meticulous investigation and experimentation, the paper explores strategies to optimize nutrient utilization, enhance system productivity, and reduce ecological footprint. The review delves into the intricate dynamics of the aquaponic ecosystem, examining the interactions between fish, plants, and microorganisms. It also provides a critical assessment of the environmental implications, including energy efficiency and waste management. The findings contribute to the advancement of aquaponic

Sustainable Aquaculture: Recent Advances and Techniques. Dhanaji W. Patil, Sandip S. Markad (Eds.)

DOI: 10.1201/9781779641113-14

technology, offering insights that can guide the development of more sustainable and resilient agricultural systems. This research emphasizes the pivotal role of aquaponics in addressing the dual challenges of food security and environmental conservation in the 21st century.

14.1 INTRODUCTION

The increasing global population and corresponding surge in food requirements pose substantial challenges to traditional agricultural practices. The continuous rise in the human population, particularly in developing regions, underscores the pressing need to produce more food to meet the mounting demand. This surge in demand places significant pressure on conventional agricultural methods that have been the backbone of food production for centuries. The world's population is expanding at an unprecedented rate, with estimates projecting significant growth in the coming decades. As more individuals inhabit the planet, the demand for food escalates proportionally. Fulfilling the nutritional requirements of the surging human population will be one of the utmost challenges that the world has to face, with projections suggesting that the world's human population is estimated to reach 10 billion by 2050. In order to meet the supplementary demand for food owing to the almost 30% surge in population, global food production has to be augmented by 50% (FAO, 2017).

Rising incomes and urbanization further contribute to changing dietary habits, with a shift towards higher consumption of resource-intensive products such as meat and dairy. This change amplifies the overall demand for agricultural products (FAO, 2017; Vermeulen et al., 2020). Traditional agriculture, characterized by soil-based cultivation, extensive use of chemical fertilizers, and reliance on large-scale irrigation, faces immense pressure to keep pace with the escalating demand. Conventional farming practices are often constrained by limited arable land, water scarcity, and environmental concerns associated with excessive use of agrochemicals (DeLonge et al., 2016; Joshi et al., 2022; Zerssa et al., 2023). Additionally, climate change introduces further unpredictability, affecting crop yields and amplifying the challenges faced by traditional agriculture (FAO, 2018; Joshi et al., 2022; Vasdravanidis et al., 2022).

Inefficient resource utilization, coupled with the need to preserve biodiversity and minimize environmental impact, underscores the necessity for innovative and sustainable approaches to ensure long-term food security.

Addressing global food demand necessitates a paradigm shift towards resource-efficient, resilient, and sustainable food production practices. The expansion of agriculture and allied sectors is considered the most efficient approach for achieving food security and curbing poverty (Joshi et al., 2022). Alternative systems, such as aquaponics, hydroponics, and other forms of controlled-environment agriculture, present opportunities to revolutionize food production and meet the challenges posed by the growing population, changing dietary patterns, and climate change (Conijn et al., 2018).

14.2 ENVIRONMENTAL AND SUSTAINABILITY CHALLENGES IN AGRICULTURE AND CONVENTIONAL FARMING

The pursuit of sustainability in agriculture is confronted by a myriad of challenges, stemming from the depletion of natural resources and the environmental impact associated with conventional farming practices. As societies strive to balance current needs with the obligation to protect the environment for future generations, addressing these sustainability challenges becomes paramount.

1. **Chemical Inputs:** The widespread use of chemical fertilizers and pesticides in conventional agriculture contributes to soil and water pollution. Runoff from fields carries these chemicals into water bodies, posing threats to aquatic ecosystems and, in some cases, human health.
2. **Soil Erosion and Degradation:** Intensive agricultural practices, including extensive monoculture and improper land management, contribute to the erosion and degradation of soil. This compromises soil fertility, reducing its ability to support healthy plant growth and sustainable agriculture (Rizal et al., 2018).
3. **Greenhouse Gas Emissions:** Traditional farming practices, such as rice cultivation and livestock farming, contribute significantly to the emission of greenhouse gases. The potent greenhouse gases that contribute to climate change include methane arising from livestock and nitrous oxide from fertilizers (Rizal et al., 2018).
4. **Water Scarcity:** Conventional farming, often reliant on extensive irrigation, contributes to the depletion of water resources. The

excessive use of water in agriculture can lead to aquifer depletion, reduced water tables, and, in some cases, the over-extraction of water from rivers and lakes (Zerssa et al., 2023).

5. **Deforestation:** Clearing land for agriculture, particularly for livestock grazing and crop cultivation, results in deforestation. This loss of natural habitats contributes to the decline of biodiversity and exacerbates climate change.
6. **Loss of Biodiversity:** Traditional farming practices may lead to the transformation of natural ecosystems into fields of monoculture, which ultimately results in biodiversity loss. This loss affects the resilience of ecosystems and disrupts the delicate balance necessary for sustainable agriculture.

Addressing sustainability challenges requires a holistic approach that considers the environmental, social, and economic dimensions of agriculture. Transitioning towards more sustainable practices, such as agroecology, precision farming, and alternative cultivation methods like aquaponics, is essential to mitigate the adverse effects of conventional farming and build a resilient and sustainable food system.

14.3 EMERGENCE OF AQUAPONICS

The term 'aquaponics' is derived by combining the word 'aqua' from aquaculture and 'ponics' from hydroponics (Marklin et al., 2013). The emergence of aquaponics represents a pioneering approach to agriculture, seamlessly integrating aquaculture and hydroponics to create a mutually beneficial and sustainable system (Rinehart, 2010). This innovative method holds the promise of addressing key challenges in traditional farming while offering a pathway to sustainable food production.

Aquaponics is a closed recirculatory agricultural system integrating hydroponics (the soilless cultivation of plants) with aquaculture (the cultivation of aquatic organisms, typically fish) (Rizal et al., 2018). This integration is accomplished by the continuous circulation of water to facilitate the recycling of nutrients and waste from one system to another. The waste produced in the aquaculture system becomes a nutrient source for plants in the hydroponics system, while plants extract these nutrients for their growth and help in the purification and filtration of water (Joshi et al., 2022; Rizal et al., 2018; Zou et al., 2016). Aquaponic systems operate as recirculating ecosystems, emphasizing the conservation of water and nutrients. Fish waste, containing ammonia, serves as a natural fertilizer for plants. Beneficial bacteria within the system convert ammonia into nitrate, which plants absorb for growth, thereby cleansing the water for the fish (Wongkiew et al., 2017).

The most common fish species cultivated in aquaponics include catfish, tilapia, carps, perch, salmon, barramundi, trout, Murray cod, largemouth bass, etc., while vegetables such as mint, pepper, broccoli, chilies, watercress, beets, carrot, radish, and leafy vegetables are commonly used in aquaponics (Hao et al., 2020; Love et al., 2015; Mishra et al., 2020). The plant and fish species-specific growth has been observed by Buzby et al. (2016); Knaus and Palm (2017); Mamat et al. (2016); Ogah et al. (2020); Oladimeji et al. (2020); Purwandari et al. (2017).

14.3.1 OVERVIEW OF AQUAPONICS

Aquaponics is a recirculatory food production system wherein the concept of circular economy and biomimetic natural systems are used to minimize waste as well as inputs for sustainable economic returns (Obirikorang, 2021). Aquaponics is an innovative sustainable food production practice that combines hydroponics (soilless agriculture) with aquaculture (fish farming). In this integrated arrangement, the nitrogenous waste generated by aquatic animals (fish) serves as a source of nutrients for plants, whereas the plants utilize such waste as nutrients for their growth, thereby purifying and filtering water for fish culture. This closed-loop ecosystem creates a mutually beneficial environment, where the interactions between fish, plants, and beneficial microorganisms contribute to a more efficient and eco-friendly approach towards food production (Rizal et al., 2018). The aquaponics system is depicted in Figure 14.1.

14.3.2 BASIC PRINCIPLES OF AQUAPONICS

The basic principles of aquaponics revolve around creating a sustainable and closed-loop ecosystem where fish, plants, and beneficial microorganisms interact harmoniously (Rizal et al., 2018). Three fundamental principles define the essence of aquaponics:

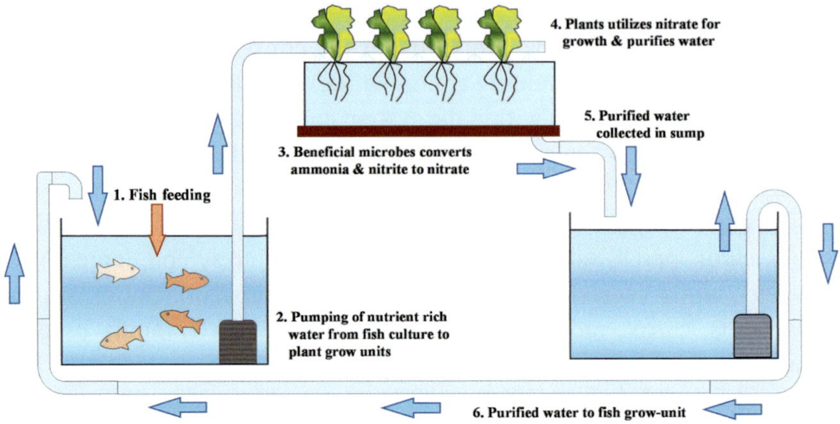

FIGURE 14.1 Diagrammatic representation of aquaponic system.

1. **Nutrient Circulation:** The nutrient recirculation system involves the cycling of essential nutrients, primarily ammonia and nitrate, between the aquatic animals (usually fish) and the cultivated plants. Fish produce waste, primarily in the form of ammonia, through respiration and excretion. Beneficial bacteria in the system convert ammonia into nitrate through a process known as nitrification. Nitrate-rich water is then circulated to the hydroponic growing beds where plants can uptake these nutrients for growth. This creates a self-sustaining and nutrient-rich environment for plant growth, minimizes the need for external fertilizers, promotes resource efficiency, and establishes a symbiotic relationship between fish and plants, ensuring optimal conditions for both (Suguna, 2021; Vasdravanidis et al., 2022).
2. **Biological Filtration:** This involves the use of microbial communities to convert harmful compounds in fish waste, particularly ammonia, into less toxic forms that can be utilized by plants. This process occurs in specialized biofilters within the aquaponic system. Ammonia, a byproduct of fish metabolism, is initially released into the water. Beneficial bacteria, such as *Nitrosomonas* and *Nitrobacter*, colonize the biofilter media (Krastanova et al., 2022). These bacteria convert ammonia into nitrite and then further into nitrate, a nutrient readily absorbed by plants. This process maintains water quality by preventing the accumulation of toxic

ammonia, facilitates the conversion of fish waste into a usable nutrient form for plants, and enhances the overall health of the aquatic environment by reducing harmful compounds (Okomoda et al., 2023; Suguna, 2021).

3. **Plant Uptake:** This involves the absorption and utilization of nutrients by the cultivated plants in the hydroponic growing beds. This process not only supports plant growth but also serves to purify and cleanse the water for the fish. Plants uptake nutrients, including nitrate, through their roots in the hydroponic beds. The nutrients are utilized by plants for essential physiological processes, promoting growth. Cleaned and nutrient-depleted water is then recirculated back to the fish tanks. Plant uptake acts as a natural filtration system, removing excess nutrients from the water. It supports healthy plant growth, ensuring a continuous supply of fresh produce and contributes to the overall balance and stability of the aquaponic ecosystem (Krastanova et al., 2022; Rizal et al., 2018).

Understanding and optimizing these basic principles are crucial for the successful design, operation, and sustainability of aquaponic systems. By leveraging the symbiotic relationships between fish, plants, and microorganisms, aquaponics offers a holistic and efficient approach to food production.

14.3.3 COMPONENTS OF AN AQUAPONIC SYSTEM

The components of an aquaponic system work in concert to create a sustainable and interconnected environment where fish, plants, and microorganisms thrive. At the core of aquaponics is a symbiotic relationship between fish and plants. Fish waste, rich in nutrients, provides an organic fertilizer for the plants. In return, the plants act as natural biofilters, removing toxins from the water and creating a healthier environment for the fish (Kloas et al., 2015). The integration of aquaculture and hydroponics optimizes nutrient cycling, creating a harmonious balance between the nutrient needs of plants and the waste generated by fish. Thus, it is essential for achieving a balanced and productive aquaponic system.

1. **Fish Tanks:** These are the primary habitat for aquatic species within the aquaponic system. Various fish species, such as tilapia, catfish, or perch, are commonly chosen based on their adaptability to aquaponic environments. They provide a living environment for fish to grow and thrive. They also serve as a source of fish waste, rich in nutrients, which becomes a vital component of the aquaponic nutrient cycle (Love et al., 2015) (Figure 14.2).

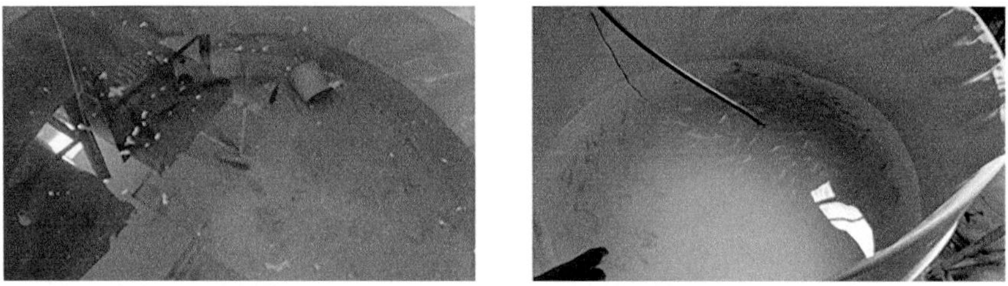

FIGURE 14.2 Fish tanks.

2. **Plant Grow Units:** Grow beds are containers or troughs where plants are cultivated without the use of soil. They are typically filled with a soilless medium such as gravel, perlite, or expanded clay pellets. They act as the location for plant cultivation, utilizing nutrient-rich water from the fish tanks (Love et al., 2015). They provide space for the root systems of plants to access and absorb essential nutrients. They serve as a natural biofilter, purifying the water by assimilating nutrients and removing potential contaminants (Figure 14.3).

FIGURE 14.3 Plant grow units.

3. **Pumps and Plumbing:** These systems are crucial for the circulation of water within the aquaponic setup. They facilitate the movement of water between the fish tanks and the hydroponic grow beds. They circulate water from fish tanks to grow beds, ensuring plants receive the necessary nutrients, and facilitate the flow of water back to fish tanks after nutrient uptake by plants. They maintain water circulation and aeration, promoting oxygenation for both fish and plants (Ibrahim et al., 2023; Joshi et al., 2022).

4. **Beneficial Microorganisms:** These including bacteria, play a vital role in the nitrogen cycle within the aquaponic system. These microorganisms colonize the biofilter media and contribute to nutrient conversion. They convert ammonia produced by fish into nitrite and then further into nitrate. They facilitate the breakdown of organic matter, contributing to the overall health of the system. They contribute to the establishment of a stable and balanced microbial community (Love et al., 2015; Okomoda et al., 2023).

These components work synergistically to create a closed-loop ecosystem where fish waste is converted into nutrients for plant growth, and plants, in turn, contribute to water purification. The efficient circulation of water and the presence of beneficial microorganisms are key to maintaining the balance and sustainability of the aquaponic system (Krastanova et al., 2022).

14.3.4 AQUAPONIC DESIGNS

The aquaponic systems are mainly classified according to the type of hydroponic system being used. These types primarily include Deep Water Culture (DWC), Nutrient Film Technique (NFT), and Media-Based Growing Bed (MBGB) (Hao et al., 2020; Krastanova et al., 2022). Though the names of these system types may vary, the basic concept and design used are similar.

1. **Deep Water Culture (DWC):** It is the simpler form of aquaponic system that can be used in large-scale commercial aquaponics for growing plant species like lettuce, basil, and other leafy plants. It consists of plant grow units with plants on Styrofoam sheets, which float over nutrient-rich water media (Mishra et al., 2020). This system facilitates easier cleaning with a higher rate of nitrate

removal. However, this system requires an additional setup for biofiltration, aeration, and a large volume of water (Hao et al., 2020) (Figure 14.4).

FIGURE 14.4 Plant grow units used in deep water culture.

2. **Nutrient Film Technique (NFT):** These systems lend higher efficiency in terms of water utilization and potential for customization of design as per requirements, at the cost of higher expenses (Goddek et al., 2015). The horizontally laid plastic pipes are used as plant grow units for growing vegetables in this system. Water is pumped from the biofiltration system into NFT pipes at a predefined water flow rate to create a shallow stream of nutrient-rich water flowing along the bottom. The pipe comprises a number of holes along the top into which plants are placed to grow (Mishra et al., 2020) (Figure 14.5).

FIGURE 14.5 Plant grow units used in nutrient film technique.

3. **Media-Based Growing Bed (MBGB):** It is one of the most popular designs of aquaponic systems used for small-scale aquaponics. It is suitable for beginners in the aquaponics system as they are space-saving, cost-effective, and have a very simple design. In MBGB, the media is used in the grow beds for filtration and to support the roots of the plants (Mishra et al., 2020). These systems are found to be difficult to maintain and clean (Figure 14.6).

FIGURE 14.6 Plant grow units used in media-based growing bed.

14.4 AQUAPONICS POTENTIAL FOR SUSTAINABLE FOOD PRODUCTION

14.4.1 OVERALL RESOURCE EFFICIENCY IN AQUAPONICS

Aquaponics has the potential to significantly reduce water usage compared to traditional soil-based agriculture. The closed-loop nature of the system minimizes nutrient runoff and maximizes resource utilization. Resource efficiency is a key attribute of aquaponic systems, harnessing the principles of nutrient recycling and minimizing reliance on external fertilizers. This sustainable agricultural practice optimizes the use of resources, addressing challenges associated with traditional farming methods. Aquaponics offers several advantages that make it a sustainable and resource-efficient method of agriculture. The key benefits include resource efficiency, reduced environmental impact, and the ability for year-round production.

1. **Water Conservation by Recirculation:** Aquaponic systems are highly water-efficient compared to traditional soil-based agriculture, where water may be lost through runoff or evaporation. The closed-loop nature of the system allows for the recirculation of water between fish tanks and hydroponic grow beds, minimizing overall water consumption (Krastanova et al., 2022; Rizal et al.,

2018; Vasdravanidis et al., 2022). The controlled environment of aquaponics allows for precise water management. Techniques such as optimizing pH levels, monitoring nutrient concentrations, and implementing water-saving technologies contribute to efficient water use. It is an effective means of growing food as it requires only 10% of the water and no use of chemical fertilizers compared to traditional farming methods (Joshi et al., 2022; Sharma et al., 2018; Suguna, 2021). The reduced water usage in aquaponics makes it more resilient to drought conditions, a significant advantage in regions prone to water scarcity (Okomoda et al., 2023).

2. **Optimizing Space Utilization:** Aquaponic systems can be implemented vertically or in controlled environments, optimizing space utilization. This is especially valuable in urban areas or regions with limited arable land (Krastanova et al., 2022; Okomoda et al., 2023; Rizal et al., 2018).
3. **Energy Efficiency and Optimized Nutrient Cycling:** While energy is required for maintaining water circulation and aeration, aquaponic systems can be designed for energy efficiency. The reduced need for mechanical nutrient application and irrigation, as seen in traditional farming, contributes to overall energy savings. The symbiotic relationship between fish and plants in aquaponics results in efficient nutrient cycling. Fish waste is converted into nutrients for plants, reducing the need for external fertilizers and enhancing resource utilization (Joshi et al., 2022; Rizal et al., 2018). Nutrients circulate within the system, creating a self-sustaining environment. This closed-loop approach minimizes nutrient loss, making the most of the resources available, as indicated by near zero-waste discharge (Okomoda et al., 2023; Rizal et al., 2018).

The emphasis on resource efficiency in aquaponics not only addresses the challenges of resource depletion and environmental impact associated with traditional agriculture but also contributes to the development of sustainable and economically viable farming practices.

14.4.2 REDUCED ENVIRONMENTAL IMPACT

Aquaponics provides notable environmental benefits, offering a sustainable alternative to traditional agriculture. By minimizing the need for chemical fertilizers and preventing chemical and nutrient runoff and soil degradation, aquaponics mitigates the environmental impact associated with traditional farming practices.

1. **Chemical-Free Cultivation:** Aquaponic systems minimize or eliminate the need for synthetic fertilizers and pesticides. Traditional farming often relies on chemical inputs, leading to the contamination of soil and water through runoff (Love et al., 2015; Okomoda et al., 2023).

2. **Reduction in Nutrient Runoff:** Nutrient runoff from fields, a common issue in traditional agriculture, can lead to water pollution and eutrophication of water bodies. In aquaponics, the closed-loop nutrient cycle minimizes the risk of nutrient runoff, mitigating negative impacts on aquatic ecosystems (Rizal et al., 2018).

3. **Economic Benefits:** By reducing reliance on external fertilizers, aquaponic systems offer economic benefits to farmers. The closed-loop nature of nutrient cycling minimizes the ongoing expenses associated with purchasing and applying synthetic fertilizers (Rizal et al., 2018; Suguna, 2021). This cycle of water involving aquaculture wastewater reuse (Endut et al., 2011) prevents the environmental pollution caused by traditional fisheries (Hao et al., 2020). Aquaponics has lower maintenance as labor work of weeding is not required for the crops; it requires only about half an hour of human monitoring for system operation (Liang and Chien, 2013).

4. **Mitigation of Soil and Water Pollution:** Conventional agricultural practices, particularly those involving extensive plowing and monoculture, contribute to soil erosion. Aquaponics, being a soil-less system, avoids such issues and helps prevent the degradation of fertile topsoil. Aquaponics relies on the natural nutrient cycle created by fish waste and beneficial microorganisms. This reduces or eliminates the use of chemical fertilizers and prevents nutrient runoff. By doing so, aquaponics helps mitigate soil erosion and water pollution, contributing to a healthier and more sustainable environment (Krastanova et al., 2022; Okomoda et al., 2023).

5. **Lower Greenhouse Gas Emissions:** Aquaponic systems generally produce fewer greenhouse gas emissions compared to traditional farming practices, particularly those associated with livestock. Methane emissions from enteric fermentation and manure management are reduced in aquaponics (Ibrahim et al., 2023).

6. **Biodiversity Conservation:** Traditional agriculture often involves clearing large areas of land, leading to habitat destruction and biodiversity loss. The controlled and compact nature of aquaponic systems reduces the need for extensive land use, contributing to biodiversity conservation.

14.4.3 YEAR-ROUND PRODUCTION

Controlled indoor environments in aquaponic systems enable year-round production, decoupling food production from seasonal variations (Suguna, 2021). This continuous production enhances overall system productivity, providing a consistent supply of fresh produce.

1. **Controlled Indoor Environments:** Aquaponic systems can be implemented indoors, allowing for precise control over environmental factors such as temperature, light, and humidity. This enables year-round production, decoupled from seasonal variations (Rizal et al., 2018).

2. **Climate Resilience:** The ability to operate in controlled environments makes aquaponics more resilient to external climate conditions, ensuring a consistent and reliable supply of fresh produce regardless of external weather patterns (Okomoda et al., 2023).

3. **Increased Crop Turnover and Overall System Productivity:** The continuous and controlled nature of aquaponics allows for faster growth rates and shorter crop cycles, leading to increased crop turnover and higher overall productivity throughout the year (Okomoda et al., 2023; Suguna, 2021).

4. **Enhanced Plant and Fish Growth:** The interconnected relationships within an aquaponic system lead to high nutrient cycling efficiency. Plants grown in aquaponic systems often exhibit

accelerated growth rates compared to traditional soil-based cultivation. Fish waste, a byproduct of aquaculture, becomes a valuable resource for hydroponic plant cultivation (Love et al., 2015). The continuous availability of nutrients in the form of nutrient-rich water and the absence of soil-borne diseases contribute to the overall health and productivity of plants (Rizal et al., 2018). The clean and well-oxygenated water provided by plant roots contributes to the well-being and growth of fish, as observed in rainbow trout culture. The elimination of ammonia and other waste compounds ensures a healthy aquatic environment (Vasdravanidis et al., 2022). Oladimeji et al. (2020) reported about 5- and 11-times higher pumpkin production and yield in the aquaponic system as compared to irrigated and non-irrigated land, respectively. The fish yield in the aquaponic system was also found to be 29% and 75% higher compared to the recirculatory and static aquaculture systems, respectively.

Aquaponics, with its resource-efficient practices, reduced environmental impact, and capacity for year-round production, represents a sustainable and innovative approach to agriculture. As the world faces challenges related to population growth, water scarcity, and climate change, aquaponics stands out as a promising solution contributing significantly to global food security, sustainability, and environmental conservation.

14.5 CHALLENGES AND CONSIDERATIONS

While aquaponics offers numerous advantages, it also presents unique challenges and considerations that require careful attention for successful implementation and sustained productivity. Key challenges include system balancing, species selection, and operational management.

14.5.1 SYSTEM BALANCING

1. **Achieving Equilibrium:** Maintaining a harmonious balance among fish, plants, and microorganisms is crucial for the success of an aquaponic system. Fluctuations in nutrient levels, pH, or water temperature can disrupt this equilibrium, affecting the health of both fish and plants (Kloas et al., 2015; Okomoda et al., 2023; Suguna, 2021).

2. **Optimizing Nutrient Cycling:** Striking the right balance in nutrient cycling is essential (Okomoda et al., 2023). Overloading the system with fish waste may lead to excessive nutrient levels, impacting plant health, while insufficient nutrient supply can hinder plant growth (Vasdravanidis et al., 2022). The feed rate ratio for leafy vegetables is 40-50 $g/m^2/day$, and the fruiting vegetables require 50-80 $g/m^2/day$. The omnivorous fish, such as tilapia, common carp, and Magur (*Clarias batrachus*), are fed with 32% protein feed. While carnivorous fish like snakeheads are fed with 40-42% protein feed (Suguna, 2021).
3. **Microbial Community Dynamics:** The health and activity of beneficial microorganisms, particularly in the biofilter, play a vital role in nutrient conversion. Maintaining the right conditions for microbial communities is a continuous challenge (Okomoda et al., 2023).

14.5.2 SPECIES SELECTION

1. **Compatibility:** Selecting compatible fish and plant species is crucial for the success of the system. Some fish may have specific environmental requirements that need to align with the needs of the chosen plant species. Tilapia, carps, and catfish are highly suitable for aquaponics in tropical regions. Nile tilapia with leafy vegetables like lettuce, spinach, and cabbage are good crops for aquaponic systems (Suguna, 2021).
2. **Growth Rates:** Different fish and plant species have varying growth rates. Ensuring a balance where neither fish nor plants outgrow the system is essential for sustained productivity (Buzby et al., 2016).
3. **Temperature Tolerance:** Matching the temperature requirements of both fish and plants is important. Some fish species may prefer warmer water, while certain plants thrive in slightly cooler conditions.

Environmental parameters such as temperature and humidity, pH, and mineral concentrations should be optimum for species growth. Maintaining an optimum water temperature of 22–24°C, pH in the range of 5.6–7.3, and DO of 3–10 mg/L for fish and crops is necessary. The other levels of tolerance are: alkalinity of 50–250 mg/L, CO_2 of 0–30 mg/L, hardness of 50–350 mg/L, salinity of 0–10 ppt, and nitrite concentrations of 0–0.8 mg/L (Krastanova et al., 2022).

14.5.3 OPERATIONAL MANAGEMENT

1. **Parameter Monitoring:** Regular monitoring and adjustment of key parameters such as pH, temperature, dissolved oxygen, and nutrient levels are essential for system health. Fluctuations in these parameters can impact the well-being of fish and the uptake of nutrients by plants (Krastanova et al., 2022; Rizal et al., 2018; Suguna, 2021).
2. **Disease Prevention:** Disease management is crucial in aquaponics, as disease outbreaks in fish can have cascading effects on plant health and overall system stability. Implementing preventive measures and prompt responses to signs of illness are essential (Okomoda et al., 2023; Rizal et al., 2018; Suguna, 2021).
3. **Skill and Knowledge Requirements:** Successful aquaponic operation requires a deep understanding of the interdependencies within the system. Skilled management is necessary to troubleshoot issues and make informed adjustments (Ibrahim et al., 2023; Okomoda et al., 2023).

Addressing these challenges involves ongoing research, monitoring, and adaptive management. Aquaponic practitioners must continually refine their understanding of system dynamics and implement best practices to ensure long-term success. Despite these challenges, the potential benefits of aquaponics, when managed effectively, make it a compelling and sustainable method of agriculture.

14.6 SUGGESTIONS AND FUTURE RECOMMENDATIONS

Future research recommendations for aquaponics include understanding microbial ecology and its role in system stability and nutrient cycling;

exploring crop-specific nutrient requirements within aquaponic systems; and optimizing conditions for diverse plant species. Future studies should also explore the resilience of aquaponic ecosystems under varying conditions, including changes in climate and external stressors. To further advance the field, future research endeavors should investigate microbial ecology, crop-specific nutrient requirements, economic viability, and scalability. Additionally, examining the resilience of aquaponic ecosystems under varying conditions and exploring community-based models and the integration of technology—such as automation and sensor-based monitoring—will contribute to a more comprehensive understanding of the technology's potential. Research should focus on the economic viability and scalability of aquaponic systems, assessing their potential for widespread adoption and commercial success.

14.7 CONCLUSION

The present study provides valuable insights into the intricate dynamics of aquaponic ecosystems. The focus on nutrient cycling and environmental impact accentuates the commitment to developing sustainable agricultural practices that address the growing global demand for food while minimizing environmental degradation. Sustainable agricultural practices within aquaponic systems involve optimizing resource utilization, minimizing environmental impact, and fostering long-term resilience. Aquaponics demonstrates a reduction in water usage, improved energy efficiency, and effective waste management, contributing to sustainability and a reduced ecological footprint, thereby minimizing environmental impact. The research underscores aquaponics as a resource-efficient method of agriculture, requiring less water and energy compared to traditional farming practices. Aquaponics will eventually emerge as a successful and environmentally friendly alternative, mitigating the impact of agriculture on water resources, reducing pollution, imparting climate resilience, and boosting food production.

KEYWORDS

- **aquaponics**
- **climate resilience**
- **ecosystem dynamics**
- **environmental degradation**
- **hydroponic grow beds**
- **nutrient cycling**
- **sustainability**

REFERENCES

Buzby, K. M., Waterland, N. L., Semmens, K. J., & Lin, L. S. (2016). Evaluating aquaponic crops in a freshwater flow-through fish culture system. *Aquaculture*, *460*, 15–24.

Conijn, J. G., Bindraban, P. S., Schroder, J. J., & Jongschaap, R. E. E. (2018). Can our global food system meet food demand within planetary boundaries? *Agriculture, Ecosystems & Environment*, *251*, 244–256.

DeLonge, M. S., Miles, A., & Carlisle, L. (2016). Investing in the transition to sustainable agriculture. *Environmental Science & Policy*, *55*, 266–273.

Endut, A., Jusoh, A., Ali, N., Wan Nik, W. N. S., & Hassan, A. (2009). Effect of flow rate on water quality parameters and plant growth of water spinach (*Ipomoea aquatica*) in an aquaponic recirculating system. *Desalination and Water Treatment*, *5*, 19–28.

FAO. (2017). *The Future of Food and Agriculture–Trends and Challenges*. Rome. https://www.fao.org/3/i6583e/i6583e.pdf (accessed on 10th July 2025).

FAO. (2018). *The Future of Food and Agriculture: Alternative Pathways to 2050*. Food and Agriculture Organization of the United Nations, Rome, Italy. https://www.fao.org/3/I8429EN/i8429en.pdf (accessed on 10th July 2025).

Goddek, S., Delaide, B., Mankasingh, U., Ragnarsdottir, K. V., Jijakli, H., & Thorarinsdottir, R. (2015). Challenges of sustainable and commercial aquaponics. *Sustainability*, *7*(4), 4199–4224.

Hao, Y., Ding, K., Xu, Y., Tang, Y., Liu, D., & Li, G. (2020). States, trends, and future of aquaponics research. *Sustainability*, *12*(18), 7783.

Ibrahim, L. A., Shaghaleh, H., El-Kassar, G. M., Abu-Hashim, M., Elsadek, E. A., & Alhaj Hamoud, Y. (2023). Aquaponics: A sustainable path to food sovereignty and enhanced water use efficiency. *Water*, *15*, 4310.

Joshi, E., Agarwal, P., Sasode, D. S., Khambalkar, P., Bordoloi, P., Ginger, D., & Joshi, N. (2022). Aquaponics: An innovative sustainable food production farming system. In *Recent Advances in Agricultural Science and Technology for Sustainable India Part-II* (pp. 72–78). Mahima Research Foundation and Social Welfare, Varanasi, India.

Kloas, W., Roman, G., Daniela, B., Johannes, G., Henrik, M., Uwe, S., Georg, S., Johanna, S., Martin, T., Bernd, W., Sven, W., Andrea, Z., & Bernhard, R. (2015). A new concept for aquaponic systems to improve sustainability, increase productivity, and reduce environmental impacts. *Aquaculture Environment Interactions*, *7*, 179–192.

Knaus, U., & Palm, H. W. (2017). Effects of the fish species choice on vegetables in aquaponics under spring-summer conditions in northern Germany (Mecklenburg Western Pomerania). *Aquaculture*, *473*, 62–73.

Krastanova, M., Sirakov, I., Ivanova-Kirilova, S., Yarkov, D., & Orozova, P. (2022). Aquaponic systems: Biological and technological parameters. *Biotechnology & Biotechnological Equipment*, *36*(1), 305–316.

Liang, J. Y., & Chien, Y. H. (2013). Effects of feeding frequency and photoperiod on water quality and crop production in a tilapia-water spinach raft aquaponics system. *International Biodeterioration and Biodegradation*, *85*, 693–700.

Love, D. C., Fry, J. P., Li, X., Hill, E. S., Genello, L., Semmens, K., & Thompson, R. E. (2015). Commercial aquaponics production and profitability: Findings from an international survey. *Aquaculture*, *435*, 67–74.

Mamat, N. Z., Shaari, M. I., & Abdul Wahab, N. A. A. (2016). The production of catfish and vegetables in an aquaponic system. *Fisheries and Aquaculture Journal*, *7*(4), 5–7.

Marklin, R. W., Mathison, M., Mayer, B., Nagurka, M. L., Vikram, C., & Schabelski, J. (2013). Aquaponics: A sustainable food production system that provides research projects for undergraduate engineering students. *Mechanical Engineering Faculty Research Publications*, *12*.

Mishra, R. P., Das, S. K., Kumar, N., & Mishra, J. P. (2020). Sustainable aquaponics system and its challenges: A review. *International Journal of Current Microbiology and Applied Sciences*, *9*(5), 577–588.

Obirikorang, K. A., Sekey, W., Gyampoh, B. A., Ashiagbor, G., & Asante, W. (2021). Aquaponics for improved food security in Africa: A review. *Frontiers in Sustainable Food Systems*, *5*, 705549.

Ogah, S. I., Kamarudin, M. S., Nurul Amin, S. M., & Puteri Edaroyati, M. W. (2020). Biological filtration properties of selected herbs in an aquaponic system. *Aquaculture Research*, *51*(5), 1771–1779.

Okomoda, V. T., Oladimeji, S. A., Solomon, S. G., Olufeagba, S. O., Ogah, S. I., & Ikhwanuddin, M. (2023). Aquaponics production system: A review of historical perspective, opportunities, and challenges of its adoption. *Food Science & Nutrition*, *11*, 1157–1165.

Oladimeji, A. S., Okomoda, V. T., Olufeagba, S. O., Solomon, S. G., Abol-Munafi, A. B., Alabi, K. I., Ikhwanuddin, M., Martins, C. O., Umaru, J. A., & Hassan, A. (2020). Aquaponics production of catfish and pumpkin: Comparison with conventional production systems. *Food Science and Nutrition*, *8*, 2307–2315.

Purwandari, Y., Effendi, H., & Wardiatno, Y. (2017). The use of gouramy (*Osphronemus goramy*) rearing wastewater for growing romaine lettuce (*Lactuca sativa* L. var. *longifolia*) in aquaponic system. *Asian Journal of Microbiology, Biotechnology and Environmental Sciences*, *19*(2), 359–366.

Rinehart, L. (2010). *Aquaponics—Integration of Hydroponics with Aquaculture*. ATTRA—National Sustainable Agriculture Information Service.

Rizal, A., Dhahiyat, Y., Zahidah, Y., Andriani, Y., Handaka, A. A., & Sahidin, A. (2018). The economic and social benefits of an aquaponic system for the integrated production of fish and water plants. *IOP Conference Series: Earth and Environmental Science*, *137*, 012098.

Sharma, P. K., Kumar, J. S. S., & Anand, S. (2018). Aquaponics: A boon for income generation in water deficient areas of India like Rajasthan. *International Journal of Fisheries and Aquatic Studies*, *6*(6), 170–173.

Suguna, T. (2021). Aquaponics systems: Future food production system. *International Journal of Current Microbiology and Applied Sciences*, *10*(11), 397–406.

Vasdravanidis, C., Alvanou, M. V., Lattos, A., Papadopoulos, D. K., Chatzigeorgiou, I., Ravani, M., Liantas, G., Georgoulis, I., Feidantsis, K., & Ntinas, G. K. (2022). Aquaponics as a promising strategy to mitigate impacts of climate change on rainbow trout culture. *Animals*, *12*, 2523.

Vermeulen, S. J., Park, T., Khoury, C. K., & Bene, C. (2020). Changing diets and the transformation of the global food system. *Annals of the New York Academy of Sciences*, *1478*, 3–17.

Wongkiew, S., Hu, Z., Chandran, K., Lee, J. W., & Khanal, S. K. (2017). Nitrogen transformations in aquaponic systems: A review. *Aquacultural Engineering*, *76*, 9–19.

Zerssa, G. W., Hailemariam, M., & Tadele, K. T. (2023). Improving the sustainability of agriculture: Challenges and opportunities. In S. Lousada (Ed.), *Land-Use Management–Recent Advances, New Perspectives, and Applications*.

Zou, Y., Hu, Z., Zhang, J., Xie, H., Liang, S., Wang, J., & Yan, R. (2016). Attempts to improve nitrogen utilization efficiency of aquaponics through nitrifiers addition and filler gradation. *Environmental Science and Pollution Research*, *23*, 6671–6679.

CHAPTER 15

Implications of Fish Cell Lines and Cell Culture Systems to Attain Sustainability in Aquaculture

GOWHAR IQBAL,[1,*] MODI KIRAN PIYUSHBHAI,[2] and ARYA SINGH[1]

[1]*ICAR–Central Institute of Fisheries Education, Panch Marg, Mumbai, Maharashtra, India*

[2]*Institute of Fisheries Post Graduate Studies, Tamil Nadu Dr. J. Jayalalithaa Fisheries University, OMR Campus, Vaniyanchavadi, Tamil Nadu, India*

*Corresponding author

ABSTRACT

Aquaculture, a crucial component of world food security, confronts sustainability issues pertaining to resource usage and environmental effect. In order to achieve sustainability goals, this chapter aims to explore potential impacts of employing cell lines and cell culture systems in aquaculture. Adoption of cell culture technologies presents opportunities to address problems related to traditional aquaculture methods, such as disease outbreaks, habitat degradation, and overfishing. Cell culture systems provide fine control over growth conditions, reducing the amount of resources used and waste generated. This study examines the development, challenges, and potential for scaling of cell culture technologies in aquaculture. It also covers the socioeconomic and regulatory aspects that affect how cell lines are incorporated into aquaculture procedures. All things considered, the combination of cell lines and cell culture systems

Sustainable Aquaculture: Recent Advances and Techniques. Dhanaji W. Patil, Sandip S. Markad (Eds.)

DOI: 10.1201/9781779641113-15

has significant potential to improve aquaculture's resilience and sustainability, supporting the long-term sustainability of seafood production while reducing its negative environmental effects.

15.1 INTRODUCTION

Fish cell lines (FCL) have become valuable tools in transgenic, reproductive, toxicological, drug development, genetic, environmental, and virology research in aquaculture. These cell lines—established from over 700 cell lines derived from various fish species (for example: *Trichogaster lalius* [dwarf gourami], *Paralichthys olivaceus* [Japanese flounder], *Geophagus proximus*, and *Astyanax bimaculatus*) and tissues—offer a platform for genetic manipulation and transgene expression (Furo et al., 2023; Jeong and Kim, 2023). They offer certain advantages compared to mammalian cell culture, such as wide temperature tolerance, resistance to hypoxic conditions, and genomic similarity to humans (e.g., *Danio rerio*, ~80%), making them ideal for interdisciplinary applications (Geyer et al., 2023). Cells are isolated from various tissues of fish, and based on that, specific cell lines are produced, such as embryonic stem cell lines and germ cell lines (primordial germ cells, gonadal germ stem cells) (Hong et al., 2011).

15.2 APPLICATIONS OF FISH CELL LINE

15.2.1 USES OF FISH CELL LINES (FCL) IN TRANSGENIC RESEARCH

In the past few years, there has been an urgent need for the development of transgenic fishes to improve various commercial traits (growth, meat quality, disease resistance) in order to meet global demand. Accordingly, the production and utilization of fish cell lines for transgenic studies has drastically increased. FCL play a vital role in investigating basic fish biology and molecular biomarker development. Wang (2018) demonstrated that techniques such as recombinant baculovirus delivery have been successfully employed to transduce fish cells (*Mylopharyngodon piceus* bladder/fin/kidney [MPB/MPF/MPK]; *Oryzias latipes* spermatogonia [SG3]; and *Danio rerio* embryonic fibroblast [ZF4] cells), enabling the creation of stable transgenic cell lines. Moreover, methods such as electroporation have been explored for mass foreign gene transfer—three main steps include introduction, expression, and germ line transmission of transgenes—into fish eggs and sperm, facilitating the generation of transgenic fish expressing foreign genes (Inoue et al., 2022). Additionally, continuous fibroblast cell lines derived from fish muscle tissue have been utilized in transgenic and genetic manipulation studies, demonstrating their utility through genetic stability confirmation and growth potential assessment. These diverse approaches highlight the significance of fish cell lines in advancing transgenic research in aquaculture (Kumar et al., 2016).

15.2.2 USES OF FISH CELL LINES (FCL) IN REPRODUCTIVE RESEARCH

In aquaculture, increasing production requires improving the quality of both fish spermatozoa and oocytes, which indirectly enhances fertilization and survival in culture practices. Various researchers have transplanted germ cells in fish, showing that surrogate broodstock can be produced in aquaculture systems. For the first time, rainbow trout seedlings were produced through intraperitoneal transplantation of PGCs. Xu (2015) established and characterized a marine FCL using ovary tissue of *Verasper moseri* (barfin flounder) for transgenic research and cytotechnological manipulation of this species. Numerous authors have reported the use of fish cell lines in reproductive research, as described in Table 15.1.

15.2.3 USES OF FISH CELL LINES (FCL) IN TOXICOLOGICAL AND ENVIRONMENTAL RESEARCH

To assess the potential threat of chemicals, traditional animal models (rats, mice, rabbits, and fish) have long been used. However, these models are expensive, time-consuming, and require ethical clearance, among other limitations. Hence, the establishment and development of various cell lines from different fish tissues/cells offer an economically viable and unique screening model for toxicological and environmental assessment. These methods are rapid and provide insights into chemical impacts at both the cellular level (e.g., antioxidant enzyme activity, cell death rate,

TABLE 15.1 Fish Cell Lines (FCL) Used in Reproductive Research

Fish Species	Cell Line Name	Tissue	Usage of FCL	References
Cynoglossus semilaevis (half-smooth tongue sole)	CSGC	Testis	Host–pathogen interaction; understanding sex determination system	Zhang et al. (2011)
Cynoglossus semilaevis pseudomale (half-smooth tongue sole)	CSPMG	Gonad cells	For sex reversal, determination, and gonad development in flatfish	Sun et al. (2015)
Fugu rubripes	TSOC1 and TSOC2	Ovarian	To research sexually transmitted infections, the relationship between oogonia cells and primordial germ cells, and virus–host cell interaction	Zheng et al. (2015)
Paralichthys olivaceus (olive flounder)	POSC	Testis	Gene expression, spermatogenesis studies, Sertoli cell–germ cell interaction, growth and development, and testis mechanism	Limin et al. (2016)
Oryzias latipes (transgenic medaka)	mis-GFP transgenic cell lines	Gonadal somatic cell lines	Proliferation and meiosis of germ cells induced through this cell line	Kawabe et al. (2020)
Epinephelus coioides (grouper)	SSCs	Spermatological	Genetic engineering, disease-related research, and germ cell transfer	Zhong et al. (2022)

morphological assessment) and the molecular level (e.g., gene/protein expression, comet assay) using various parameters.

In developmental toxicology, alternative models such as whole embryo culture, the mouse embryonic stem cell test, and the *Danio rerio* (zebrafish) assay (embryo/larvae) have been developed. All four models showed comparable relative potencies as observed in *in-vivo* studies, with the zebrafish assay being particularly effective (Lee et al., 2012). Numerous authors have used fish cell lines in toxicological and environmental studies, as described in Table 15.2.

15.2.4 USES OF FISH CELL LINES (FCL) IN DRUG DEVELOPMENT

Fish cell lines (FCL) in pharmaceutical and drug research are gaining the attention of researchers and the pharmaceutical industry because traditional animal testing presents many challenges. In contrast, cell lines offer several advantages: easy transportation, low maintenance cost, ability to mimic certain disease conditions, no need to sacrifice animals, rapid growth rate, higher susceptibility to bacterial and viral infections, transgene expression, reproducible results, and support for drug–drug interaction studies to assess combination effects of drugs (Jassim, 2023).

To investigate the effects of various viruses—including red sea bream iridovirus (RSIV), infectious spleen and kidney necrosis virus (ISKNV), hirame rhabdovirus, and spring viraemia virus—a cell line known as the dwarf gourami (*Trichogaster lalius*) caudal fin cell line (DGF) was established. Researchers reported that the FCL shows persistence with ISKNV infection and is useful for understanding the pathogenesis of ISKNV in fish (Jeong and Kim, 2023). A heart-derived goldfish cell line (GH), when infected with viral hemorrhagic septicemia virus (VHSV), infectious hematopoietic necrosis virus (IHNV), spring viremia of carp virus (SVCV), infectious pancreatic necrosis virus (IPNV), channel catfish virus (CCV), or grass carp reovirus (GCRV), demonstrated its value as a tool for viral pathogenesis research (Hongli et al., 2017). Studied fish cell lines (FCL) in drug development are shown in Table 15.3.

TABLE 15.2 Fish Cell Lines Used for Toxicological and Environmental Studies

Cell Line Name	Tested Toxic Chemical	Outcome of the Study	References
Rainbow trout intestinal cell line (RTgutGC)	Copper sulfate	This cell line could be used to predict gastrointestinal permeability of metals	Langan et al. (2017)
	16 hydrophobic and volatile chemicals	Used to assess biotransformation or chemical transfer to find the non-lethal concentration for downstream application	Schug et al. (2019)
Centropomus viridis brain and heart cell lines	Xenobiotic (glyphosate)	Can be used to investigate cellular and molecular mechanisms of toxicants	Soriano et al. (2021)
Gilthead seabream brain-derived cell line (SaB-1)	Nanoplastics combined with metals arsenic and methylmercury	Showed alteration in oxidative stress and increased toxicity of both metals	Fernandez et al. (2021)
Danio rerio fibroblast-like cells (ZF4) and hepatic cells	Individual and mixtures of neodymium (Nd^{3+}), gadolinium (Gd^{3+}), and ytterbium (Yb^{3+})	Mixtures of lanthanides showed a synergistic effect on cell viability	Fleurbaix et al. (2022)
Oncorhynchus mykiss epithelial cells (RTgill-W1)		Less sensitive compared to zebrafish cell lines	
Atlantic salmon gill epithelial cell line (ASG-10)	Benzocaine	ASG-10 line has metabolic capacity to biotransform xenobiotics in the gills through induction of CYP1A expression	Ivanova et al. (2023)
Rainbow trout liver cell line (RTL-W1)	Anionic compounds: pentachlorophenol (PCP), diclofenac (DCF), tecloftalam (TT), and benzotriazole-tert-butyl-hydroxyl-phenyl propanoic acid (BHPP)	This FCL can be used in the future to predict bioconcentration and biomagnification of anionic compounds	Balk et al. (2023)

TABLE 15.3 Fish Cell Lines (FCL) in Drug Development

Cell Line Name	Viral/Bacterial Assessed	Outcome of the Study	References
Grass goldfish (*Carassius auratus*)	Transfection using Lipofectamine LTX and Xfect	This cell line is useful for artificial fish meat production and studies on myogenesis-related functional genes	Li et al. (2021)
Japanese flounder (*Paralichthys olivaceus*)	Bohle virus (BIV), viral hemorrhagic septicemia virus (VHSV), hirame rhabdovirus (HIRRV), infectious hematopoietic necrosis virus (IHNV), and lymphocystis disease virus (LCDV)	This FCL will help study immune genes and mechanisms in fish for disease prevention and treatment	Yang et al. (2022)

15.2.5 USES OF FISH CELL LINES (FCL) IN GENE-EDITING AND GENETIC ENGINEERING TECHNOLOGY

Genetically modified fish cell lines hold significant potential for various biotechnological and clinical purposes. The introduction of CRISPR (clustered regularly interspaced palindromic repeats)–Cas9 technology has revolutionized the field of gene editing (Iqbal et al., 2023). By harnessing CRISPR–Cas9, it becomes possible to generate enhanced fish cell lines, which in turn could greatly aid aquaculture biotechnological research, particularly in studying fish diseases. Utilizing genome-editing technology to genetically modify cell lines could significantly increase the transfection efficiency of fish cell lines, enabling their effective use in virus production for vaccine development. Although most applications of this approach have involved gene editing in mammalian cell lines, its use in fish cell lines is still in its early stages due to low transfection efficiency (Dehler et al., 2016).

Knockout cells or animals are generated when CRISPR–Cas9 is co-expressed with a gRNA specific to the target gene. Gene knockouts reveal a gene's function by altering its expression in a cell (Iqbal et al., 2023). Over the past few years, numerous genetic methodologies have been explored and applied to enhance the growth rate of aquatic species. The utilization of genetic editing tools such as TALENs, CRISPR–Cas9, and transgenic technology has contributed to expanding our understanding of somatic growth regulation in fish (Tao et al., 2021). The CRISPR–Cas genome-editing platform can be used *in vivo* in zebrafish to introduce site-specific insertion/deletion (indel) mutations (Hwang et al., 2013). Researchers have conducted investigations on various fish species to genetically modify genes for diverse objectives.

The advent of CRISPR–Cas9 technology marked a revolutionary breakthrough in molecular biology, offering a precise and efficient method for generating specific mutations within the genomes of various living cells. Since its inception, diverse gene-editing approaches using CRISPR–Cas9 have been extensively employed in mammalian cell lines, while comparatively limited information exists regarding genetic manipulation in fish cell lines. Recently, genome-editing technology has been applied to knock out *MSTN* in numerous fish species, spanning from model to marine fish, as evidenced by studies conducted by Chisada et al. (2011), Kim et al. (2019), Kishimoto et al. (2019), and Lee et al. (2009).

Regional Fish Co., Ltd., in collaboration with Kyoto University, Kinki University, Japan's Ministry of Health, Labor and Welfare, and the Ministry of Agriculture, Forestry and Fisheries, successfully engineered the genome-edited "Madai" red sea bream. By employing CRISPR gene-editing technology to disrupt a gene responsible for inhibiting muscle growth, the genome-edited fish demonstrated a 17 percent increase in skeletal muscle mass. This resulted in the edible portion of the fish being 1.2 to 1.6 times larger than that of conventional varieties, accompanied by a noteworthy enhancement in feed utilization efficiency, estimated at around 14 percent (Kishimoto et al., 2018).

15.2.6 USE OF CELL LINES IN FISH HEALTH MANAGEMENT

Fish cell lines have proven to be a valuable resource for researching a wide range of important aquaculture-related topics, including fish growth, disease, reproduction, genetics, and biotechnology (Table 15.4). Furthermore, fish cell lines serve as excellent *in vitro* models for research in immunology, pathology, and toxicity.

TABLE 15.4 Examples of Widely Used Fish Cell Cultures (Sobhana, 2015)

SL. No.	Cell Line	Fish and Tissue of Origin
1	RTG-2	Rainbow trout gonad
2	CHSE-214	Chinook salmon embryo
3	FHM	Flathead minnow – caudal peduncle
4	BB	Brown bullhead – caudal peduncle
5	CAR	Goldfish fin
6	EPC	Common carp – *Epithelioma papulosum cyprini*
7	SSN-1	Snakehead fin

The isolation and characterization of viruses is the most commonly used application of fish cell lines. The most often utilized fish cell lines for the identification and diagnosis of intracellular fish infections are BF-2, RTG-2, EPC, FHM, CHSE-214, and CCO. The extensive use of antibiotics and other substances that present a serious risk to the aquatic environment, along with the economic loss caused by fish sickness, has made it one of the most essential concerns for sustainable production practices in aquaculture. Fish cell culture offers a lot of promise for developing methods and instruments for aquaculture disease control. *In vitro* models that employ cell culture techniques and experimental systems enable a deeper comprehension of the intricate relationships underpinning disease outbreaks and their progression through the dissection of the connections between aggressors and hosts (Villena et al., 2003). Fish cell lines may be useful for researching disease causes, creating diagnostic assays, and creating medications and vaccines to combat fish diseases.

It has been found that the fish cell line model is helpful in identifying viral infections, and appropriate measures must be taken to protect the key aquaculture species' health. One of the most sensitive methods for identifying the major pathogens posing viral illness in many of the fish species and other organisms is the isolation of viruses using fish cell lines. Therefore, the utilization of fish cell lines for such objectives is necessary for the development of control methods to stop the viral illness from spreading. Table 15.5 provides a detailed inventory of fish cell lines utilized in virus susceptibility research. Investigation into viral diseases in significant aquaculture species has been made easier by the isolation and propagation of viruses using fish cell culture, which has accelerated virological studies. One of the foundations of a cell culture-based virus surveillance system is the propagation of viruses in the system.

TABLE 15.5 Some Fish Cell Lines That are Susceptible to Viruses

SL. No.	Fish Cell Line	Tissue/Species of Origin	Susceptible to Virus	References
1.	CFF	*Pristolepis fasciatus*	Tilapia lake virus (TiLV)	Behera et al. (2018)
2.	GF cells (grunt fin)	Fin of *blue striped grunt*	Red sea bream iridovirus	Clem et al. (1961)
3.	SSF-1	Fin of *gilt-head seabream*	Lymphocystis disease virus	Pérez-Prieto et al. (1999)
4.	SF	*Lates calcarifer*	Nodavirus	Chang et al. (2001)
5.	SISK	Kidney of *Lates calcarifer*	Nodavirus	Sarath Babu et al. (2013)
6.	GS-1	Spleen of *Epinephelus coioides*	Fish ranavirus	Huang et al. (2018)

When examining a unique virus that would not normally proliferate in any regular cell line, highly specialized cell lines are employed. Bluegill fry (BF-2), Chinook salmon embryo (CHSE-214), *Epithelioma papulosum cyprini* (EPC), fathead minnow (FHM), rainbow trout gonad (RTG-2), and SAF-1 are among the fish cell lines that have demonstrated susceptibility to a number of commonly found viruses, including infectious pancreatic necrosis (IPN), VHSV, IHNV, IPNV, SVC, koi herpesvirus (KHV), and channel catfish virus (CCV), which have negatively impacted a number of aquaculture species (Lorenzen et al., 1999). Developing control strategies to stop the spread of viral infections would be beneficial, given our understanding of viral infection transmission between the two most significant aquaculture species: snakehead fish and mandarin fish. Using fish cell cultures or cell lines, viral pathogenesis and host–pathogen interactions can be investigated.

15.2.7 USES OF FISH CELL LINES (FCL) IN VACCINE DEVELOPMENT

A dependable and durable substitute for current vaccine production methods is cell-culture-based technology. A significant number of live fish are needed for the vaccine's development and potency tests. When producing and evaluating the effectiveness of fish vaccines, fish cell culture might be utilized instead of whole live fish. Based on Dhar et al. (2014), the primary application of cell lines such as Vero, Madin-Darby canine kidney (MDCK), and chicken embryo fibroblasts (CEFs) has been in the manufacture of viral vaccines. Compared to mammalian cell lines, transfection efficiency is lower in fish cell lines. Recombinant protein and other products cannot be produced from low-transfected cell lines. The transfection efficacy of mammalian cell lines was increased to 100% with the aid of the optimal cell type and procedure combination (Li et al., 2017). There are very few experiments being conducted utilizing fish cell lines to generate viral vaccines.

The viruses that were grown in grunt fin (GF) cells were used to create the formalin-inactivated RSIV vaccine (Genzel et al., 2015). For iridovirus and NNV protection, a number of inactivated or attenuated fish viral vaccines have been produced (Oh et al., 2016), and some of these have been brought to the commercialization level (Sato et al., 2010). To produce vaccines against megalocytivirus, betanodavirus, herpesvirus, and aquareovirus, further work is needed to create specialized cell lines that promote the growth of these viruses, as there are currently few cell lines available for their replication. Modern vaccine technology has many uses for fish cell cultures, including recombinant and DNA/RNA particle vaccinations. While several are presently undergoing trials to create vaccinations, only a few fish cell lines have been used in the viral propagation process that led to the invention of vaccines and diagnostics.

It is estimated that industrial fishing practices and marine capture activities have significantly reduced ocean biomass by as much as 80% (Myers and Worm, 2003). This decline, compounded by the effects of global warming on marine environments, poses a serious threat to wild fish populations (Funk and Brown, 2009). While some view aquaculture's emergence as a positive ecological and economic development (Tidwell and Allan, 2001), others argue that its benefits have been exaggerated and that it fails to address the underlying pressures on wild fish populations. The significance of genetic modification and enclosed aquaculture systems has led to the emergence of an innovative idea known as cellular aquaculture, which involves producing fish from cultured cells. Aquaculture has the potential to reduce the pressure on wild-caught fisheries. But that might not be the case. Fish farming and fishing activities have altered natural environments because carnivorous farmed fish frequently feed on wild-caught fish (Naylor et al., 2000). Aquaculture has been recognized as a viable means of providing food for the world's expanding population; but, in order to accomplish this, less wild fish would need to be used as aquaculture feed, since a large portion of farmed fish are fed fishmeal that comes from wild fisheries (Merino et al., 2012). Consequently, the expansion of aquaculture coincides with an increase in wild fish capture, thus perpetuating the strain on marine ecosystems.

15.3 CELL-BASED AQUACULTURE FOR SUSTAINABLE AQUACULTURE GROWTH

Cellular agriculture involves the creation of agricultural products using cell cultures instead of whole plants or animals. The issues of traditional animal husbandry, such as those related to public health, the environment, and animal welfare, have drawn increasing attention to this technique as a potential solution. The concept of cellular aquaculture was initially explored by NASA in a research program focused on producing edible muscle protein from *goldfish* cells for long-duration space missions (Rubio et al., 2019). A deeper comprehension of myogenesis within muscle cell and tissue culture is crucial for fully harnessing the potential of cellular aquaculture (Benjaminson et al., 2002). The idea of creating seafood from farmed fish tissues and cells is also beginning to take hold as a viable strategy to deal with issues related to marine capture and industrial aquaculture. Marine animal cell production for cell-based seafood could be scaled up by leveraging biomedical engineering advances, such as closed-system bioreactor technology that produces cells from terrestrial animals. Muscle cell lines have been developed from marine as well as freshwater fish species (Anchelin et al., 2011).

Efforts have been undertaken to develop crustacean tissue cell lines, aiming to enhance their isolation and maintenance methods. These efforts have involved utilizing short-term cultures derived from several isolations of primary cells, as documented by Rinkevich in 2005. While short-term cultures suffice for laboratory-scale investigations, they do not meet the demands of long-term objectives associated with mass production and commercialization in cell-based seafood applications. To achieve its full potential, the advancement of cell-based seafood research necessitates a deeper understanding of the culture of fish muscle cells and tissues. Further exploration is needed into tailored serum-free media formulations suitable for fish cell culture, and the development of bioreactor designs specifically optimized to meet the requirements of fish cells for industrial manufacturing. Studies focusing on the molecular mechanisms of cell-based systems will furnish essential research data for the production of fish through cellular methods. Certain inquiries into harvested muscle tissues from freshwater and marine fish offer intriguing perspectives on the feasibility of developing a muscle cell culture system (Powell et al., 1989). Research needs to delve deeper into the biology of fish muscle cells, including their growth mechanisms, nutrient requirements, and cellular signaling pathways. Therefore, it is imperative to conduct research focusing on the sustained culture of these cells over extended periods to facilitate progress toward achieving these goals.

Lab-grown meat presents a promising avenue for sustainable growth in aquaculture, offering solutions to critical challenges faced by traditional seafood production. Cultivating fish meat from cell cultures minimizes the environmental footprint associated with large-scale fishing and aquaculture operations. Moreover, it mitigates the strain on wild fish populations, conserving marine ecosystems and promoting biodiversity. As technology continues to advance, lab-grown seafood holds the potential to revolutionize the aquaculture industry, ensuring a sustainable and ethically sound food supply for future generations while addressing pressing environmental and ethical concerns (Rubio et al., 2019) (Figure 15.1).

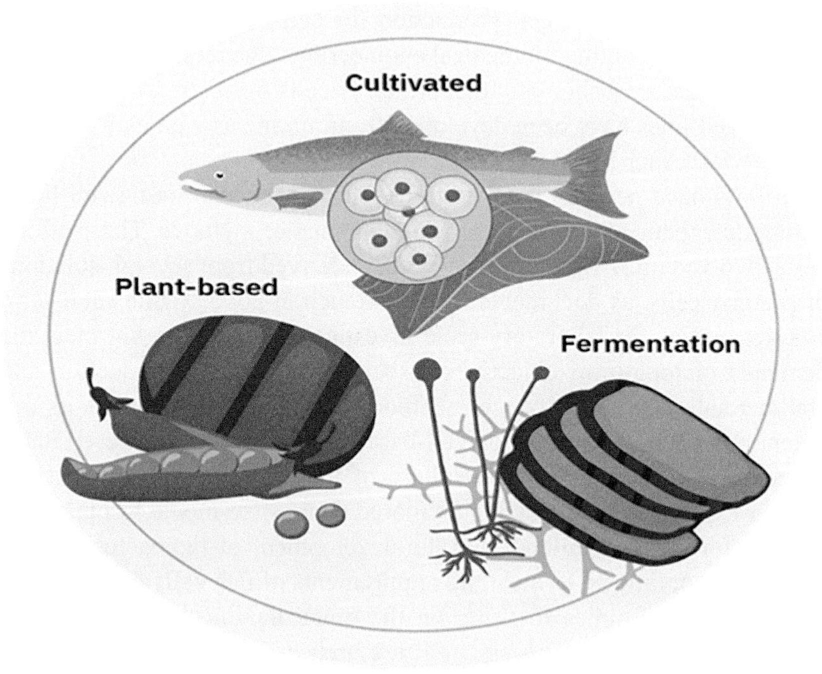

FIGURE 15.1 Alternative proteins.

Source: Adapted from Good Food Institute.

15.3.1 ALTERNATIVE PROTEINS

Cell-based fish meat has the potential to contribute significantly to sustainable aquaculture for several reasons:

1. **Reduced Environmental Impact:** Traditional aquaculture often involves practices that can harm the environment, such as habitat destruction, pollution, and the use of antibiotics and chemicals. Cell-based fish meat production can mitigate many of these issues by reducing the need for large-scale fishing operations and minimizing pollution associated with fish farming.

2. **Conservation of Marine Ecosystems:** Overfishing is a major threat to marine ecosystems and biodiversity. By producing fish meat from cells in a controlled environment, cell-based aquaculture can help relieve pressure on wild fish populations and promote the conservation of marine ecosystems.
3. **Resource Efficiency:** Cell-based fish meat production requires fewer resources than traditional aquaculture, including land, water, and feed. This efficiency can help alleviate some of the strain on natural resources and reduce the environmental footprint of seafood production.
4. **Animal Welfare:** Because cell-based fish meat doesn't require rearing and killing live animals, it lessens the suffering associated with standard fishing and aquaculture methods.
5. **Quality and Safety:** It is possible to manufacture cell-based fish meat in a regulated setting free from pollutants and pathogens that are typically present in traditional aquaculture environments. This can result in a safer and higher-quality product for consumers.
6. **Mitigation of Illegal, Unreported, and Unregulated (IUU) Fishing:** Cell-based fish meat can provide a sustainable alternative to fish products sourced from illegal, unreported, and unregulated fishing operations, helping to combat this destructive practice.

Overall, cell-based fish meat holds great promise for promoting sustainability in aquaculture by offering a more environmentally friendly, resource-efficient, and ethical alternative to conventional fish farming and fishing practices. A unique approach to seafood production and a fascinating prospect for cellular agriculture is the production of seafood from marine cell cultures. As technology develops and becomes more scalable and affordable, it has the potential to significantly contribute to satisfying the rising demand for seafood worldwide while minimizing the detrimental effects on marine ecosystems and biodiversity (Rubio et al., 2019).

15.4 CONCLUSIONS

Fish cell lines have shown promising outcomes in several important areas of aquaculture *in vitro* research, such as virology, nutrition and metabolism, vaccine development, and transgenic fish production. In contemporary biological research, the use of fish cell culture is growing. Research on fish cell culture, however, faces numerous difficulties, including contamination and misidentification. Fish cell lines will be used far more frequently to address fish disease, genetics, and biotechnological interventions as aquaculture expands globally. A unique approach to seafood production and a fascinating prospect for cellular agriculture is the production of seafood from marine cell cultures. The use of fish cell lines in the development of aquaculture vaccines and other cell-derived products has not yet been thoroughly investigated. Thus, it is essential to scale up fish cell culture systems to fully realize the potential benefits of using fish cell cultures in cell-based aquaculture.

KEYWORDS

- **cell culture**
- **cell lines**
- **cellular-aquaculture**
- **CRISPR/Cas9**
- **disease diagnosis**
- **gene editing**
- **vaccine**

REFERENCES

Anaguiven, A., Avalos-Soriano, A., García-Gasca, A., & Yáñez-Rivera, B. (2021). The development and evaluation of brain and heart cell lines from a marine fish for use in xenobiotic-induced cytotoxicity testing. *ATLA–Alternatives to Laboratory Animals*, *49*(4), 147–156. https://doi.org/10.1177/02611929211052565 (accessed on 10th July 2025).

Behera, B. K., Pradhan, P. K., Swaminathan, T. R., Sood, N., Paria, P., Das, A., Verma, D. K., Kumar, R., Yadav, M. K., Dev, A. K., & Parida, P. K. (2018). Emergence of tilapia lake virus associated with mortalities of farmed Nile tilapia *Oreochromis niloticus* (Linnaeus 1758) in India. *Aquaculture*, *484*, 168–174.

Bonetta, L. (2005). The inside scoop—Evaluating gene delivery methods. *Nature Methods*, *2*(11), 875–883.

Chang, S. F., Ngoh, G. H., Kueh, L. F. S., Qin, Q. W., Chen, C. L., Lam, T. J., & Sin, Y. M. (2001). Development of a tropical marine fish cell line from Asian seabass (*Lates calcarifer*) for virus isolation. *Aquaculture*, *192*(2–4), 133–145.

Clem, L. W., Moewus, L., & Sigel, M. M. (1961). Studies with cells from marine fish in tissue culture. *Proceedings of the Society for Experimental Biology and Medicine*, *108*, 762–766.

Dhar, A. K., Manna, S. K., & Allnutt, F. C. T. (2014). Viral vaccines for farmed finfish. *VirusDisease*, *25*, 1–17.

Fleurbaix, E., Parant, M., Maul, A., & Cossu-Leguille, C. (2022). Toxicity of lanthanides on various fish cell lines. *Ecotoxicology*, *31*(7), 1147–1157. Available from: https://doi.org/10.1007/s10646-022-02574-y (accessed on 10th July 2025).

Furo, I. D. O., Nogueira, L. S., de Sousa, R. P. C., Oliveira, G. C. S., da Silva, D. M. D. S., Malaquias, A. C., & de Oliveira, E. H. C. (2023). New parameters for the *in vitro* development of cell lines from fish species. *bioRxiv*, pp. 2023–05.

Genzel, Y. (2015). Designing cell lines for viral vaccine production: Where do we stand? *Biotechnology Journal*, *10*(5), 728–740.

Gerhard, F., Balk, P., Hollender, J., & Schirmer, K. (2023). Investigating the bioaccumulation potential of anionic organic compounds using a permanent rainbow trout liver cell line. *Social Science Research Network*, *174*, 107798. https://doi.org/10.2139/ssrn.4308054 (accessed on 10th July 2025).

Geyer, N., Kaminsky, S., Confino, S., Livne, Z. B. M., Gothilf, Y., Foulkes, N. S., & Vallone, D. (2023). Establishment of cell lines from individual zebrafish embryos. *Laboratory Animals*, *57*(5), 518–528.

González-Fernández, C., Díaz Baños, F. G., Esteban, M. Á., & Cuesta, A. (2021). Functionalized nanoplastics (NPs) increase the toxicity of metals in fish cell lines. *International Journal of Molecular Sciences*, *22*(13), 7141. https://doi.org/10.3390/IJMS22137141 (accessed on 10th July 2025).

Hong, N., Li, Z., & Hong, Y. (2011). Fish stem cell cultures. *International Journal of Biological Sciences*, *7*(4), 392.

Huang, S. M., Kuo, S. T., Kuo, H. C., & Chang, S. K. (2018). Assessment of fish iridoviruses using a novel cell line GS-1, derived from the spleen of orange-spotted grouper *Epinephelus coioides* (Hamilton) and susceptible to ranavirus and megalocytivirus. *Journal of Veterinary Medical Science*, *80*(11), 1766–1774.

Hwang, W. Y., Fu, Y., Reyon, D., Maeder, M. L., Tsai, S. Q., Sander, J. D., Peterson, R. T., Yeh, J. J., & Joung, J. K. (2013). Efficient genome editing in zebrafish using a CRISPR-Cas system. *Nature Biotechnology*, *31*(3), 227–229.

Inoue, K., & Yamashita, S. (2022). The techniques using electroporation to generate transgenic fish. In *Transgenic Animals* (pp. 129–132). CRC Press.

Iqbal, G., Dar, S. A., Quyoom, N., Malik, M. A., Gul, S., & Mir, S. A., et al. (2023). CRISPR/Cas9-gene editing technology. *World Journal of Aquaculture Research and Development*, *3*(1), 1017.

Iqbal, G., Quyoom, N., Singh, L. S., Ganpatbhai, A. V. K., Bhat, N. M., Gul, S., Malik, M. A., Mohanty, A., Mir, S. A., & Dar, S. A. (2023). Genome editing technology in fishes. *Current Journal of Applied Science and Technology*, *42*(23), 20–26. https://doi.org/10.9734/cjast/2023/v42i234170 (accessed on 10th July 2025).

Ivanova, L., Kruse, C., & Solhaug, A. (2023). Atlantic salmon gill epithelial cell line (ASG-10) as a suitable model for xenobiotic biotransformation. *Metabolites*, *13*(6), 771. https://doi.org/10.3390/metabo13060771 (accessed on 10th July 2025).

Jassim, Z. E. (2023). Cell line culture in pharmaceutical development and application: A review. العراقية للصيدلة, *0*(0), 17–22. Available from: https://doi.org/10.33899/iphr.2023.138155.1023 (accessed on 10th July 2025).

Jeong, Y., & Kim, K. I. (2023). A new cell line derived from the caudal fin of the dwarf gourami (*Trichogaster lalius*) and its susceptibility to fish viruses. *Biology*, *12*(6), 829. https://doi.org/10.3390/biology12060829 (accessed on 10th July 2025).

Jing, H., Lin, X., Xu, L., Gao, L., Zhang, M., Wang, N., & Wu, S. (2017). Establishment and characterization of a heart-derived cell line from goldfish (*Carassius auratus*). *Fish Physiology and Biochemistry*, *43*(4), 977–986. Available from: https://doi.org/10.1007/S10695-017-0345-4 (accessed on 10th July 2025).

Kawabe, T., Kariya, H., Hara, S., Shirozu, T., Shiraishi, E., Mukai, K., Yazawa, T., Inoue, S., & Kitano, T. (2020). Transcriptional regulation of Müllerian inhibiting substance (MIS) and establishment of a gonadal somatic cell line using mis-GFP transgenic medaka (*Oryzias latipes*). *Frontiers in Endocrinology*, *11*, 578885.

Langan, L. M., Harper, G. M., Owen, S. F., Purcell, W. M., Jackson, S. K., & Jha, A. N. (2017). Application of the rainbow trout derived intestinal cell line (RTgutGC) for ecotoxicological studies: Molecular and cellular responses following exposure to copper. *Ecotoxicology*, *26*(8), 1117–1133. https://doi.org/10.1007/S10646-017-1838-8 (accessed on 10th July 2025).

Lee, H. Y., Inselman, A. L., Kanungo, J., & Hansen, D. K. (2012). Alternative models in developmental toxicology. *Systems Biology in Reproductive Medicine*, *58*(1), 10–22.

Li, C., Fu, X., Lin, Q., Liu, L., Liang, H., Huang, Z., & Li, N. (2017). Autophagy promoted infectious kidney and spleen necrosis virus replication and decreased infectious virus yields in CPB cell line. *Fish & Shellfish Immunology*, *60*, 25–32.

Lorenzen, E., Carstensen, B., & Olesen, N. J. (1999). Inter-laboratory comparison of cell lines for susceptibility to three viruses: VHSV, IHNV and IPNV. *Diseases of Aquatic Organisms*, *37*(2), 81–88.

Oh, S. Y., Kim, W. S., Oh, M. J., & Nishizawa, T. (2016). Multiplication rate of red seabream iridovirus (RSIV) in rock bream *Oplegnathus fasciatus* at different fish rearing temperatures. *Fish Pathology*, *51*(4), 194–198.

Peng, L., Zheng, Y., You, F., Wu, Z., Zou, Y., & Zhang, P. (2016). Establishment and characterization of a testicular Sertoli cell line from olive flounder *Paralichthys olivaceus*. *Chinese Journal of Oceanology and Limnology*, *34*(5), 1054–1063.

Perez-Prieto, S. I., Rodriguez-Saint-Jean, S., Garcia-Rosado, E., Castro, D., Alvarez, M. C., & Borrego, J. J. (1999). Virus susceptibility of the fish cell line SAF-1 derived from gilt-head seabream. *Diseases of Aquatic Organisms*, *35*(2), 149–153.

Sarath Babu, V., Abdul Majeed, S., Nambi, K. S. N., Taju, G., Madan, N., Sundar Raj, N., & Sahul Hameed, A. S. (2013). Comparison of betanodavirus replication efficiency in ten Indian fish cell lines. *Archives of Virology*, *158*, 1367–1375.

Sato, A., & Okamoto, N. (2010). Induction of virus-specific cell-mediated cytotoxic responses of isogeneic ginbuna crucian carp, after oral immunization with inactivated virus. *Fish & Shellfish Immunology*, *29*(3), 414–421.

Schug, H., Maner, J., Hülskamp, M., Begnaud, F., Debonneville, C., Berthaud, F., Gimeno, S., & Schirmer, K. (2019). Extending the concept of predicting fish acute toxicity *in vitro* to the intestinal cell line RTgutGC. *ALTEX–Alternatives to Animal Experimentation*, *37*(1), 37–46. Available from: https://doi.org/10.14573/ALTEX.1905032 (accessed on 10th July 2025).

Sobhana, K. S. (2015). Development of marine fish cell lines and stem cell lines: Applications in mariculture and marine biodiversity.

Sun, A., Chen, S. L., Gao, F. T., Li, H. L., Liu, X. F., Wang, N., & Sha, Z. X. (2015). Establishment and characterization of a gonad cell line from half-smooth tongue sole *Cynoglossus semilaevis* pseudomale. *Fish Physiology and Biochemistry*, *41*, 673–683.

Villena, A. J. (2003). Applications and needs of fish and shellfish cell culture for disease control in aquaculture. *Reviews in Fish Biology and Fisheries*, *13*, 111–140.

Wang, Q., Fang, J., Pan, Q., Wang, Y., Xue, T., Li, L., & Chen, T. (2018). Efficient and stable delivery of multiple genes to fish cells by a modified recombinant baculovirus system. *International Journal of Molecular Sciences*, *19*(12), 3767.

Xu, X., Fan, T., Jiang, G., & Yang, X. (2015). Establishment and characterization of a new marine fish cell line from ovary of barfin flounder (*Verasper moseri*). *Journal of Ocean University of China*, *14*, 1105–1110.

Yang, Y., Ren, Y., Zhang, Y., Wang, G., He, Z., Liu, Y., Cao, W., Wang, Y., Chen, S. L., Fu, Y., & Hou, J. (2022). A new cell line derived from the spleen of the Japanese flounder (*Paralichthys olivaceus*) and its application in viral study. *Biology*, *11*(12), 1697. https://doi.org/10.3390/biology11121697 (accessed on 10th July 2025).

Zhang, B., Wang, X., Sha, Z., Yang, C., Liu, S., Wang, N., & Chen, S. L. (2011). Establishment and characterization of a testicular cell line from the half-smooth tongue sole, *Cynoglossus semilaevis*. *International Journal of Biological Sciences*, *7*(4), 452.

Zheng, D., Zhang, B., Song, W., Pan, L., Ma, C., & Xiao, G. (2015). Initiation of two ovarian cell lines from *Fugu rubripes* (Temminck et Schlegel). *Acta Oceanologica Sinica*, *34*, 98–102.

Zhong, C., Tao, Y., Liu, M., Wu, X., Yang, Y., Wang, T., Meng, Z., Xu, H., & Liu, X. (2022). Establishment of a spermatogonial stem cell line with potential of meiosis in a hermaphroditic fish, *Epinephelus coioides*. *Cells*, *11*(18), 2868.

CHAPTER 16

Application of Fish Genetics and Biotechnology in Aquatic Health Management

VIVEK R. TANDEL,[1,*] D. T. VAGHELA,[1] RITIKA A. TANDEL,[2] KISHAN SAKARIYA,[3] and JANSHI TANDEL[4]

[1]*Department of Aquatic Environment Management, College of Fisheries Science, Kamdhenu University, Veraval, Gujarat, India*

[2]*Department of Fish Processing Technology, College of Fisheries Science, Kamdhenu University, Veraval, Gujarat, India*

[3]*Department of Aquaculture, College of Fisheries Science, Kamdhenu University, Veraval, Gujarat, India*

[4]*Department of Fisheries Resource Management, College of Fisheries Science, Kamdhenu University, Veraval, Gujarat, India*

*Corresponding author

ABSTRACT

Fish genetics and biotechnology are pivotal in revolutionizing aquatic health management, providing groundbreaking solutions for sustainable aquaculture and conservation practices. Leveraging genetic tools, researchers can pinpoint and selectively propagate robust fish strains endowed with heightened disease resistance, thereby bolstering the well-being of entire populations and diminishing the reliance on antibiotics. The integration of biotechnological methods, such as molecular diagnostics, enables swift and precise disease identification, facilitating

Sustainable Aquaculture: Recent Advances and Techniques. Dhanaji W. Patil, Sandip S. Markad (Eds.)

DOI: 10.1201/9781779641113-16

timely interventions to curtail the proliferation of pathogens. Furthermore, gene editing technologies play a crucial role in crafting transgenic fish with desirable traits, amplifying aquaculture productivity while curbing environmental repercussions. The amalgamation of these advancements not only fortifies the overall health of aquatic ecosystems but also fosters the sustainable management of fisheries. This comprehensive approach ensures the enduring vitality of global aquatic resources, contributing to the preservation of biodiversity and the livelihoods of communities dependent on fisheries. In essence, the synergy between fish genetics and biotechnology stands as a beacon for responsible and forward-thinking aquatic health management in the pursuit of a resilient and sustainable aquatic future.

16.1 INTRODUCTION

Aquaculture, fisheries, and the food business can all benefit from the use of biotechnology as a powerful instrument for sustainable progress. Scientists are exploring biotechnological methods to enhance the production of seafood in response to increasing consumer demand and the decline of natural marine environments. This has spurred the rapid growth of aquaculture as a burgeoning area of animal research. In this context, biotechnology empowers scientists to discern and amalgamate desirable traits in fish and shellfish, aiming to boost yield and enhance the quality of seafood. Studies are in progress to determine which genes are linked to increased synthesis of fish growth factors naturally occurring in the environment and the innate defense mechanisms marine creatures use to fend off microbial diseases. Even though contemporary biotechnology is already making a big difference, aquaculture development faces enormous obstacles as a result of it (Mayekar et al., 2013).

Aquaculture might be greatly improved and demand could be met by using contemporary biotechnology to increase aquatic species output. The utilization of genetic modification and biotechnology holds immense potential for enhancing both the quality and quantity of fish cultivated in aquaculture. With the increasing demand for aquaculture, biotechnology emerges as a crucial solution to address this growing need. Similar to other biotech-enhanced foods, aquaculture products will undergo rigorous regulation before being approved for sale. Moreover, biotech aquaculture offers environmental benefits. When effectively combined with other technologies, biotechnology can significantly contribute to meeting the needs of a burgeoning and urbanizing population in the coming millennium. However, the successful development and application of biotechnology necessitate a robust foundation of research and understanding across various disciplines such as biology, genetics, and pathology. Without sustained investment in basic research, the full potential of these technologies cannot be realized. It is imperative that biotechnological projects remain integrated within their scientific context for successful outcomes (Tanekhy, 2020).

16.2 FISH GENETICS IN AQUATIC HEALTH MANAGEMENT

Aquaculture production is growing rapidly worldwide, but a significant portion of the organisms utilized in this sector are not the result of structured breeding and enhancement efforts. The biological characteristics of aquatic organisms present numerous opportunities for swift genetic advancement, especially with the advent of novel genetic and genomic technologies that allow for the implementation of improvement programs across a wider range of species.

Dunham thoroughly examined the history of fish genetics in 2004, focusing particularly on traits such as growth, maturation, environmental adaptability, and disease resistance. Features such as external fertilization and high fecundity contribute to fish genetic improvement by facilitating the collection and fertilization of a large number of gametes under controlled settings. Large family sizes in breeding schemes enable strong selection pressures and increase genetic parameter estimations, bringing advances to market faster. Many aquatic species still retain traits from their wild ancestors, showcasing considerable genetic diversity and demonstrating medium to high heritability across various traits. Aquatic organisms can be altered by their surroundings, and one genotype can have several different phenotypes. Single-sex populations can be produced by carefully breeding sex-reversed individuals or manipulating phenotypic sex through the use of sex hormones during the sexually labile period. This greatly increases output. Furthermore, manipulating daylength can affect features such as the onset of sexual maturity (McAndrew and Napier, 2011).

While new "-omic" technologies like proteomics and genomics have been employed in studying numerous species, their practical applications in real-world scenarios remain limited. Within the larger framework of developing profitable and efficient aquaculture, genetic applications are essential—especially when it comes to producing feed inputs of superior quality that represent noteworthy advances for the sector.

16.3 BIOTECHNOLOGY IN AQUATIC HEALTH MANAGEMENT

As biotechnology has evolved and become more sophisticated, its application in aquaculture has become more widespread, aiming to enhance success rates and promote sustainability. This section explores various biotechnological applications in aquaculture, including genetic modification to customize species for different environments or market demands. It also discusses the incorporation of algae-infused commercially formulated feed for its high unsaturated fatty acids, as well as the integration of biosensors to monitor the welfare of cultured animals in real time. Disease management strategies such as vaccinations are also examined to enhance species' resistance against pathogens. These advancements are crucial in commercial aquafarming, where high-density species are confined, posing challenges to natural improvements.

Biotechnological interventions have become indispensable in the aquaculture sector, enhancing its efficiency and cost-effectiveness. The section also addresses controversial issues and sustainability concerns related to the broader adoption of biotechnology in aquaculture, reflecting public awareness. Biotechnology has significantly contributed to the advancement of aquaculture, agriculture, fisheries, and human health, providing the capacity to create new genes and genotypes in fish, plants, and animals. Although biotechnology's application in the fisheries and aquaculture industry is relatively recent, it holds immense potential to boost fish production by emphasizing rapid growth, immunity, and efficient harvest, potentially revolutionizing fish farming and contributing to biodiversity conservation and productivity enhancement.

16.4 APPLICATIONS OF BIOTECHNOLOGY AND FISH GENETICS IN AQUATIC HEALTH MANAGEMENT

16.4.1 BIOTECHNOLOGY IN FISH BREEDING

Fish reproduction can be accelerated most effectively with the use of gonadotropin-releasing hormone (GnRH), a biotechnological tool.

Bhattacharya et al. (2002) highlighted the pivotal significance of GnRH as the primary controller and central initiator of the reproductive cascade across all vertebrates. Originally discovered in the hypothalamus of pigs and lambs, GnRH is a decapeptide known for its ability to stimulate the pituitary production of follicle-stimulating hormone (FSH) and luteinizing hormone (LH). In most placental mammals, including humans, a single form of GnRH has been identified as the sole neuropeptide responsible for triggering the release of LH and FSH. However, in non-mammalian species, except for guinea pigs, researchers have identified structural variations in twelve different GnRH variants. Among these, seven or eight distinct varieties have been identified in various fish species (Halder et al., 1991).

The latest purification and characterization of GnRH were conducted by Carolsfeld et al. (2000). The structural differences and their biological functions have served as a basis for synthesizing a variety of chemical equivalents. One notable example of such an analog is the salmon GnRH analog, which is widely used in fish breeding and is sold commercially around the world under the brand name *Ovaprim*. The development of GnRH technology is credited with making induced fish breeding a success.

16.4.2 TRANSGENESIS

The term transgenesis, often known as transgenics, describes the stable maintenance, transmission, and expression of exogenous genes or DNA within the host genome. In the context of aquaculture, this technology offers a substantial chance to improve or change the genetic characteristics of fish, mollusks, and crustaceans that are commercially important. Zhu et al. (1985) in China are recognized for producing the first transgenic fish when they documented the temporary expression of potential transgenics. It is important to note, however, that they did not provide any molecular proof of the transgene's incorporation. Since then, several fish species have benefited from the technique's successful application, which has significantly advanced transgenesis research (Figure 16.1).

With the advent of transgenic technology, the urgency to produce sterile offspring to prevent transgenic stocks from breeding with wild populations has heightened. Technological advancements have broadened the scope of options for generating fish that are either sterile or whose reproductive

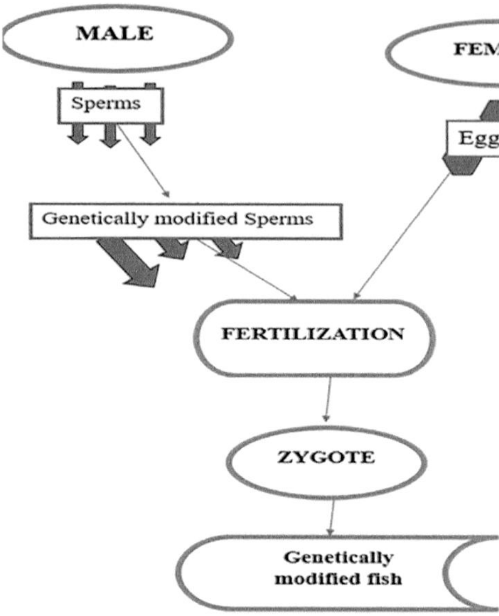

FIGURE 16.1 Revolution of transgenic fish.

behavior can be meticulously regulated using inducible promoters. This is a highly valuable development that enables transgenic populations to grow and reproduce at optimal levels while preventing escape and reproduction. Furthermore, a recent focus of transgenic fish research has been on improving fish resistance to cold temperatures (Fletcher et al., 2001). Few fish can withstand water temperatures considerably below 0–1°C, making coldwater temperatures a major stressor for many species. This is often a significant issue for aquaculture in colder regions. It is intriguing to observe that certain marine teleost species possess elevated serum levels of glycoproteins known as antifreeze glycoproteins (AFGP) or antifreeze proteins (AFP), ranging from 10 to 25 mg/ml. These proteins effectively decrease the freezing temperature by preventing the formation of ice crystals (Mayekar et al., 2013).

16.4.3 GENETIC HYBRIDS

The quality of the stock is the primary factor in producing high-grade fishery products. A strong and healthy stock directly affects the overall yield value by reducing the possibility of inherited illnesses or impairments.

Technology had little impact in the early phases of aquafarming; selection programs were the primary method of choosing stock. These systems often relied on phenotypes—or observable characteristics—of individuals or pedigree data (Abdelrahman et al., 2017). To align with Charles Darwin's revolutionary theory, the standard practice was to favor those with the best reproductive traits. By applying the theories of survival of the fittest and natural selection, the goal was to produce offspring with inherited desirable features. Undoubtedly, genetic hybridization is indispensable for enhancing the survival rates of aquatic species cultivated in controlled environments. However, this approach has also sparked considerable debate. Concerns arise when genetically modified organisms are unintentionally released or manage to stray into the wild. Escapees from aquaculture environments may interbreed with native species already inhabiting the area, intensifying competition for resources such as food and shelter. Within this context, the potential interaction between native and genetically modified species has become a topic of ongoing discussion (Dunham, 2023).

16.4.4 CHROMOSOME ENGINEERING

The breeding of cultured fish species has extensively utilized chromosome sex manipulation techniques, as outlined by Lakra and Das in 1998. These techniques encompass the production of polyploidy, such as triploidy and tetraploidy, as well as the alteration of uniparental chromosome inheritance, including gynogenesis and androgenesis. These methods play a vital role in advancing fish breeding by providing a rapid means of achieving gonadal sterilization, enhancing sex control, improving hybrid viability, and facilitating cloning.

In most vertebrates, including fish, the typical chromosomal configuration is diploid, characterized by two complete sets of chromosomes in somatic cells. Polyploid individuals possess one or more additional sets of chromosomes, resulting in triploids having three sets and tetraploids having four sets, and so forth. Among these manipulations, induced triploidy is widely acknowledged and effective for producing sterile fish in aquaculture and fisheries management. The methods employed to induce triploidy and manipulate other chromosome sets in fish, along with the applications of these biotechnologies in aquaculture and fisheries management, are thoroughly documented (Chourrout, 1987) (Figure 16.2).

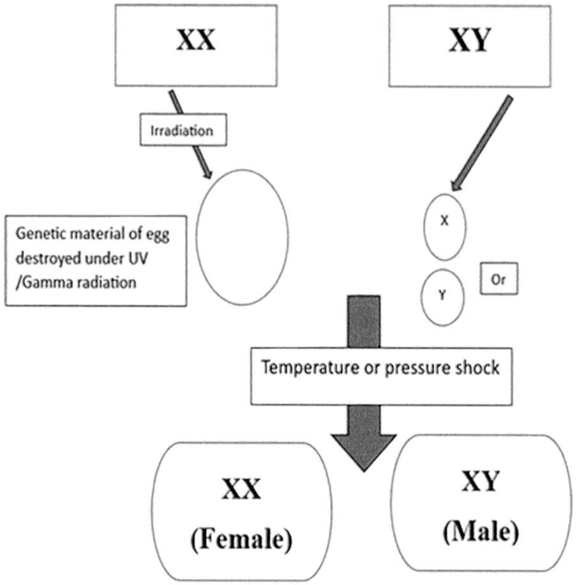

FIGURE 16.2 Chromosome manipulation in fish.

The application of tetraploid breeding lines shows potential for aquaculture, providing a straightforward method to produce ample sterile triploid fish through straightforward crosses between tetraploids and diploids, as highlighted by Chourrout et al. in 1986. However, despite the successful induction of tetraploidy in several finfish species, it is important to acknowledge that the viability of tetraploids has often been limited in various instances, as noted by Rothbard et al. in 1997.

16.4.5 HYBRIDIZATION

The extensive literature on interspecific and intergeneric hybridization in fish underscores the potential use of hybrids in aquaculture, as emphasized by Bartley et al. in 2001. The significance of hybrids is evident on a national scale in various countries. For example, in Israel, the hybrid *O. niloticus* × *Oreochromis aureus* is prized for its skewed male sex ratio and tolerance to cold temperatures, as noted by Hulata et al. in 1993. Similarly, in the USA, the sunshine bass (*Morone chrysops* × *Morone saxatilis*) has exhibited faster growth and superior overall culture characteristics compared to either parental species under commercial culture conditions. However, despite numerous reported hybrids, only a few have been successfully cultured for prolonged periods, as discussed by Smith in 1988. This limited success is often attributed to the added complexities in production or challenges associated with reintroducing hybrids back into the parental species, which may result in the loss of desirable traits.

16.4.6 GENETIC MARKERS

Liu and Cordes (2004) have conducted comprehensive reviews on molecular markers in aquatic organisms. Their research utilizes two types of markers: Type 1, which are genes with known functions, and Type 2, consisting of anonymous DNA segments. Type 2 markers—more especially, microsatellite markers—have so far shown to have a considerable impact. Numerous microsatellite loci, characterized by short tandem repeat DNA sequences, have been discovered in various aquatic species, including *carp*, *tilapia*, *sea bass*, *sea bream*, and *salmonids*.

Genetic markers serve multiple purposes in managing both cultivated and wild stocks. One notable application is determining the origin of farm stock that escaped, as seen in *Atlantic salmon*, *rainbow trout*, and *Atlantic cod*, as highlighted by Glover in 2010. Additionally, these markers are crucial for detecting and characterizing potential gene introgression from farm escapees into wild *salmon* populations (Skaala et al. 2006).

16.4.7 BIOSENSOR IN WATER MANAGEMENT

Maintaining water quality is a paramount concern in aquaculture, necessitating constant and vigilant monitoring. The quality of water in the cultivation environment is a critical factor that significantly influences the success or failure of the entire aquaculture industry and ultimately determines the survival of the cultured species. Water quality degradation is a major problem in many countries, mostly due to contaminated outflows from domestic sewage, agricultural runoff, and industrial waste. Contaminated water sources frequently contain pathogenic microbes, toxic compounds from pesticides and herbicides at increased concentrations, and hazardous heavy metals. These pollutants eventually show up as abnormal pH levels and critically low dissolved oxygen (DO) levels in the water quality measures. Some organisms may be able to adapt and survive in such harsh conditions, though unavoidably with negative consequences. Examples of these impacts include food contamination in humans as a result of toxins in the food chain biomagnifying and bioaccumulating, and reproductive dysfunction in which men may exhibit feminization (Tyler et al., 1998).

16.4.8 GENETIC MAPPING

The genetic linkage maps established by Danzmann and Gharbi in 2007 facilitate the identification of Quantitative Trait Loci (QTL), specific genome regions harboring genes that influence significant traits. According to Korol et al. (2007), data on QTL for species like *common carp*, *tilapia*, and *Atlantic salmon* are continually expanding. However, many of these maps still require refinement to achieve higher precision, as noted by Hayes et al. (2006). Nonetheless, researchers have expedited the identification of QTL linked to resistance to infectious pancreatic necrosis (IPN) by leveraging significant differences in recombination rates between genders. These identified QTLs have been successfully integrated into commercial *salmon* breeding programs, as highlighted by Houston et al. (2008).

The rapid progress in identifying expressed sequence tags (EST) is opening new possibilities for Type 1 marker mapping. This advancement enables the study of gene expression in stocks under different developmental or environmental conditions, providing insights into active genes and factors affecting gene expression. The provision of a radiation hybrid (RH) mapping panel has made EST mapping more manageable. Senger et al. (2006) reported the formation of the first panel in *gilthead sea bream Sparus auratus*, a commercially relevant species. The development of comparable panels for other species, such as *sea bass*, is still underway.

16.5 CONCLUSION

The fusion of fish genetics and biotechnology into aquatic health management represents a paradigm shift in enhancing the sustainability and resilience of aquaculture. Genetic markers, such as Quantitative Trait Loci (QTL), and the ongoing mapping of expressed sequence tags (EST) provide researchers with the tools to identify and understand crucial traits related to disease resistance and overall health in aquatic species. This integration enables targeted breeding programs and precision management strategies aimed at bolstering the health and viability of aquaculture stocks. This invaluable knowledge serves as the cornerstone for developing targeted breeding programs meticulously designed to bolster disease resistance within populations. Furthermore, the arsenal of biotechnological tools enables precise and efficient disease diagnostics, propelling advancements in vaccine development. Notably, the incorporation of beneficial traits into breeding programs through biotechnology amplifies the genetic robustness of aquatic species. These innovative strategies collectively contribute to proactive and strategic management of aquatic health, providing effective solutions to combat challenges posed by diseases and environmental stressors in aquaculture. The overarching outcome is the fortification of a more resilient, sustainable, and adaptive aquaculture industry.

KEYWORDS

- **aquaculture**
- **aquatic health management**
- **biotechnology**
- **disease resistance**
- **expressed sequence tags**
- **fish genetics**
- **fortification**
- **genetic markers**
- **sustainable aquaculture**

REFERENCES

Abdelrahman, H., ElHady, M., Alcivar-Warren, A., Allen, S., Al-Tobasei, R., Bao, L., & Chappell, J. (2017). Aquaculture genomics, genetics and breeding in the United States: Current status, challenges, and priorities for future research. *BMC Genomics*, *18*(1), 191.

Bartley, D. M., Rana, K., & Immink, A. J. (2000). The use of inter-specific hybrids in aquaculture and fisheries. *Reviews in Fish Biology and Fisheries*, *10*, 325–337.

Battacharya, S., Dasgupta, S., Datta, M., & Basu, D. (2002). Biotechnology input in fish breeding. *Indian Journal of Biotechnology*, *1*, 29–38.

Carolsfeld, J., Powell, J. F., Park, M., Fischer, W. H., Craig, A. G., Chang, J. P., & Sherwood, N. M. (2000). Primary structure and function of three gonadotropin-releasing hormones, including a novel form, from an ancient teleost, herring. *Endocrinology*, *141*(2), 505–512.

Chourrout, D. (1987). Genetic manipulation in fish: Review of methods. In *Proc. World Symp. on Selection, Hybridization, and Genetic Engineering in Aquaculture* (pp. 111–126). Heeneman.

Chourrout, D., Chevassus, B., Krieg, F., Happe, A., Burger, G., & Renard, P. (1986). Production of second-generation triploid and tetraploid rainbow trout by mating tetraploid males and diploid females—Potential of tetraploid fish. *Theoretical and Applied Genetics*, *72*, 193–206.

Danzmann, R. G., & Gharbi, K. (2007). Linkage mapping in aquaculture species. In *Aquaculture Genome Technologies* (p. 139–167).

Dunham, R. A. (2023). *Aquaculture and Fisheries Biotechnology: Genetic Approaches* (Vol. 13). CABI.

Fletcher, G. L., Hew, C. L., & Davies, P. L. (2001). Antifreeze proteins of teleost fishes. *Annual Review of Physiology*, *63*(1), 359–390.

Flourizel, I., Akpevwe, O., & Gabriel, O. (2023). Aquaculture advancement in biotechnology: Review. *International Journal for Research in Agricultural and Food Science*, *9*.

Glover, K. A. (2010). Forensic identification of fish farm escapees: The Norwegian experience. *Aquaculture Environment Interactions*, *1*(1), 1–10.

Halder, S., Sen, S., Bhattacharya, S., Ray, A. K., Ghosh, A., & Jhingran, A. G. (1991). Induced spawning of Indian major carps and maturation of a perch and a catfish by murrel gonadotropin releasing hormone, pimozide and calcium. *Aquaculture*, *97*(4), 373–382.

Hayes, B. J., Gjuvsland, A., & Omholt, S. (2006). Power of QTL mapping experiments in commercial Atlantic salmon populations, exploiting linkage and linkage disequilibrium and effect of limited recombination in males. *Heredity*, *97*(1), 19–26.

Houston, R. D., Haley, C. S., Hamilton, A., Guy, D. R., Tinch, A. E., Taggart, J. B., & Bishop, S. C. (2008). Major quantitative trait loci affect resistance to infectious pancreatic necrosis in Atlantic salmon (*Salmo salar*). *Genetics*, *178*(2), 1109–1115.

Hulata, G., Wohlfarth, G. W., Karplus, I., Schroeder, G. L., Harpaz, S., Halevy, A., & Kavessa, M. (1993). Evaluation of *Oreochromis niloticus* × *O. aureus* hybrid progeny of different geographical isolates, reared under varying management regimes. *Aquaculture*, *115*(3–4), 253–271.

Korol, A., Shirak, A., Cnaani, A., & Hallerman, E. M. (2007). Detection and analysis of quantitative trait loci (QTL) for economic traits in aquatic species. In *Aquaculture Genome Technologies* (p. 169).

Lakra, W. S., & Das, P. (1998). Genetic engineering in aquaculture. *Indian Journal of Animal Sciences*.

Liu, Z. J., & Cordes, J. F. (2004). DNA marker technologies and their applications in aquaculture genetics. *Aquaculture*, *238*(1–4), 1–37.

Mayekar, T. S., Salgaonkar, A. A., Koli, J. M., Patil, P. R., Chaudhari, A., Pawar, N., & Kapse, P. (2013). Biotechnology and its applications in aquaculture and fisheries. *Aquafind-Aquat Fish Database.*

McAndrew, B., & Napier, J. (2011). Application of genetics and genomics to aquaculture development: Current and future directions. *The Journal of Agricultural Science, 149*(S1), 143–151.

Rajee, O., & Alicia, T. K. M. (2019). Biotechnological application in aquaculture and its sustainability constraints. *International Journal of Advanced Biotechnology and Research, 10*(3), 1.

Rothbard, S., Shelton, W. L., Kulikovsky, Z., Rubinshtein, I., Hagani, Y., & Moav, B. (1997). Chromosome set manipulations in the black carp. *Aquaculture International, 5*, 51–64.

Senger, F., Priat, C., Hitte, C., Sarropoulou, E., Franch, R., Geisler, R., & Galibert, F. (2006). The first radiation hybrid map of a perch-like fish: The gilthead seabream (*Sparus aurata* L.). *Genomics, 87*(6), 793–800.

Skaala, Ø., Wennevik, V., & Glover, K. A. (2006). Evidence of temporal genetic change in wild Atlantic salmon, *Salmo salar* L., populations affected by farm escapees. *ICES Journal of Marine Science, 63*(7), 1224–1233.

Smith, T. I. (1988). Aquaculture of striped bass and its hybrids in North America. *Aquaculture Magazine, 14*(1), 40–49.

Tanekhy, M. (2020). Applications of biotechnology in fish. *Egyptian Journal of Animal Production, 57*(Suppl. Issue), 157–161.

Tyler, C., Jobling, S., & Sumpter, J. P. (1998). Endocrine disruption in wildlife: A critical review of the evidence. *Critical Reviews in Toxicology, 28*(4), 319–361.

Zhu, Z., He, L., & Chen, S. (1985). Novel gene transfer into the fertilized eggs of gold fish (*Carassius auratus* L. 1758). *Journal of Applied Ichthyology, 1*(1), 31–34.

CHAPTER 17

Exploring Biochar Production Techniques and Their Applications in Aquaculture: A Pathway to Sustainable Development

JORA KIRANKUMAR,[1,*] KAILASH VADHER,[1] DABHI RAJKUMAR,[1] PATHAN IMRAN,[1] and MAYUR BHADARKA[2]

[1]*Department of Aquaculture, College of Fisheries Science, Kamdhenu University, Veraval, Gujarat, India*

[2]*Department of Aquatic Environment Management, College of Fisheries Science, Kamdhenu University, Veraval, Gujarat, India*

*Corresponding author

ABSTRACT

Biochar, a carbon-rich material from thermochemical biomass conversion, offers multifaceted sustainability and efficiency benefits in aquaculture. This review synthesizes growing research on its application, covering production methods (pyrolysis, gasification, hydrothermal carbonization), techniques, and outcomes. Biochar significantly improves water quality and nutrient cycling, enhancing ecosystem resilience. It functions as a sustainable alternative to activated carbon, mitigating carbon emissions. Applications include improving soil health, supporting waste management, acting as a feed additive, neutralizing acid, and mitigating global warming. Specifically, biochar adsorbs excess nutrients (ammonia, nitrite, nitrate, phosphate), stabilizes nitrogen/phosphorus retention, reduces leaching, and fosters beneficial microbial biofilms for pollutant degradation.

Sustainable Aquaculture: Recent Advances and Techniques. Dhanaji W. Patil, Sandip S. Markad (Eds.)

DOI: 10.1201/9781779641113-17

Bacterial inoculation further boosts its bioremediation capacity. Analysis of current research and case studies underscores biochar's promise as a cost-effective, environmentally friendly solution for increasing aquaculture productivity while minimizing ecological footprints. Biochar technology represents an innovative and sustainable approach for aquaculture wastewater treatment.

17.1 INTRODUCTION

The aquaculture sector has undergone substantial growth, yielding 179 million metric tons (MMT) of fish in 2018, and forecasted to escalate to 214 MMT by 2030, as per FAO projections. Despite this expansion, the industry faces hurdles in effectively managing resources and maintaining sustainability. Consequently, there's a recognized necessity to reform the existing cultural systems and approaches to tackle these issues, as highlighted by Abdel-Tawwab et al. (2019). In addition, the sludge or waste material contains harmful nitrogen-based substances that originate from leftover feed and fish excrement. These elements frequently contribute to the development of eutrophication in bodies of water (Lananan et al., 2014). The sustainability of the aquaculture industry is significantly challenged by the use of ingredients, notably fish meal and oil, which are extensively employed in aquafeed production due to their high demand.

In addition to accounting for over 50% of the input in aquaculture, the feed also functions as a significant source of pollution, with more than 75% being released into the environment. In this particular situation, the researchers have successfully addressed the challenges to a certain extent to provide efficient solutions. An example of addressing this challenge is through the adoption of resource recycling technologies like biofloc (Debbarma et al., 2021, 2022), which efficiently manage water quality and optimize feed utilization within the aquaculture system. In recent times, there has been a growing emphasis on employing biochar derived from waste and biomass materials for wastewater treatment in aquaculture (Mahari et al., 2022). Biochar has also been promoted for improving bacterial communities in aquaponics systems and for animal feed production (Man et al., 2020). Therefore, biochar emerges as a feasible solution for mitigating harmful waste while also offering resources for aquatic users.

For most of us, the word *biochar* is not familiar. The name comes from the Greek *bios*, which means "life," and *char*, which refers to the charcoal that is made when biomass is carbonized. The earliest Amerindian villages in the Amazon region are referred to as *Terra Preta de Índio* locally. They developed a method of creating dark earth called biochar through slash-and-char techniques (Joseph and Lehmann, 2009). The depletion of trace fuels, the rising costs of oil, and the growing environmental issues on a global scale are demanding the advancement and utilization of alternative energy sources as a viable replacement for conventional fossil fuels. Biomass is a term used to describe the collective organic materials derived from living matter or a combination of organic and inorganic substances. It has a diverse range of sources, including plants, animals, animal excrement, sludge, and waste wood (Tripathi et al., 2016). Biochar is a form of charcoal commonly utilized for enhancing soil quality. The primary difference between biochar and charcoal is their production methods. Biochar is exclusively manufactured using the modern pyrolysis technique, while charcoal can be produced using either the traditional or modern methods.

Currently, the concept of biochar can be defined as a term that intentionally produces black carbon. Its main objective is to effectively manage carbon to mitigate climate change. Furthermore, biochar is employed to improve soil health and yield favorable agricultural outcomes. Its versatility as an organic material extends to diverse sectors, including aquaculture, where it has demonstrated promising results (Lan et al., 2016; Schmidt et al., 2019).

There is a large body of literature on the utilization of biochar in numerous fields. Unfortunately, there is presently no comprehensive paper focusing only on biochar preparation methods and its use in the aquaculture sector. Regarding the aquaculture sector's potential for pollution and its reliance on natural resources, further study into the potential utilization of biochar ingredients is required. This research relies on previously reported studies. As a result, in this brief analysis, we aimed to investigate the prospective applications of biochar in aquaculture for the benefit of scholars and scientists.

17.2 WHAT IS BIOCHAR?

Biochar is a carbon-rich substance similar to charcoal. It is produced by burning agricultural and forest waste organic materials, as well as biomass, at high temperatures and low oxygen levels (Liang et al., 2021; Zhang et al., 2019). Biochar is generated by a method known as pyrolysis, in which biomass (such as wood, manure, or leaves) is burned in the absence of oxygen, generating oil and gas as byproducts (Grammelis et al., 2018). It is 70% carbon, with the remaining elements being hydrogen, oxygen, and nitrogen.

Biochar is a black material having small pores and a delicate texture. It is lightweight, has a big surface area, and has a pH that makes it suitable for usage in soil. Biochar is applied to agricultural soil to increase its quality. Biochar is a stabilized biomass that can be mixed into soil to intentionally change the soil's atmosphere, thereby improving agricultural yield and lowering pollution (Rawat et al., 2019). The qualities of biochar vary based on the raw material (biomass) utilized and the processing conditions.

17.3 THE SUBSTANCE UTILIZED FOR THE PRODUCTION OF BIOCHAR

Biochar can be made from a variety of feedstocks, including grass, coconut hulls, bamboo, sugarcane bagasse, cow dung, wood chips, rice husk, wheat straw, cassava rhizome, and other agricultural leftovers (Dao et al., 2013; Ronsse et al., 2013). According to recent findings, biochar created from carbonizing organic wastes can be utilized as a substitute that not only contributes to soil carbon sequestration but also alters its physicochemical and biological properties.

However, the priority is on the use of "true wastes" to reduce delays to local carbon and nutrient recycling systems. Some of the raw materials that are currently being used on a commercial scale include tree bark, wood chips, agricultural residues such as nutshells, straw, and rice hulls, as well as grass and various organic wastes such as alcohol producers' wheat, bagasse from sugarcane production, mill waste, chicken litter, dairy manure, sewage sludge, and paper sludge.

17.4 BIOCHAR MANUFACTURING PROCEDURES

The International Biochar Project defines biochar as a solid substance formed by the thermochemical degradation of biomass in an oxygen-limited environment. Biochar is formed in a solid state through dry carbonization, pyrolysis, or gasification of biomass. It can also be generated as a slurry by hydrothermal carbonization (HTC) of biomass under pressure (Ahamad et al., 2014). Depending on the temperature, biomass is converted into a variety of products such as charcoal, biofuel, and other bio-based products by thermochemical processes such as pyrolysis, gasification, HTC, and torrefaction (Bridgwater, 2012; Kumar and Gupta, 2009). Biochar, produced by the thermochemical decomposition of biomass, serves a dual benefit: it decreases carbon emissions while also acting as a long-term alternative to activated carbon and other carbon-based materials (Cha et al., 2016).

Biochar is a key product created from biomass after extracting small amounts of condensable liquid (bio-oil) and non-condensable gases (syngas). Several types of pyrolysis units or reactors function on the same principle in terms of oxygen availability, but their heating rates, pressure conditions, and residence times may differ, affecting the proportions of the final products (Bridgwater, 2012; Kumar and Gupta, 2009). Historically, pyrolysis engineers have attempted to reduce the generation of char, which was regarded as a bothersome and low-value waste component. As a result, there has been a major increase in research focused on fast pyrolysis, which aims to maximize the generation of condensable oil, and combustion, which is similar to fast pyrolysis but involves a controlled supply of oxygen. These methods primarily produce heavy oil or gas, with only minor amounts of char (Sohi et al., 2010).

17.4.1 PYROLYSIS

Pyrolysis, a thermal process that breaks down biomass in an oxygen-limited atmosphere, is commonly used to produce biochar. This leads to the formation of gas and liquid byproducts, as well as charcoal leftovers. Pyrolysis is performed utilizing a variety of heating methods (as shown in Table 17.1) and equipment designs at temperatures ranging from 450°C to 700°C, resulting in significant differences in yields of different product fractions (Bartoli et al., 2019; Sakhiya et al., 2020). Pyrolysis technology is heavily influenced by residence time, pyrolytic temperature (e.g., slow and fast pyrolysis processes), pressure, adsorbent size, and heating rate and method (e.g., pyrolysis initiated by fuel combustion, electrical heating, or microwaves) (Nartey and Zhao, 2014). When compared to biochar produced by traditional pyrolysis, the biochar produced via microwave pyrolysis is of higher quality. According to Kumar et al. (2020), it has a larger pore capacity and surface area. Pyrolysis can be divided into two categories based on the heating rate and residence period.

17.4.1.1 SLOW PYROLYSIS

The slow pyrolysis process is becoming more and more popular because it can produce a lot of biochar and can be done on a small scale (Hodgson et al., 2016; Kim et al., 2020). Slow pyrolysis, another name for the standard carbonization process, is heating biomass slowly over a long period of time—usually several days—in order to make biochar (Sohi et al., 2010). Using heat produced by the partial burning of the feed, external heaters, or hot gas recirculation, slow pyrolysis takes place at atmospheric pressure. Several types of reactors, including screw pyrolyzers and agitated drum and rotating kilns, have been utilized in this slow pyrolysis method to produce biochar. This process has been used to produce charcoal for ages.

17.4.17.2 FAST PYROLYSIS

Under these conditions, biochar is produced with a residence time of less than 10 seconds and a high heating rate exceeding 200 K/min. Rapid pyrolysis is a thermal conversion process that needs to occur at moderate temperatures (between 500°C and 1000°C), with short residence times (less than 2 seconds), and rapid heating rates (more than 2°C/s). According to some (Bridgwater, 2012; Duman et al., 2011), bio-oil made from biomass can be used as a feedstock for the synthesis of chemicals and as a source of energy. Fast pyrolysis produces more bio-oil and less biochar than slow pyrolysis, whereas slow pyrolysis produces more biochar. This is the primary distinction between the two pyrolysis methods.

17.4.2 GASIFICATION

As a method of converting biomass, gasification includes partial combustion for a short period (usually 10 to 20 seconds) at temperatures between 600°C and 1200°C (Kambo and Dutta, 2015). According to Qambrani et al. (2017), the carbon source, such as biomass or other organic materials, decomposes to produce hydrogen (H_2), carbon dioxide (CO_2), and possibly small hydrocarbons like methane (CH_4). During gasification, a regulated amount of oxidizing agent is introduced at extreme temperatures (over 700C) to convert biomass into a mostly gaseous mixture (syngas including CO, H_2, CO_2, CH_4, and smaller amounts of higher hydrocarbons).

One of the most critical parameters that directly affect gasification reactions is temperature. Higher temperatures cause a rise in H_2, CO, and carbon production and a fall in CO_2, CH_4, hydrocarbons, and tar (Taba et al., 2012). About 10% by weight of biomass is typically converted into biochar during gasification. When gasification occurs, oxidizing agents such as oxygen, air, steam, or combinations of these gases can be utilized. Figure 17.1 illustrates that syngas produced through air gasification has low heating values of 4–7 MJ/Nm^3, while syngas from steam gasification has higher heating values of 10–14 MJ/Nm^3. By gasification, the biochar output averages around 10% of the biomass weight.

In gasification, oxidizing agents may include oxygen, air, steam, or combinations of these gases. While steam gasification yields syngas with greater heating values of 10 to 14 MJ/Nm^3, air gasification provides syngas with lower heating values, ranging from 4 to 7 MJ/Nm^3 (Figure 17.1).

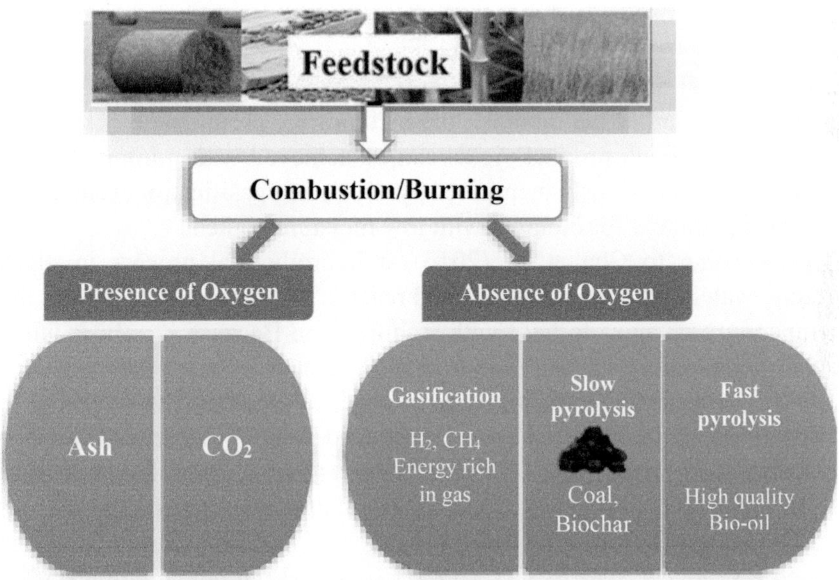

FIGURE 17.1 Schematic representation of biochar production pathways from various biomass feedstocks under different thermal processes.

Source: Adapted from Lan et al., 2021

Note: Combustion in the presence of oxygen produces ash and carbon dioxide (CO_2), whereas thermochemical conversion in the absence of oxygen—such as gasification, slow pyrolysis, and fast pyrolysis—yields biochar, syngas, and bio-oil in varying proportions depending on the processing method.

The following stages describe the gasification mechanism:

i. **Oxidation/Combustion:** Energy for gasification is primarily obtained from the oxidation and combustion of gasification agents. Water, carbon dioxide (CO_2), and carbon monoxide (CO) are produced when these agents react with the combustible components inside the gasifier.

ii. **Drying:** This stage involves the complete removal of moisture from the biomass without energy recovery. The moisture level varies among biomass types. Drying is especially important when processing biomass with high moisture content.

17.4.3 HYDROTHERMAL CARBONIZATION

In 1913, Bergius discovered the HTC technique, which involves the formation of coal-like material from cellulose. In addition to being referred to as HTC, wet torrefaction, and subcritical water treatment, this process is also known as artificial coalification (Chi et al., 2020). Depending on the moisture content of the biomass, the methods for producing biochar vary in efficiency. High product yields can be achieved with minimal energy loss using the previously mentioned pyrolysis and gasification techniques, which are suitable for biomass with low moisture content.

According to Cha et al. (2016), a hydrothermal process involves adding water and biomass to a closed reactor. After a specified period, the temperature changes to help with stabilization. Because it only requires a low temperature of 180°C to 250°C (Lee et al., 2018), this approach appears to be a cost-effective way to produce biochar. The temperature affects how different products are produced. HTC biochar is formed at temperatures lower than 250°C (Zhang et al., 2017). Hydrothermal liquefaction is the process that yields bio-oil at temperatures ranging from 250°C to 400C. Hydrothermal gasification is the process that produces gaseous products, such as CO, CO_2, H_2, and CH_4, at temperatures higher than 400°C (Khorram et al., 2016).

Depending on the characteristics of the feedstock, reaction temperature, and pressure, the char production of biomass via non-hydrothermal carbonization (NHTC) at low temperatures can vary from 30% to 60%. HTC uses water to produce biochar, which could make it cheaper for feedstocks that contain significant amounts of moisture.

17.4.4 TORREFACTION

A relatively new method for creating biochar is called torrefaction. Biomass is subjected to a variety of decomposition processes (Khorram et al., 2016; Wang et al., 2017; Yaashikaa et al., 2019) and a relatively long residence time (20–120 minutes) in order to remove carbon dioxide, moisture, and oxygen using inert atmospheric air in the absence of oxygen, at temperatures ranging from 200C to 300°C.

There are three main ways to carry out the torrefaction process:

i. **Steam Torrefaction:** Biomass is heated with steam for approximately ten minutes at a temperature not exceeding 260°C.

ii. **Wet Torrefaction (HTC):** Biomass reacts with water at temperatures between 180°C and 260°C, with a residence period ranging from five to 240 minutes.

iii. **Oxidative Torrefaction:** The final phase to attain the required temperature involves treating biomass with oxidizing agents such as combustion gases to generate heat energy (Figure 17.2).

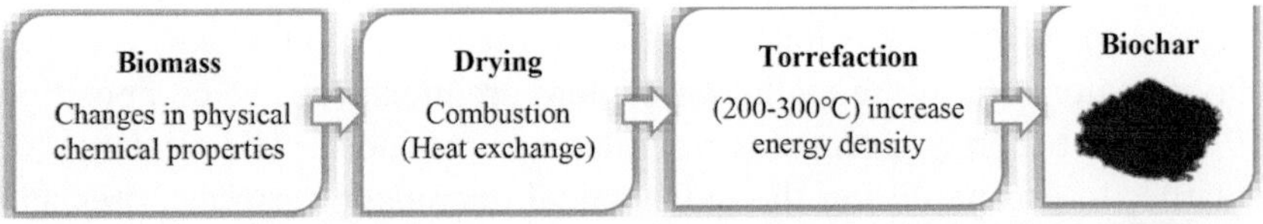

FIGURE 17.2 Schematic overview of the torrefaction process showing sequential stages from biomass through drying, heat treatment (200–300°C), and final biochar production.

Source: Adapted from Wang et al., 2020

17.5 CHARACTERISTICS OF BIOCHAR

According to Akhil et al. (2021), biochar shows several key characteristics that are essential to evaluating its efficacy in eliminating toxic substances and analyzing its uses. Among the essential characteristics of biochar are:

- pH;
- Pore composition;
- Surface area;
- Stability.

17.5.1 SURFACE AREA

Generally, it is preferable to have a larger surface area for the biochar. Conditions such as temperature, time, and the kind of feedstock utilized have a significant impact on surface area. Temperature is the primary factor influencing biochar's surface area (Akhil et al., 2021). Volatile chemicals are released, and hydrogen ions and oxygen-carrying functional groups are lost when the temperature rises from 300°C to 700°C during the biochar formation process. This substantial increase in temperature results in the biochar's surface area increasing from 3.6 to 82.9 m^2/g. A large specific surface area improves the efficiency of pollution removal and assists in the adhesion of heavy metals (Son et al., 2018).

17.5.2 PORE COMPOSITION

Evaluating the pore properties of biochar is dependent on temperature. The total pore volume of the biochar increases from 0.5 to 0.6 cc/g when the temperature increases from 400°C to 500°C. According to Yakout (2017), this temperature rise may cause pore blockages through melting. Jia et al. (2018) noted that biochar's well-organized nanoscale pores make it easier for heavy metal ions to enter the material. In biochars, volatiles play a major role in the evolution of pore architecture. The composition of biochar becomes porous as volatiles are produced. The tendency of volatiles to escape during thermochemical degradation causes pores and gaps to emerge on the surface of the biochar (Zhang et al., 2017).

17.5.3 PH

One crucial aspect of biochar is its pH. Based on pH values, biochars can be categorized as basic or acidic. The pH range of biochars is typically 4.0–12.0. The pH increases with temperature, moving closer to an alkaline level due to increased volatilization. Biochar produced via pyrolysis at 600°C showed a basic pH of 9.0, indicating that the surface became alkaline at higher temperatures. As long as there is electrostatic interaction between the heavy metal and the biochar, a higher pH promotes greater heavy metal adsorption. Higher pH levels increase negative charges, thereby facilitating heavy metal adsorption. However, in some cases, a high pH may inhibit metal adsorption by dissolving heavy metals into the solution (Jia et al., 2018; Yakout, 2017).

17.5.4 STABILITY

The most notable quality of biochar is its outstanding stability. The pyrolysis process, pH levels, and feedstock used all significantly impact this stability. Raising the temperature during pyrolysis promotes the creation of more stable biochar. According to Yang et al. (2018), biochar produced at 600°C was found to be more stable than that produced at 250°C, 350°C, and 450°C.

Some weight reduction indicates that pyrolysis performed in the presence of oxygen does not affect the stability of the biochar. However, optimized pyrolysis enhances the stability of biochar, making it beneficial for several applications, including as an additive (Das et al., 2017).

17.6 FACTORS DETERMINING THE QUALITY OF BIOCHAR

The main factors affecting the characteristics of biochar include temperature, particle size, heating rate, and feedstock. These parameters directly influence the yield of biochar, though not necessarily its quality (Yaashikaa et al., 2019). According to Libra et al. (2011), the properties of the feedstock vary depending on its maturity and climate, including its chemical composition, particle size, pH, volatile content, moisture content, and energy content.

The presence of lignin is crucial when selecting a feedstock. A feedstock with a high lignin content produces biochar with a higher concentration of aromatic carbon and greater porosity (Leng and Huang, 2018). In addition to feedstock, the temperature at which biochar is produced is critical for determining its structure and physicochemical properties. The breakdown of biomass components such as lignin, hemicellulose, and cellulose is highly temperature-dependent. Temperature also influences the release of volatile chemicals and the formation and transformation of intermediate compounds. Harvey et al. (2012) stated that hydroxyl groups oxidize at temperatures below 500°C, while methylene groups undergo dehydrogenation at temperatures above 500°C. Residence time, which may vary from a few minutes to several days, is another important factor in biochar production. It significantly affects the severity of reaction conditions, product yield, material composition, and overall quality (Kumar et al., 2020).

The distribution of micro- and macropores, surface area, and char forms can all be impacted by residence duration. The physical properties of biochar are influenced by shrinkage, elemental realignment, and structural melting throughout the production process. Furthermore, both the characteristics of biochar and the total yield of char are significantly affected by moisture content. The amount of moisture present affects the charring process and is a common factor in the creation of activated carbon.

17.7 BENEFITS OF BIOCHAR

The ecosystem is affected by the increasing intensity of fish farming systems in both direct and indirect ways. Untreated effluent from many aquaculture facilities is released into open water bodies, causing problems such as eutrophication. Additionally, there are concerns regarding the use of sustainable feed to meet the growing demand for marine seafood oil and meal. Furthermore, the presence of antibiotic residues in fish meat and the environment poses serious health risks to humans (Figure 17.3).

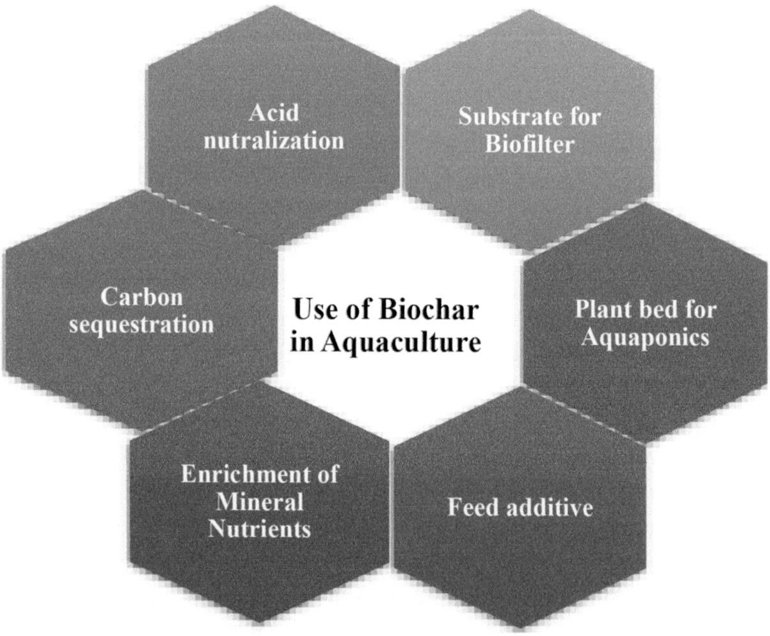

FIGURE 17.3 Benefits of biochar in the aquaculture industry.

17.7.1 SEWAGE TREATMENT

Fish farms have various applications for biochar. It helps reduce the accumulation of heavy metals in fish organs. For this purpose, dust and rice hull biochar are particularly effective. As an adsorbent, biochar has been widely used to remove nutrients, organic pollutants, and toxic metals from wastewater. It is commonly used as a water filtration medium, especially for drinking water and aquaculture systems. According to Xiang et al. (2020), biochar technology presents a new, cost-effective, and environmentally sustainable method for treating wastewater. Slurry biochar offers an efficient and eco-friendly adsorption solution for utilizing sewage effluent waste. It can remove microscopic pollutants from treated wastewater and drinking water at rates of 35% to 97% and 67% to 99%, respectively (Regkouzas and Diamadopoulos, 2019).

As sustainable substitutes for conventional bio-media, char and the mineral zeolite can effectively reduce nitrate levels in aquaculture effluents (Paul and Hall, 2021). The addition of biochar reduced soil NO_3^- levels in all fertilizer treatments by 16%. Biochar improves soil quality by enhancing the efficiency of conventional chemical fertilizers, reducing their demand, and increasing crop yields. By increasing NH_4^+, NO_3^-, and PO_4^{3-} concentrations in water, biochar (derived from sugarcane bagasse) supports primary productivity and stabilizes fluctuations in alkalinity. Without significantly altering the sediment, it also increases the essential macronutrient potassium (K) in inland saline water (ISW) (Raul et al., 2021). According to Khalil et al. (2018), ammonia from aquaculture ponds can be successfully removed using biochar.

17.7.2 POND WATER QUALITY MANAGEMENT

Water quality management is a crucial aspect of aquaculture, requiring both investment and ongoing monitoring. Limited evidence suggests that biochar can be effectively utilized in water quality regulation. A study by Deng et al. (2020) found that biochar derived from spent mushroom substrate (SMS) through pyrolysis at 400°C is highly effective in eliminating ammonia from water and both ammonia and hydrogen sulfide (H_2S) from sediment. The use of biochar can also raise the pH of both sediment and water, suggesting its effectiveness in addressing pond acidity issues (Lin et al., 2021). Achieng and Shikuku (2020) reported that copper ions, known as hazardous heavy metals, can be absorbed from water sources by biochar made from fish skins through a gradual combustion process at 600°C.

17.7.3 ANTIBIOTIC REMEDIATION

The excessive use of antibiotics—including fluoroquinolone, levofloxacin, and tetracycline—and their accumulation in water bodies has become a serious environmental and health risk. Levofloxacin may be removed from a suspension using biochar made from wood pieces and rice grains (Yi et al., 2016). Subsequently, tetracycline can be efficiently removed from a liquid solution or absorbed using biochar derived from rice husks (Fan et al., 2018). Biochar made from fish bones has also proven effective in absorbing tetracycline (Modenes et al., 2021). The efficiency of bamboo-derived biochar in adsorbing ofloxacin and enrofloxacin—two fluoroquinolone antibiotics—was investigated in synthetic wastewater. Wang et al. (2015) reported that the elimination of these antibiotics was over 99% through adsorption. Biochar produced from yak dung can also remove tetracyclines and fluoroquinolones from water (Wu et al., 2018).

17.7.4 AQUACULTURE FARMS UTILIZING BIOCHAR AS A REGULATING ASPECT

Biochar can be utilized in biofloc technology (BFT) systems that rely on tilapia as a solid carbon material to mitigate the harmful impacts of nitrogen compounds. It presents a viable option as a carbon source in BFT applications. The fish anesthetic MS-222 can be removed from water or recirculating aquaculture systems (RAS) using biochar made from agricultural residues such as eucalyptus bark, nut and walnut shells, peach stones, fruit seeds, and olive fruit waste (Ferreira et al., 2016; Mopoung et al., 2020). This is possible even though the anesthetic has a strong affinity for the aqueous phase.

In aquaculture production systems, biochar improves soil quality, enhances crop yields, improves water quality, reduces greenhouse gas emissions from soil, reduces nutrient leaching and soil acidity, and decreases fertilizer and irrigation needs (Bhutia and Bhutia, 2021).

Mangrove-derived biochar has also proven effective in reducing nitrogen release in shrimp pond sediments. A significant decrease in cumulative water volume and nitrogen absorption from sediment was observed when 8% of this biochar was applied. Additionally, the abundance of nitrifying and denitrifying bacteria in the sediment increased in the presence of biochar (Be et al., 2021).

17.7.5 BIOCHAR AS FISH FEED

Biochar is rich in nutrients. The microalgae generated during its production retain high protein and carbohydrate levels, making it suitable as an animal feed supplement (Viegas et al., 2021). According to Man et al. (2020), certain studies have shown that the growth of specific fish species improves with biochar supplementation, as measured by weight gain, protein efficiency ratio, specific growth rate (SGR), and feed conversion ratio (FCR).

Thu et al. (2010) reported that feeding Japanese flounder 0.5% bamboo-derived biochar improved weight gain, SGR, FCR, and protein efficiency ratio. Brown trout (*Salmo trutta*) fed 0.2–0.3 kg of biochar per kg of feed also showed increased body weight, SGR, and FCR (Khaki et al., 2017).

According to Quaiyum et al. (2014), striped catfish fed a diet supplemented with 2% bamboo biochar showed improved body weight, survival rate, FCR, and SGR. In another study, 1% rice husk–derived biochar in striped catfish feed led to a 36% higher growth rate (Lan et al., 2016). Mabe et al. (2018) noted that varying concentrations of bamboo biochar fed to juvenile common carp in China resulted in measurable improvements in overall health and quality. Blood markers such as glucose, triglycerides, high-density lipoprotein (HDL), aspartate aminotransferase, total protein, and total cholesterol all showed significant enhancement.

17.7.6 SYNTHESIS OF SUSTAINABLE POND FERTILIZER

To optimize profitability, India uses large quantities of fertilizer in both agriculture and aquaculture. However, this overuse has resulted in wastewater high in phosphorus. As an alternative, phosphorus-rich leftover biochar can be used as a biofertilizer, providing a more economical and environmentally friendly solution than chemical fertilizers. Due to its high water-retention capacity and biodegradability, biochar can also be applied in various settings, including lawns, gardens, and highway medians (Arun et al., 2020).

17.7.7 ACID NEUTRALIZATION

According to FT-IR analysis, biochar maintains an alkaline pH and contains calcium, magnesium, and carbonate ions. It can serve as an alternative to reduce the need for lime application in aquaculture ponds due to its strong buffering capacity (Schmidt and Wilson, 2012).

17.7.8 REDUCTION IN SEEPAGE AND LEACHING LOSS OF NUTRIENTS

Applying biochar to the bottom of ponds can reduce seepage by enhancing water retention. Inland saline aquaculture ponds in Haryana, Punjab, and Rajasthan experience significant seepage losses. As a result, large volumes of saline groundwater are extracted for aquaculture use. In addition, biochar's high anion exchange capacity allows it to retain nitrate ions produced during nitrification. Preventing nitrate leaching is critical for avoiding groundwater contamination and ensuring public health and safety.

17.7.9 USE AS PLANT BED IN AQUAPONICS SYSTEM

Biochar beads serve as effective plant substrates in aquaponics systems. They retain nutrients from wastewater and promote the symbiotic association of mycorrhizal fungi with plant roots, which enhances mineral and nutrient uptake from water.

17.7.10 SUBSTRATE FOR BIOFILTER

Biochar can adsorb pesticides and heavy metals, making it a viable substrate for biofilters and effluent treatment systems in shrimp farms. Its ability to support diverse microbial associations and biofilm formation further enhances its utility in water purification (Nguyen, 2015).

17.7.11 CARBON SEQUESTRATION

Biochar contributes to climate change mitigation by sequestering carbon and reducing greenhouse gas emissions from agricultural fields. It has the potential to capture approximately 12% of annual anthropogenic CO_2-equivalent emissions. The use of biochar in aquaculture systems also reduces emissions, making it a dual-purpose tool. However, further research is needed to explore its full potential across other applications.

17.7.12 ENHANCEMENT OF MINERAL NUTRIENTS

The mineral nutrient profile of inland saline pond water differs from that of brackish water or seawater. Deficiencies in potassium (with seawater requiring at least 50%) and imbalances in the calcium-to-magnesium ratio (minimum 1:2.25 for shrimp) can result in muscle deformities in shrimp. Biochar derived from banana peduncle, sugarcane bagasse, and other agricultural wastes contains abundant potassium, calcium, and magnesium, making it a valuable mineral supplement (Dahlawi et al., 2018).

17.8 CONCLUSION

Biochar is a sustainable platform for carbon-based material production and waste valorization. It can be produced using various methods depending on the intended application. Different types of biomass materials—such as agricultural crop residues, forestry residues, sewage sludge, manure, and solid organic municipal waste—can be utilized for biochar production. Although many studies have focused on diverse aspects of biochar application, its use as a catalyst and for the removal of heavy metals from wastewater has primarily been explored on a small scale.

Therefore, it is crucial to conduct extensive research to gain deeper insights into the long-term sustainability and economic viability of biochar. Additionally, it is important to introduce new technologies in the field of biochar production—including conversion methods, machinery, and activation techniques—to support further advancements in the development of biochar-based catalysts.

KEYWORDS

- **agricultural crop residues**
- **aquaculture**
- **biochar**
- **biomass**
- **catalysts**
- **long-term sustainability**
- **pyrolysis**
- **valorization**

REFERENCES

Abakari, G., Luo, G., Meng, H., Yang, Z., Owusu-Afriyie, G., & Kombat, E. O. (2020). The use of biochar in the production of tilapia (*Oreochromis niloticus*) in a biofloc technology system-BFT. *Aquacultural Engineering*, *91*, 102–123.

Abdel-Tawwab, M., Monier, M. N., Hoseinifar, S. H., & Faggio, C. (2019). Fish response to hypoxia stress: Growth, physiological, and immunological biomarkers. *Fish Physiology and Biochemistry*, *45*(3), 997–1013.

Achieng, G. O., & Shikuku, V. O. (2020). Adsorption of copper ions from water onto fish scales derived biochar: Isothermal perspectives. *Journal of Materials and Environmental Science*, *11*(11), 1816–1827.

Ahmad, M., Rajapaksha, A. U., Lim, J. E., Zhang, M., Bolan, N., & Mohan, D. (2014). Biochar as a sorbent for contaminant management in soil and water: A review. *Chemosphere*, *99*, 19–33.

Akhil, D., Lakshmi, D., Kartik, A., Vo, D. V. N., Arun, J., & Gopinath, K. P. (2021). Production, characterization, activation and environmental applications of engineered biochar: A review. *Environmental Chemistry Letters*, 1–37.

Arun, J., Gopinath, K. P., Vigneshwar, S. S., & Swetha, A. (2020). Sustainable and eco-friendly approach for phosphorus recovery from wastewater by hydrothermally carbonized microalgae: Study on spent bio-char as fertilizer. *Journal of Water Process Engineering*, *38*, 101567.

Bartoli, M., Giorcelli, M., Rosso, C., Rovere, M., Jagdale, P., & Tagliaferro, A. (2019). Influence of commercial biochar fillers on brittleness/ductility of epoxy resin composites. *Applied Sciences*, *9*(15), 3109.

Be, S., Vinitnantharat, S., & Pinisakul, A. (2021). Effect of mangrove biochar residue amended shrimp pond sediment on nitrogen adsorption and leaching. *Sustainability*, *13*, 7230. https://doi.org/10.3390/su13137230 (accessed on 10th July 2025).

Bhutia, T. P., & Bhutia, R. N. (2021). Waste reduction cum utilization: An approach for sustainable aquaculture. *Food and Scientific Reports*, *2*(7), 34–37.

Bolan, N., Hoang, S., Beiyuan, J., Gupta, S., Hou, D., & Karakoti, A. (2021). Multifunctional applications of biochar beyond carbon storage. *International Materials Reviews*. https://doi.org/10.1080/09506608.20217.1922047 (accessed on 10th July 2025).

Bridgwater, A. V. (2012a). Review of fast pyrolysis of biomass and product upgrading. *Biomass and Bioenergy*, *38*, 68–94.

Bridgwater, A. V. (2012b). Review of fast pyrolysis of biomass and product upgrading. *Biomass and Bioenergy*, *38*, 68–94. https://doi.org/10.1016/j.biombioe.20117.017.048 (accessed on 10th July 2025).

Castiglioni, M., Rivoira, L., Ingrando, I., Meucci, L., Binetti, R., & Fungi, M. (2022). Biochars intended for water filtration: A comparative study with activated carbons of their physicochemical properties and removal efficiency towards neutral and anionic organic pollutants. *Chemosphere*, *288*, 132538.

Cha, J. S., Park, S. H., Jung, S. C., Ryu, C., Jeon, J. K., & Shin, M. C. (2016). Production and utilization of biochar: A review. *Journal of Industrial and Engineering Chemistry*, *40*, 1–15.

Czernik, S., & Bridgwater, A. (2004). Overview of applications of biomass fast pyrolysis oil. *Energy & Fuels*, *18*(2), 590–598.

Dao, T. T., Canh, N. T., Trach, N. X., & Preston, T. R. (2013). Effect of different sources of biochar on growth of maize in sandy and feralite soils. *Livestock Research for Rural Development*, *25*.

Das, O., Kim, N. K., Sarmah, A. K., & Bhattacharyya, D. (2017). Development of waste-based biochar/wool hybrid bio composites: Flammability characteristics and mechanical properties. *Journal of Cleaner Production*, *144*, 79–89.

Debbarma, R., Biswas, P., & Singh, S. K. (2021). An integrated biomarker approach to assess the welfare status of *Ompok bimaculatus* (Pabda) in biofloc system with altered C/N ratio and subjected to acute ammonia stress. *Aquaculture (Amsterdam, Netherlands)*, *545*, 737184.

Debbarma, R., Meena, D. K., Biswas, P., Meitei, M. M., & Singh, S. K. (2022). Portioning of microbial waste into fish nutrition via frugal biofloc production: A sustainable paradigm for greening of environment. *Journal of Cleaner Production*, *334*, 130246.

Deng, B., Shi, Y., Zhang, L., Fang, H., Gao, Y., & Luo, L. (2020). Effects of spent mushroom substrate-derived biochar on soil CO_2 and N_2O emissions depend on pyrolysis temperature. *Chemosphere*, *246*, 125608. https://doi.org/10.1016/j.chemosphere.2019.125608 (accessed on 10th July 2025).

Duman, G., Okutucu, C., Ucar, S., Stahl, R., & Yani, J. (2017). The slow and fast pyrolysis of cherry seed. *Bioresource Technology*, *102*(2), 1869–1878.

Fan, S., Wang, Y., Li, Y., Wang, Z., Xie, Z., & Tang, J. (2018). Removal of tetracycline from aqueous solution by biochar derived from rice straw. *Environmental Science and Pollution Research*, *25*, 29529–29540. https://doi.org/10.1007/s11356-018-2976-0 (accessed on 10th July 2025).

FAO. (2018). *State of the World Fisheries and Aquaculture*. Rome: Food and Agriculture Organization of the United Nations, 1–227.

Ferreira, C. I., Calisto, V., Otero, M., Nadais, H., & Esteves, V. I. (2016). Comparative adsorption evaluation of biochars from paper mill sludge with commercial activated carbon for the removal of fish anesthetics from water in recirculating aquaculture systems. *Ecological Engineering*, *74*, 76–83.

Grammelis, P., Margaritis, N., & Kourkoumpas, D. S. (2018). Provided for non-commercial research and educational use. Not for reproduction, distribution or commercial use. https://doi.org/10.1016/B978-0-12-809597-3.00445-4 (accessed on 10th July 2025).

Gwenzi, W., Chaukura, N., Noubactep, C., & Mukome, F. N. D. (2017). Biochar-based water treatment systems as a potential low-cost and sustainable technology for clean water provision. *Journal of Environmental Management*, *197*, 732–749.

Harvey, O. R., Herbert, B. E., Kuo, L. J., & Louchouarn, P. (2012). Generalized two-dimensional perturbation correlation infrared spectroscopy reveals mechanisms for the development of surface charge and recalcitrance in plant-derived biochars. *Environmental Science & Technology*, *46*, 10641–10650.

Jia, M., Wang, F., Bian, Y., Stedtfeld, R. D., Liu, G., & Yu, J. (2018). Sorption of sulfamethazine to biochars as affected by dissolved organic matters of different origin. *Bioresource Technology*, *248*, 36–43.

Kambo, H. S., & Dutta, A. (2015). A comparative review of biochar and hydrochar in terms of production, physic-chemical properties and applications. *Renewable and Sustainable Energy Reviews*, *45*, 359–378.

Kameyama, K., Miyamoto, T., Iwata, Y., & Shiono, T. (2016). Influences of feedstock and pyrolysis temperature on the nitrate adsorption of biochar. *Soil Science and Plant Nutrition*, *62*(2), 180–184. https://doi.org/10.1080/00380768.2015.1136553 (accessed on 10th July 2025).

Khaki, N. D., Malcevschi, A., Voccia, A., & Marzano, F. (2017). Interaction of dietary biochar (black carbon) on the growth performance and survival rate of early-stage larvae of brown trout (*Salmo trutta*). *Aquaculture Europe*, Dubrovnik, Croatia.

Khalil, A., Sergeevich, N., & Borisova, V. (2018). Removal of ammonium from fish farms by biochar obtained from rice straw: Isotherm and kinetic studies for ammonium adsorption. *Adsorption Science & Technology*, *36*(5–6), 1294–1309.

Khorram, M. S., Zhang, Q., Lin, D., Zheng, Y., Fang, H., & Yu, Y. (2016). Biochar: A review of its impact on pesticide behavior in soil environments and its potential applications. *Journal of Environmental Sciences*, *44*, 269–279.

Kumar, A., Saini, K., & Bhaskar, T. (2020). Hydochar and biochar: Production, physicochemical properties and techno-economic analysis. *Bioresource Technology*, 123442. Available from: https://doi.org/10.1016/j.biortech (accessed on 10th July 2025).

Kumar, S., & Gupta, R. B. (2009). Biocrude production from switchgrass using subcritical water. *Energy & Fuels*, *23*(10), 5151–5159.

Lan, N. T., Anto, S., Ahamed, T. S., Kumar, S. S., Shanmugam, S., Samuel, M. S., Mathimani, T., Brindhadevi, K., & Pugazhendhi, A. (2021). A review on biochar production techniques and biochar-based catalyst for biofuel production from algae. *Fuel*, *287*, 1194117. https://doi.org/10.1016/j.fuel.2020.119411 (accessed on 10th July 2025).

Lan, T. T., Preston, T. R., & Leng, R. A. (2016). Feeding biochar or charcoal increased the growth rate of striped catfish (*Pangasius hypophthalmus*) and improved water quality. *Livestock Research for Rural Development*, *28*, 84.

Lananan, F., Hamid, S. H. A., Din, W. N. S., Khatoon, H., Jusoh, A., & Endut, A. (2014). Symbiotic bioremediation of aquaculture wastewater in reducing ammonia and phosphorus utilizing effective microorganism (EM-1) and microalgae (*Chlorella* sp.). *International Biodeterioration & Biodegradation*, *95*, 127–134.

Lee, D. J., Cheng, Y. L., Wong, R. J., & Wang, X. D. (2018). Adsorption removal of natural organic matters in waters using biochar. *Bioresource Technology*, *260*, 413–416.

Lehmann, J. (2007). Bio-energy in the black. *Frontiers in Ecology and the Environment*, *5*(7), 381–387. https://doi.org/10.1890/1540-9295 (accessed on 10th July 2025).

Leng, L., & Huang, H. (2018). An overview of the effect of pyrolysis process parameters on biochar stability. *Bioresource Technology*, *270*, 627–642.

Liang, L., Xi, F., Tan, W., Meng, X., Hu, B., & Wang, X. (2021). Review of organic and inorganic pollutants removal by biochar and biochar-based composites. *Biochar*, *3*(3), 255–2817.

Libra, J. A., Ro, K. S., Kammann, C., Funke, A., Berge, N. D., & Neubauer, Y. (2011). Hydrothermal carbonization of biomass residuals: A comparative review of the chemistry, processes and applications of wet and dry pyrolysis. *Biofuels*, 71–106.

Lin, J., Cheng, A., Shiu, Y., Wong, Y., Yeh, S., & Simangunsong, T. (2021). Using the biochar produced from spend mushroom substrate to improve the environmental condition of aquaculture pond. *Aquaculture Research*, *52*, 3532–3539. https://doi.org/10.1111/are.15194 (accessed on 10th July 2025).

Mabe, L. T., Su, S., Tang, D., Zhu, W., Wang, S., & Dong, Z. (2018). The effect of dietary bamboo charcoal supplementation on growth and serum biochemical parameters of juvenile common carp (*Cyprinus carpio* L.). *Aquaculture Research*, *49*(3), 1142–1152. https://doi.org/10.1111/are.13564 (accessed on 10th July 2025).

Mahari, W. A. W., Waiho, K., Azwar, E., Fazhan, H., Peng, W., & Ishak, S. D. (2022). A state-of-the-art review on producing engineered biochar from shellfish waste and its application in aquaculture wastewater treatment. *Chemosphere*, *288*, 132559.

Man, K. Y., Chow, K. L., Man, Y. B., Mo, W. Y., & Wong, M. H. (2020). Use of biochar as feed supplements for animal farming. *Critical Reviews in Environmental Science and Technology*. https://doi.org/10.1080/10643389.2020.1721980 (accessed on 10th July 2025).

Manyuchi, M. M., Mbohwa, C., & Muzenda, E. (2018). Potential to use municipal waste biochar in wastewater treatment for nutrients recovery. *Physics and Chemistry of the Earth, Parts A/B/C*, *107*, 92–95. https://doi.org/10.1016/j.pce.2018.07.002 (accessed on 10th July 2025).

Modenes, A. N., Bazarin, G., Borba, C. E., Locatelli, P. P. P., Borsato, F. P., & Pagno, V. (2021). Tetracycline adsorption by tilapia fish bone-based biochar: Mass transfer assessment and fixed-bed data prediction by hybrid statistical-phenomenological modeling. *Journal of Cleaner Production*, *279*, 123775. https://doi.org/10.1016/j.jclepro.2020.123775 (accessed on 10th July 2025).

Mopoung, S., Udeye, V., Viruhpintu, S., Yimtragool, N., & Unhong, V. (2020). Water treatment for fish aquaculture system by biochar-supplemented planting panel system. *The Scientific World Journal*, https://doi.org/10.1155/2020/7901362 (accessed on 10th July 2025).

Nartey, O. D., & Zhao, B. (2014). Biochar preparation, characterization, and adsorptive capacity and its effect on bioavailability of contaminants: An overview. *Advances in Materials Science and Engineering*.

Nhuchhen, D. R., Afzal, M. T., Dreise, T., & Salema, A. A. (2018). Characteristics of biochar and bio-oil produced from wood pellets pyrolysis using a bench scale fixed bed, microwave reactor. *Biomass and Bioenergy*, *119*, 293–303.

Paul, D., & Hall, S. G. (2021). Biochar and zeolite as alternative biofilter media for denitrification of aquaculture effluents. *Water*, *13*(19), 2703.

Qambrani, N. A., Rahman, M. M., Won, S., Shim, S., & Ra, C. (2017). Biochar properties and eco-friendly applications for climate change mitigation, waste management, and wastewater treatment: A review. *Renewable and Sustainable Energy Reviews*, *79*, 255–273.

Quaiyum, M. A., Jahan, R., Jahan, N., Akhter, T., & Sadiqul, I. M. (2014). Effects of bamboo charcoal added feed on reduction of ammonia and growth of *Pangasius hypophthalmus*. *Journal of Aquaculture Research & Development*, *5*, 269–273. https://doi.org/10.4172/2155-9546.1000269 (accessed on 10th July 2025).

Raul, C., Bharti, V. S., Dar Jaffer, Y., Lenka, S., & Krishna, G. (2021). Sugarcane bagasse biochar: Suitable amendment for inland aquaculture soils. *Aquaculture Research*, *52*(2), 643–654.

Rawat, J., Saxena, J., & Sanwal, P. (2019). Biochar: A sustainable approach for improving plant growth and soil properties. In *Biochar—An Imperative Amendment for Soil and the Environment* (pp. 1–17). IntechOpen.

Regkouzas, P., & Diamadopoulos, E. (2019). Adsorption of selected organic micro-pollutants on sewage sludge biochar. *Chemosphere*, *224*, 8408517.

Ronsse, F., Van Hecke, S., Dickinson, D., & Prins, W. (2013). Production and characterization of slow pyrolysis biochar: Influence of feedstock type and pyrolysis conditions. *GCB Bioenergy*, *5*(2), 104–115.

Sakhiya, A. K., Anand, A., & Kaushal, P. (2020). Production, activation, and applications of biochar in recent times. *Biochar*, *2*(3), 253–285. https://doi.org/10.1007/s42773-020-00047-1 (accessed on 10th July 2025).

Schmidt, H., Hagemann, N., Draper, K., & Kammann, C. (2019). The use of biochar in animal feeding. *PeerJ*, *7*, e7373. https://doi.org/10.7717/peerj.7373 (accessed on 10th July 2025).

Sohi, S. P., Krull, E., Lopez-Capel, E., & Bol, R. (2010). A review of biochar and its use and function in soil. *Advances in Agronomy*, *105*, 47–82.

Son, E. B., Poo, K. M., Chang, J. S., & Chae, K. J. (2018). Heavy metal removal from aqueous solutions using engineered magnetic biochars derived from waste marine macro-algal biomass. *Science of the Total Environment*, *615*, 161–168.

Stylianou, M., Christou, A., Michael, C., Agapiou, A., Papanastasiou, P., & Fatta-Kassinos, D. (2021). Adsorption and removal of seven antibiotic compounds present in water with the use of biochar derived from the pyrolysis of organic waste feedstocks. *Journal of Environmental Chemical Engineering*, *9*, 105868.

Taba, L. E., Irfan, M. F., Daud, W. A. M. W., & Chakrabarti, M. H. (2012). The effect of temperature on various parameters in coal, biomass and CO-gasification: A review. *Renewable and Sustainable Energy Reviews*, *16*(8), 5584–5596.

Thu, M., Koshio, S., Ishikawa, M., & Yokoyama, S. (2010). Effects of supplementation of dietary bamboo charcoal on growth performance and body composition of juvenile Japanese flounder, *Paralichthys olivaceus*. *Journal of the World Aquaculture Society*, *41*, 255–262. https://doi.org/10.1111/j.1749-7345.2010.00365.x (accessed on 10th July 2025).

Tripathi, M., Sahu, J. N., & Ganesan, P. (2016). Effect of process parameters on production of biochar from biomass waste through pyrolysis: A review. *Renewable and Sustainable Energy Reviews*, *55*, 467–4817.

Viegas, C., Gouveia, L., & Gonçalves, M. (2021). Aquaculture wastewater treatment through microalgal biomass: Potential applications on animal feed, agriculture, and energy. *Journal of Environmental Management*, *286*, 112187.

Wang, H., Baek, K., Xue, J., Li, Y., & Beiyuan, J. (2020). Preface—Biochar and agricultural sustainability. *Journal of Soils and Sediments*, *20*, 3015–3016.

Wang, L., Barta-Rajnai, E., Skreiberg, Ø., Khalil, R., Czégény, Z., & Jakab, E. (2017). Impact of torrefaction on woody biomass properties. *Energy Procedia*, *105*, 1149–1154.

Wang, S., Bin, G., Yuncong, L., Ahmed, M., Andrew, R. Z., & Lena, Q. M. (2015). Manganese oxide-modified biochars: Preparation, characterization, and sorption of arsenate and lead. *Bioresource Technology*, *181*, 13–17. https://doi.org/10.1016/j.biortech.2015.017.044 (accessed on 10th July 2025).

Wu, H., Dong, X., & Liu, H. (2018). Evaluating fluorescent dissolved organic matter released from wetland-plant derived biochar: Effects of extracting solutions. *Chemosphere*, *212*, 638–644.

Xiang, W., Zhang, X., Chen, J., Zou, W., He, F., & Hu, X. (2020). Biochar technology in wastewater treatment: A critical review. *Chemosphere*, *252*, 126–539.

Yaashikaa, P. R., Kumar, P. S., Varjani, S. J., & Saravanan, A. (2019). Advances in production and application of biochar from lignocellulosic feedstocks for remediation of environmental pollutants. *Bioresource Technology*, *292*, 122–130.

Yakout, S. M. (2017). Physicochemical characteristics of biochar produced from rice straw at different pyrolysis temperature for soil amendment and removal of organics. *Proceedings of the National Academy of Sciences, India Section A: Physical Sciences*, *87*(2), 207–214. https://doi.org/10.1007/s40010-017-0343-z (accessed on 10th July 2025).

Yang, Y., Sun, K., Han, L., Jin, J., Sun, H., & Yang, Y. (2018). Effect of minerals on the stability of biochar. *Chemosphere*, *204*, 310–317.

Yi, S., Gao, B., Sun, Y., Wu, J., Shi, X., & Wu, B. (2016). Removal of levofloxacin from aqueous solution using rice-husk and wood-chip biochars. *Chemosphere*, *150*, 7017.

Zhang, P., Li, Y., Cao, Y., & Han, L. (2019). Characteristics of tetracycline adsorption by cow manure biochar prepared at different pyrolysis temperatures. *Bioresource Technology*, *285*, 121–348.

Zhang, X., Gao, B., Zheng, Y., Hu, X., Creamer, A. E., & Annable, M. D. (2016). Biochar for volatile organic compound (VOC) removal: Sorption performance and governing mechanisms. *Bioresource Technology*, *245*, 606–614.

Zhang, X., Gao, B., Zheng, Y., Hu, X., Creamer, A. E., & Annable, M. D. (2017). Biochar for volatile organic compound (VOC) removal: Sorption performance and governing mechanisms. *Bioresource Technology*, *245*, 606–614.

CHAPTER 18

Developing an Ecosystem Approach to Assess Aquaculture for Effectively Reaching the Sustainable Development Goals: Case Study of Whiteleg Shrimp Farming Business

RIESTI TRIYANTI* and RIZKI APRILIAN WIJAYA

Research Center for Behavioral and Circular Economics, National Research and Innovation Agency (BRIN), Jakarta, Indonesia

*Corresponding author

ABSTRACT

With close ties to terrestrial ecosystems, aquaculture is a distinct industry that includes all aquatic environments, including freshwater, brackish/estuarine, and marine. The varied aquaculture sector contributes significantly to achieving the Sustainable Development Goals (SDGs)/Agenda 2030. Aquaculture, particularly shrimp farming, is a critical component of economy; however, it faces environmental degradation and social sustainability challenges. The chapter emphasizes on use of an Ecosystem Approach to Aquaculture (EAA) as a strategic plan for improving sustainability in the shrimp growing industry. Assessment of the current state of shrimp farming practices is elaborated through in-depth exploration, emphasizing the importance of a comprehensive strategy that includes environmental, economic, and social factors depending on relation of aquaculture with certain SDG indicators. Key components of the EAA, such as the incorporation of mangrove ecosystems, intelligent water

Sustainable Aquaculture: Recent Advances and Techniques. Dhanaji W. Patil, Sandip S. Markad (Eds.)

DOI: 10.1201/9781779641113-18

management, and responsible feed sourcing, were scrutinized for their effectiveness in promoting sustainable shrimp cultivation. Furthermore, the chapter examines the role of governmental policies and the active engagement of stakeholders, including farmers and industry players, in facilitating the adoption of sustainable practices. Real-world examples and practical insights will guide and inspire policymakers, researchers, and practitioners to assess the Ecosystem Approach, fostering sustainable and resilient shrimp farming business.

18.1 INTRODUCTION

In its 1995 Code of Conduct for Responsible Fisheries (CCRF), the Food and Agriculture Organization of the United Nations (FAO-UN) defined worldwide fisheries management. The assertion is made that the management of fisheries is obligated to ensure "high quality," "comprehensiveness," and the adequate "presence" of fish resources in ample amounts for both current and future generations (Hughes, 2021). The primary concept is the development of sustainable fisheries management through the consideration of the dynamic interactions between biotic and abiotic ecosystem components as a single, integrated process, function, and framework for policy planning. To address the issues of sustainable fisheries development, one important strategy is to practice responsible fisheries management (Adrianto, 2005).

The 2030 Agenda provides humanity with a roadmap for a future that is more prosperous, just, and sustainable, with its seventeen Sustainable Development Goals (SDGs) (UN General Assembly, 2015). It aims to reduce inequality, eradicate impoverishment and malnutrition, improve nutrition and health, and build inclusive, equitable, and peaceful communities while respecting the limits of the planet. One of the most significant issues confronting humanity today is how to feed the increasing number of people a nutritious and sustainable diet. There are numerous connections between the food system and the SDGs, including those involving resources, environments, the economy, and the welfare of people (Troell et al., 2023).

Aquaculture has advanced to the point of global industrialization through the diminishing contribution of the catch fisheries subsector to the supply of fish products. With over 3 million tons higher than the previous year's record high in 2018, and a notable rise of 2.7% from 2020, the total amount of aquatic animal production through aquaculture and capture fisheries exceeded 182 million tons in 2021. Around 92 million tons (2021) of the total 182 million tons came from the aquaculture sector, while 50 million tons (91.2 million tons) came from catch fisheries. This illustrates the growing importance of aquaculture, which accounted for 4% of the entire production in the 1950s, 5% in the 1970s, 19% in the 1990s, and 44% in the 2010s. According to the 6.3 million tons generated in 2021, the most common species was the whiteleg shrimp (*Penaeus vannamei*) (Food and Agriculture Organization of the United Nations, 2024).

By geographic region, Europe, the Americas, Oceania, and Africa produced the majority of aquatic animals worldwide in 2021, with Asia contributing up to 70% of the total. With 35% of the entire production, China remained the largest producer. Indonesia and India came in second and third, with 7% and 8% of the market, respectively. In 2021, these three nations collectively supplied around half of the worldwide aquatic animal production (Food and Agriculture Organization of the United Nations, 2024).

Compared to 30 years ago, aquaculture productivity has expanded by more than six times. As an economic activity, increasing production will certainly impact the negative effects of aquaculture socially, economically, and environmentally, harming not only other sectors but also aquaculture itself. Therefore, the public view has emerged that aquaculture is an exploitative economic activity that requires high levels of energy and resources and has the potential to produce waste at every stage of the production chain. Sustainable shrimp farming development in Indonesia still focuses on increasing production and value (Table 18.1). In 2009, the Ministry of Marine Affairs and Fisheries of the Republic of Indonesia envisioned Indonesia as the largest aquaculture producer, with a production target increase of 353% from 2009 to 2015 (Rimmer et al., 2013). The National Development Planning Agency of the Republic of Indonesia made a similar statement in 2014, noting that fisheries development still focuses on the economic aspects alone.

TABLE 18.1 Achievements and Targets of Shrimp Farming Production in Indonesia

Information	Unit	Achievement (2020)	Targets (2024)	%
Production volume	Thousand tons	1028	2000	95
Production value	IDR. billion	72,000	140,000	94
Export volume	Thousand tons	327	1003	207
Export value	USD thousand	2,878,075	8,829,612	207

Source: MMAF RI (2021).

The advancement of aquaculture has improved the economy, societal welfare, and the ability to supply animal protein. However, moving forward, it will be important to anticipate potential negative effects on aquaculture growth, such as the conversion of land for habitation, a decrease in mangrove ecosystems due to pond construction, an increase in nutrient and organic matter levels in water leading to plankton population explosions, changes to benthic conditions, eutrophication of public waters, increased waste from chemicals and medications used in fish farming operations, and the emergence of an inequitable distribution of labor income, particularly among small-scale fish farmers (Lasima et al., 2021; Nunes et al., 2011). Conflicts of interest with other parties and environmental harm often impede the growth of the fishing industry. The fundamentals of responsible fish farming must be followed, as several efforts have been made to foresee potential negative effects while enhancing the beneficial impacts of fish farming operations.

Acknowledging the intricate interplay of economic, environmental, and social variables in aquaculture facilitates the formulation of more targeted approaches to tackle the challenges and opportunities confronting the sector as it expands. Management at the local, international, and national levels is required to address the social, economic, and environmental issues (Troell et al., 2023). One strategy that might be implemented is the Ecosystem Approach to Aquaculture (EAA), an environmental approach to developing fisheries. EAA is an innovative policy that integrates aquaculture activities with broader ecological considerations to promote sustainable growth by balancing environmental and social concerns (Soto et al., 2008a). Several factors can be considered when implementing the EAA: (i) improving environmental management; (ii) improving community welfare; and (iii) improving governance in aquaculture development. The application of EAA principles aims to maintain the sustainability of aquaculture activities through three dimensions: socio-economic, environmental, and governance (Brugere et al., 2019). Based on the principle of ecosystem sustainability, parameters are developed for (i) aquaculture area suitability for environmental carrying capacities; (ii) sustainability of aquaculture area management; and (iii) environmental control in aquaculture regions. In terms of equitable welfare, the parameters are (a) social aspects and (b) economic aspects. For governance principles, the parameters are (a) institutions; (b) regulations; (c) compliance/law enforcement; (d) community participation; (e) stakeholder synergy; and (f) monitoring and evaluation. These parameters include several indicators that can serve as a reference in evaluating aquaculture activities. The results of the evaluation can provide strategic information, allowing corrective actions through aquaculture development programs to focus on the parameters and indicators with the lowest values.

EAA implementation is carried out by establishing integrated social and environmental responsibility from planning to management, which is supported by cross-sector integration in resource management. With this effort, it is hoped that good participation and cooperation between stakeholders can be built and feedback obtained for current fisheries management, especially shrimp cultivation in Indonesia. The research method in this chapter combines a literature review and data collection through interviews conducted at research locations in Indonesia. The study was conducted in 2019 and 2023. The data obtained was analyzed using descriptive methods to answer the existing objectives. With this background, the chapter has the following highlights:

- Contributing aquaculture to achieving the sustainable development goals (SDGs);
- A detailed history and roadmap of the ecosystem approach to aquaculture;
- Implementation of the ecosystem approach to aquaculture in the shrimp farming business in Indonesia.

18.2 THE PRESENT AND PROSPECTIVE ROLES OF AQUACULTURE IN ACHIEVING SUSTAINABLE DEVELOPMENT GOALS

A wide range of food production techniques are used in aquaculture, and there are many opportunities to implement socially and environmentally responsible policies (Gephart et al., 2020). Aquaculture is a significant industry that improves human well-being and is becoming increasingly vital to efforts to achieve the SDGs (Troell et al., 2023). In the pursuit of achieving the Sustainable Development Goals (SDGs), aquaculture management is crucial and strategic (Figure 18.1). A number of SDGs pertaining to the aquaculture industry include reducing poverty and hunger (SDGs 1, 2), enhancing health (SDG 3), promoting responsible production and consumption to improve the environmental sustainability of land, seas, and the climate (SDGs 6, 12, 13, 14, and 15), and diminishing disparities in quality of life, gender equality, and income (SDGs 1, 5, 8, and 10).

The notable but less evident benefits of aquaculture include its capacity to produce food in urban areas (via vertical farming, aquaponics systems, and communal farming), generate energy (such as algal biomass), support technological advancements, and forge a wide range of domestic and global linkages (SDGs 7, 9, 11, and 17) (Troell et al., 2023). To help accomplish the SDGs, sustainable aquaculture resource management must consider economic, social, and environmental sustainability to the greatest extent possible. However, there will be environmental costs associated with increasing aquaculture production due to environmental challenges. A vast array of technology is used to raise approximately 460 species/groups of algae, shellfish, and finfish in freshwater, brackish, and marine habitats, leading to a wide variety of production systems and feed requirements. This diversity results in numerous negative environmental impacts, including those related to energy consumption, water use, feed inputs, genetic potential hazards, and the discharge of nutrients and pollutants (Gephart et al., 2020).

The development of aquaculture is also faced with socio-economic challenges, particularly business sustainability. The aquaculture industry is a business closely tied to both backward and forward linkages. Backward linkages include relationships with input providers, production facilities, and capital providers, while forward linkages relate to the processing and marketing industries, as well as distribution and logistics. These forward and backward relationships can absorb labor, strengthen business partnerships between actors in the fisheries industry, and help keep up with developments in digital business. Partnerships are crucial for competing in both domestic and global markets. They also provide access to markets, capital, and technological development.

However, at the implementation level, inequality is often found in partnerships within the business due to the patron-client phenomenon, which causes imperfect information. The subordinate union of two or more parties with unequal status results in a strong-weak position where roles or functions are not equally shared, leading to unbalanced partnerships. Furthermore, the partnering parties may not be prepared, and market functions may not be properly functioning. To increase the competitiveness of fishery products and overcome these challenges, business partnerships in the form of institutions are necessary to achieve the SDGs goals.

Although aquaculture presents opportunities to contribute to the majority of the SDGs, various factors influence how different aquaculture systems will perform in diverse contexts to achieve these goals. Trade-offs between elements like environmental performance and societal benefits will arise, as they do in all food systems. However, in contrast to many terrestrial animal production methods, some aquaculture systems have relatively modest environmental footprints and may even undertake environmentally restorative tasks (Troell et al., 2023).

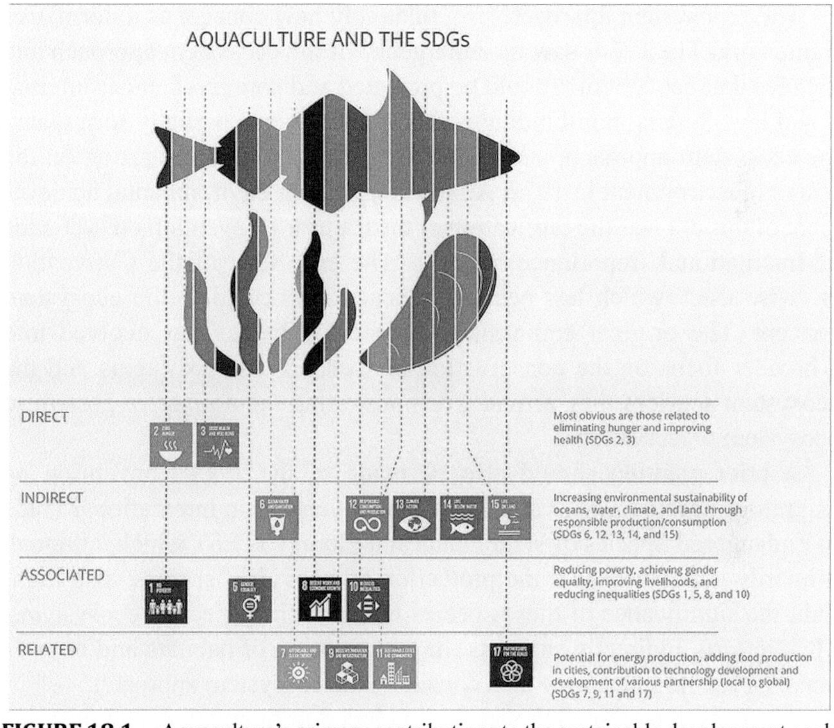

FIGURE 18.1 Aquaculture's primary contributions to the sustainable development goals in a simplified format.

Source: Adapted from Troell et al. (2023).

18.3 HISTORY AND ROADMAP ECOSYSTEM APPROACH TO AQUACULTURE

More than 40 years ago, the concept of sustainable aquaculture was first proposed, driven by predictions that fish farming would surpass other animal protein sources in the future, coupled with the stagnation of catch fisheries production (Tacon et al., 2022). Compared to a quarter of a century ago, global aquaculture production has expanded more than sixfold. As an economic activity, increasing production will mitigate the harmful impacts of aquaculture socially, economically, and environmentally, benefiting not only other industries but aquaculture itself. Therefore, it is not surprising that the public views aquaculture as an exploitative economic activity, requiring significant energy and resources and causing waste at every stage of production (Little et al., 2018).

The "ecosystem approach" is a relatively new concept as a normative framework. The 1970s saw the emergence of the ecosystem approach and the idea that ecosystems should be protected and conserved under international law. Several non-binding soft-law instruments partially formulated the ecosystem approach, starting with the Stockholm Declaration on the Human Environment in 1972. Another significant environmental achievement of this era was the ratification of the Ramsar Convention on Wetlands of International Importance in 1971. The core idea of the Convention is "wise use," which has been specifically connected to the ecosystem concept. The original emphasis on waterfowl habitat has evolved into a broader focus on the conservation of wetlands as ecosystems and the ecosystem services they provide, demonstrating the normative growth of ecosystem preservation.

A brief mention should also be made of the 1979 Convention on Migratory Species (CMS) and the 1973 Convention on International Trade in Endangered Species of Wild Fauna and Flora (CITES), which, although primarily concerned with the protection of individual species, also highlight the significance of these species within their respective ecosystems. This, in turn, indirectly supports the preservation of habitats and ecosystems. CITES practice now also considers the ecosystem approach.

An ecosystem approach was first included in international agreements in the 1980s as a normative paradigm for controlling human activity in the maritime environment. The revolutionary Convention on the Conservation of Antarctic Marine Living Resources served as the catalyst for this shift. The framework of the 1992 Convention on Biological Diversity facilitated much of the additional conceptual development at the international level. However, the ecosystem approach has an implicit foundation provided by the Law of the Sea Convention of 1982. It is said that the widespread implementation of the ecosystem approach came about due to the failure of reactive, fragmented sectors and regional techniques for environmental conservation and management. The work of the UN General Assembly has also been beneficial in reaching a consensus on the fundamental components of the ecosystem approach. In 2006, during the seventh meeting of the Unofficial Consultative Process on Oceans and the Law of the Sea (UNICPOLOS), discussions were held to solidify this approach.

There are four world organizations concerned with ecosystem-based fisheries management: the United Nations Convention on the Law of the Sea (UNCLOS), the Food and Agriculture Organization (FAO), the United Nations Environment Programme (UNEP), and the United Nations Conference on Environment and Development (UNCED). In the first period (1972–1992), these organizations focused on increasing environmental awareness. In the second period (1995–2002), they focused on harmonizing environment-based fisheries management, and from 2002 to 2019, they committed to implementing an Agreement on Indonesian Fisheries Management (Figure 18.2).

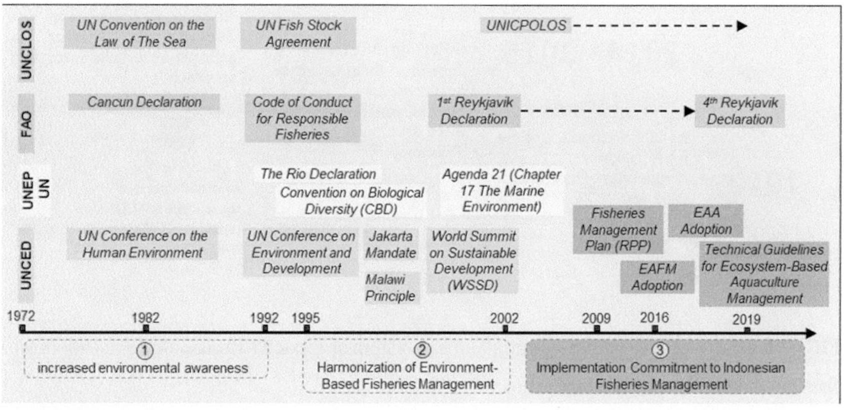

FIGURE 18.2 Historical map of the development of international instruments relevant to ecosystem-based management.

Source: Modified from various sources.

The idea of the Ecosystem Approach to Aquaculture (EAA) is an attempt to reconcile ecological and economic concerns. Aquaculture management through an ecosystem approach emphasizes four key areas of focus: (i) specific cultivation areas; (ii) an ecosystem's capacity for resilience and sustainability; (iii) dynamic interactions between environmental, biological, physical, social, and economic components; and (iv) adaptive management to changes in the environment. The understanding of the EAA concept in documents published by the FAO-UN in 2010 marked the beginning of adopting the EAA management method in cultured fisheries (Rimmer et al., 2013; Soto et al., 2008b). In 2014, EAA indicators were developed in Indonesia, and they were adopted in 2016 for cultivation management. In 2019, a policy or technical guidance for managing aquaculture areas using the ecosystem approach was issued through the Regulation of the Director General of Aquaculture Number 154/Per-DJPB/2019. An illustration of the ecosystem approach to aquaculture in Indonesia is shown in Figure 18.3.

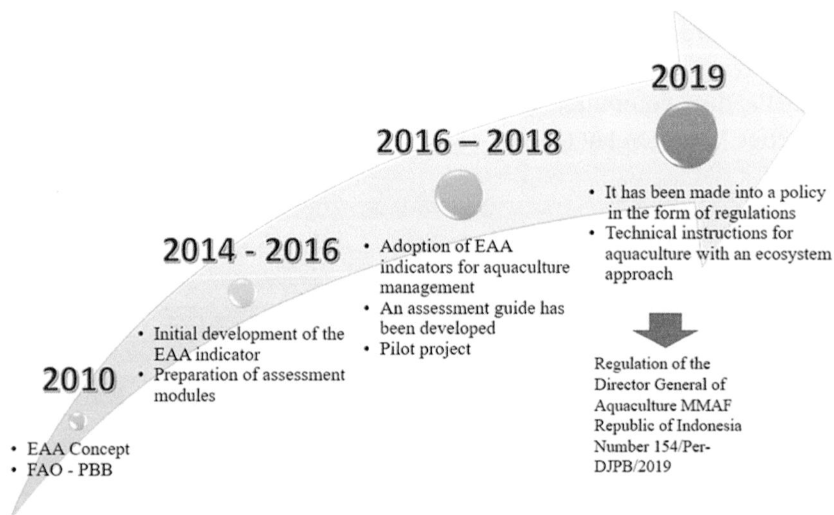

FIGURE 18.3 Roadmap for adopting an ecosystem approach to aquaculture in Indonesia.

Source: Adapted from various sources.

18.4 ECOSYSTEM APPROACH TO AQUACULTURE IMPLEMENTATION IN THE WHITELEG SHRIMP BUSINESS IN INDONESIA

In 2019, we conducted research on shrimp production and fisheries firms to examine how the EAA was being implemented in Indonesia. The research locations were in the following provinces: Aceh Besar Regency, Nangroe Aceh Darussalam Province; Pinrang Regency, South Sulawesi Province; and South Lampung Regency, Lampung Province (Figure 18.4). These three locations were selected because they serve as shrimp farming production centers in each province of Indonesia. Vannamei and tiger shrimp are the main products of traditional and semi-intensive pond shrimp farming in the three research locations. A total of 179 participants, either employees or owners of shrimp farming businesses, were involved in the study.

FIGURE 18.4 Research locations of the ecosystem approach to assess aquaculture in Indonesia.

Aquaculture managers can use the ecosystem approach as a diagnostic tool to determine which options or priorities to include in their future business strategy for aquaculture. This strategy is one way to achieve ecologically friendly and sustainable farming. The potential advantages of applying EAA to Indonesian shrimp farming companies are as follows:

- Bolstering the region's aquaculture areas;
- Boosting environmentally and socially conscious product competitiveness;
- Encouraging investment (business certainty);
- Improving environmental quality;
- Potentially lowering certification costs (regional integration);
- Ensuring aquaculture businesses' sustainability; and Providing policy feedback and management recommendations.

The need for area/cluster-based management will enable the realization of this potential.

The EAA rules and implementation guidelines we utilize are based on the 2019 Technical Rules issued by the Director General of Aquaculture under Regulation Number 154/Per-DJPB/2019. There are three principles, eleven factors, and 32 indicators. The indicators that have been prepared can be evaluated using this tool. To address any issue where one of the indicators has a low value, policy and program support are required. The tool in question is depicted in greater detail in Figure 18.5. The relevance of each indicator increases with its parameter weight. The proposed criteria and indicators are suitable for implementation in Indonesian aquaculture management.

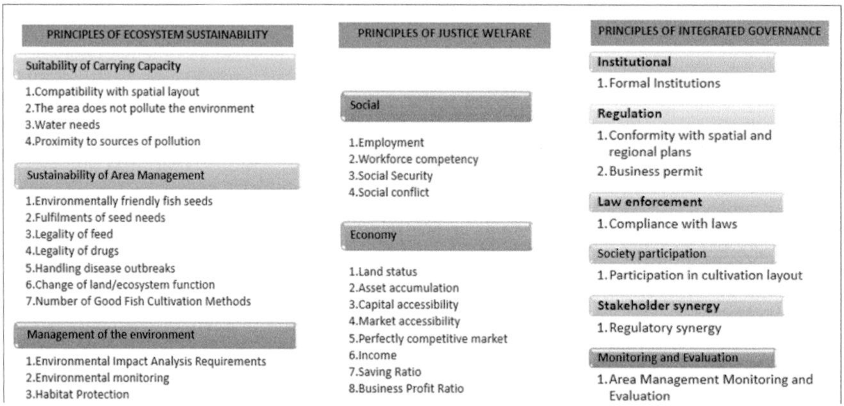

FIGURE 18.5 Principles, parameters, indicators, and weights of the ecosystem approach to aquaculture.

Source: Regulation of the Director General of Aquaculture Number 154/Per-DJPB/2019.

Managing the behavior of people, in this case, aquaculture producers, is the goal of aquaculture management. The traits of the cultivators themselves, such as age, income, education level, and prior business experience, also impact the cultivators' perceptions. The majority of growers are within the productive age range, but young individuals show little interest in becoming shrimp farmers. To sustain the agricultural industry, farmers must be replaced. Most cultivators have completed the required nine years of elementary school, and their education can influence the uptake of cultivation management policy programs and technologies. Business experience also affects the degree of adoption of cultivation practices and the implementation of management policies.

A few key points about the concept of ecosystem sustainability are as follows:

- Pond farming is often located according to the spatial arrangement established in the district/city Regional Spatial Plan and the Zoning Plan for Coastal Areas and Small Islands;

- Upstream (mainland) areas and farmer-to-farmer exchanges are the main sources of pollution for enterprises engaged in cultivation (e.g., shrimp feed);
- Many farmers still face challenges with *Cara Budidaya Ikan yang Baik* (Good Fish Farming Practices) accreditation and disease outbreaks. Outbreaks of shrimp illness have become a significant concern for farmers;
- In an effort to preserve the ecosystem, farmers have planted mangroves downstream of rivers; and
- Limited environmental monitoring and effect analyses are being conducted.

It should be mentioned that growers continue to view the concept of ecosystem sustainability as a means of increasing their operational costs. Figure 18.6 displays the findings from pond farmers' evaluations of management guidelines and standards using an ecosystem approach.

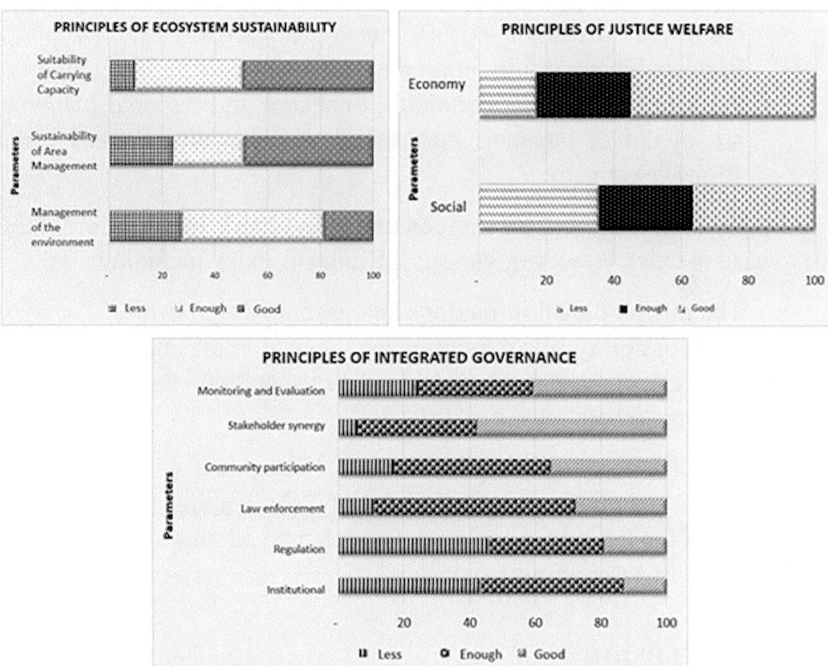

FIGURE 18.6 Indonesian Whiteleg shrimp farmers' perspectives on the ecosystem approach to aquaculture.

Source: Primary data processed.

A few key points about the equitable welfare principle are as follows:

- Since shrimp farming is still a popular export, access to capital, land, and markets remains relatively open and simple.
- Cultivators still have relatively low saving habits.
- Business actors, such as marketers or suppliers, continue to control the prices of shrimp product commodities.
- Cultivation businesses positively affect local labor absorption in each business subsystem.
- Labor competency and social security for cultivation workers remain low.

One of the aims of Law Number 27 of 2016 is to empower salt producers, fish cultivators, and fishers. When it comes to this principle, it's crucial to remember that growers still prioritize economic factors over social ones.

The integrated governance concept continues to face several obstacles. These include the following:

- No formal institution exists to manage pond fisheries.
- Farmers have not fully implemented business permit regulations.
- Regional regulations pertaining to spatial and regional planning are dynamic, meaning that zoning in aquaculture areas lacks strong legitimacy.

If aquaculture regions are not economically viable in the future, they may cease to exist. However, various advantages exist, including:

- The rarity of legal infractions against cultivation firms.
- The possibility of community engagement in management.
- Well-established laws and regulations at both the central and regional levels.

It's crucial to note that while the term "integrated with cultivation management" is simple to explain, it may be challenging for farmers to implement. If the integrated approach is implemented, this is often associated with significant transaction costs.

18.5 CONCLUSION

The goal of the Ecosystem Approach to Assess Aquaculture (EAA) policy innovation is to enhance aquaculture while creating innovative solutions that balance socio-economic and environmental aspects. EAA adoption can minimize environmental, social, and economic risks in a given area, potentially promoting sustainable development (SDGs 1, 2, 3, 5, 6, 7, 8, 9, 10, 11, 12, 13, 14, 15, and 17). The government, in collaboration with research institutes, must conduct an initial assessment (pre-assessment) as a baseline for improving/increasing the EAA's implementation status, followed by cooperation to determine the implementation strategy. Declared EAA areas that are good for achieving the targeted benefit goals must come from the government. The key to successfully enlisting the help of fish farmers and other community members to establish an EAA-based aquaculture region is the government's proactive coordination role. The following steps must be taken to implement the ecosystem approach, particularly in the shrimp cultivation industry:

- Raise local governments' awareness of the concept and application of cultivation management through an ecosystem approach as regulators;
- Educate cultivators and other business actors about the implementation of the EAA concept;
- Gain a thorough understanding of the socio-economic characteristics of each region that will implement EAA; and
- Provide regular support and oversight of shrimp pond fishery business actors.

KEYWORDS

- **aquaculture**
- **cultivation management**
- **cultivators**
- **ecosystem sustainability**
- **shrimp farming business**
- **socio-economic characteristics**
- **sustainable development goals**

REFERENCES

Adrianto, L. (2005). Implementasi Code of Conduct for Responsible Fisheries dalam Perspektif Negara Berkembang. *Indonesian Journal of International Law*, *2*(3). https://doi.org/10.17304/ijil.vol2.3.3 (accessed on 10th July 2025).

Brugère, C., Aguilar-Manjarrez, J., Beveridge, M. C. M., & Soto, D. (2019). The ecosystem approach to aquaculture 10 years on–a critical review and consideration of its future role in blue growth. *Reviews in Aquaculture*, *11*(3), 493–514. https://doi.org/10.1111/raq.12242 (accessed on 10th July 2025).

Food and Agriculture Organization of the United Nations. (2024). *Fishery and Aquaculture Statistics–Yearbook 2021*. https://doi.org/10.4060/cc9523en (accessed on 10th July 2025).

Gephart, J. A., Golden, C. D., Asche, F., Belton, B., Brugere, C., Froehlich, H. E., Fry, J. P., Halpern, B. S., Hicks, C. C., Jones, R. C., Klinger, D. H., Little, D. C., McCauley, D. J., Thilsted, S. H., Troell, M., & Allison, E. H. (2020). Scenarios for global aquaculture and its role in human nutrition. *Reviews in Fisheries Science and Aquaculture*, *29*(1), 122–138. https://doi.org/10.1080/23308249.2020.1782342 (accessed on 10th July 2025).

Hughes, A. D. (2021). Defining nature-based solutions within the blue economy: The example of aquaculture. *Frontiers in Marine Science*, *8*. https://doi.org/10.3389/fmars.2021.711443 (accessed on 10th July 2025).

Lasima, W., Supriyono, E., Junior, M. Z., Budiharsono, S., Nirmal, K., & Sukenda. (2021). The implementation strategy of Ecosystem Approach to Aquaculture (EAA) shrimp commodities in Pinrang District, South Sulawesi. *Jurnal Pengelolaan Sumberdaya Alam dan Lingkungan*, *11*(3), 504–512. https://doi.org/10.29244/jpsl.11.3.504-512 (accessed on 10th July 2025).

Little, D. C., Young, J. A., Zhang, W., Newton, R. W., Al Mamun, A., & Murray, F. J. (2018). Sustainable intensification of aquaculture value chains between Asia and Europe: A framework for understanding impacts and challenges. *Aquaculture*, *493*, 338–354. https://doi.org/10.1016/j.aquaculture.2017.12.033 (accessed on 10th July 2025).

Ministry of Marine Affairs and Fisheries. (2021). *Statistic of Marine and Fisheries Year 2021*. Pusdatin KKP. https://statistik.kkp.go.id/home.php (accessed on 10th July 2025).

Nunes, J. P., Ferreira, J. G., Bricker, S. B., O'Loan, B., Dabrowski, T., Dallaghan, B., Hawkins, A. J. S., O'Connor, B., & O'Carroll, T. (2011). Towards an ecosystem approach to aquaculture: Assessment of sustainable shellfish cultivation at different scales of space, time and complexity. *Aquaculture*, *315*(3, 4), 369–383. https://doi.org/10.1016/j.aquaculture.2011.02.048 (accessed on 10th July 2025).

Regulation of the Director General of Aquaculture Number 154/Per-DJPB/2019. https://jdih.kkp.go.id/Homedev/DetailPeraturan/6453 (accessed on 10th July 2025).

Rimmer, M. A., Sugama, K., Rakhmawati, D., Rofiq, R., & Habgood, R. H. (2013). A review and SWOT analysis of aquaculture development in Indonesia. *Reviews in Aquaculture*, *5*(4), 255–279. https://doi.org/10.1111/raq.12017 (accessed on 10th July 2025).

Soto, D., Aguilar-Manjarrez, J., Hishamunda, N., & Food and Agriculture Organization of the United Nations. (2008a). *Building an ecosystem approach to aquaculture: FAO/*

Universitat de les Illes Balears Expert Workshop, 7–11 May, 2007, Palma de Mallorca, Spain. Food and Agriculture Organization of the United Nations.

Soto, D., Aguilar-Manjarrez, J., Hishamunda, N., & Food and Agriculture Organization of the United Nations. (2008b). *Building an ecosystem approach to aquaculture: FAO/ Universitat de les Illes Balears Expert Workshop, 7–11 May, 2007, Palma de Mallorca, Spain*. Food and Agriculture Organization of the United Nations.

CHAPTER 19

Ecosystem-Based Aquaculture: Sustainable Practices for Thriving Aquatic Environments

BAMANIYA PINAK KAMLESHBHAI,[1,*] D. T. VAGHELA,[1]
ISHITA J. BAMBHANIYA,[2] and MAYUR BHADARKA[1]

[1]*Department of Aquatic Environment Management, College of Fisheries Science, Kamdhenu University, Veraval, Gujarat, India*

[2]*Department of Aquaculture, College of Fisheries Science, Kamdhenu University, Veraval, Gujarat, India*

*Corresponding author

ABSTRACT

Ecosystem-Based Aquaculture (EBA) marks a significant shift in aquaculture by integrating ecological principles to enhance sustainability and reduce environmental impacts. It acknowledges the complex relationship between aquaculture and the surrounding environment, aiming to mimic natural processes and utilize ecosystem services to support production while protecting biodiversity, water quality, and habitats. Key features of EBA include careful species selection, polyculture systems, and Integrated Multitrophic Aquaculture (IMTA). These approaches promote ecological balance and resilience, ensuring long-term productivity. Site selection and management are also critical, with attention to water flow, nutrient cycling, and habitat compatibility to create self-regulating aquaculture systems. Innovative practices like biofloc technology, mangrove restoration, and selective breeding for disease resistance are central to EBA. These not only boost production efficiency but also help restore degraded ecosystems and

Sustainable Aquaculture: Recent Advances and Techniques. Dhanaji W. Patil, Sandip S. Markad (Eds.)

DOI: 10.1201/9781779641113-19

conserve aquatic biodiversity. As global seafood demand rises, EBA offers a sustainable path forward by aligning aquaculture with environmental stewardship. By fostering a harmonious relationship between farming practices and natural ecosystems, EBA provides a resilient model that supports food security while safeguarding aquatic health and ecosystem integrity.

19.1 INTRODUCTION

19.1.1 DEFINITION AND CONCEPT

In 2006, the FAO Fisheries and Aquaculture Department acknowledged the necessity of crafting an ecosystem-centric management strategy for aquaculture to enhance the enforcement of the FAO Code of Conduct for Responsible Fisheries. FAO introduced the concept of an ecosystem approach to aquaculture (EAA), which is described as a method aimed at incorporating aquaculture into the broader ecosystem to foster sustainable growth, fairness, and the robustness of interconnected social-ecological systems (FAO, 2010).

Ecosystem-based aquaculture involves integrating environmental, social, and economic factors to promote sustainable development in aquaculture. It emphasizes the ecological significance of clean water as a primary resource and encourages practices that enhance production while minimizing reliance on ecosystem services, ensuring animal welfare, and addressing food security and waste issues. This approach underscores the importance of scientific rigor, adaptability to changing conditions, collaboration with stakeholders, and a sustained commitment to the well-being of ecosystems and communities. Additionally, it highlights the necessity of assessing the resilience of ecosystem services and conducting cost-benefit analyses in aquaculture management. Certification and sustainable seafood ranking initiatives support ecosystem services and the adoption of ecosystem-based aquaculture, although their criteria and evaluation methods may have limitations (Rector et al., 2021; Rusdi and Kodiran, 2022).

19.1.2 IMPORTANCE OF ECOSYSTEM-BASED APPROACH

The ecosystem-based approach (EBA) is crucial in aquaculture to ensure the sustainable management and conservation of marine resources. EBA takes into account the ecological, economic, societal, and cultural dynamics within marine ecosystems, aiming to maximize benefits while minimizing human impacts. Within aquaculture, EBA involves reducing human intervention, preserving biodiversity, and mitigating environmental effects (Burcu, 2023). It acknowledges the influence of human activities on aquaculture ecosystems and underscores the importance of scientific integrity, adaptability, collaboration, and long-term dedication to both ecosystems and societies (Rusdi and Kodiran, 2022). EBA also emphasizes the integration of environmental, social, and economic considerations into aquaculture management practices. Through the implementation of EBA, aquaculture can contribute essential ecosystem services, such as food provision, nutrient cycling, carbon capture, and biodiversity preservation (Zhou et al., 2022). To optimize the ecosystem services provided by mariculture, various approaches such as integrated multi-trophic aquaculture, assessments of carrying capacity, marine spatial planning, and waste treatment systems can be employed.

19.1.3 HISTORICAL CONTEXT AND DEVELOPMENT

Aquaculture boasts a rich historical trajectory, with evidence of its practice spanning millennia across various global regions. Ancient civilizations, such as those in *Hawai'i*, exemplified sustainable and integrated multi-trophic aquaculture methodologies (Dong, 2023). The advancement of aquaculture was propelled by the growing demand for seafood, alongside ornamental and religious purposes, further fueled by technological advancements during the industrial revolution. Consequently, aquaculture systems have diversified over time, encompassing extensive, semi-intensive, intensive, open, closed, and multi-trophic integrated approaches (Reto and Keller, 2022). The emergence of the ecosystem-based aquaculture paradigm has been pivotal in ensuring the sustainable progression of aquaculture, aiming to harmonize environmental, social, and economic dimensions (Rusdi and Kodiran, 2022). Spatial planning and zoning have gained prominence within aquaculture practices, particularly in light of climate change concerns, strategically aligning the location, type of aquaculture, and stakeholders' interests. Recognizing the ecological significance of clean water stands as a fundamental tenet for the advancement of ecologically sustainable aquaculture practices.

19.2 UNDERSTANDING AQUATIC ECOSYSTEMS

19.2.1 COMPONENTS OF AQUATIC ECOSYSTEMS

Aquatic ecosystems encompass a diverse array of elements, comprising plankton (both phytoplankton and zooplankton), macrophytes (large aquatic plants), and a plethora of animal species, ranging from fish and birds to insects, mammals, and various plants. Aquaculture ecosystems, functioning as a subset of industrial ecosystems, share similarities with natural aquatic ecosystems but exhibit human intervention and diminished biodiversity. Moreover, aquatic ecosystems serve as habitats for a diverse array of organisms, encompassing bacteria, fungi, algae, invertebrates, plants, and fish. The junction between terrestrial ecosystems and inland waters assumes significance in the global carbon cycle, where external organic matter influences microbial and higher-trophic-level processes. Classification of the aquatic environment hinges on parameters such as depth, water flow, salinity, and distinctive features, housing a myriad of organisms like algae, aquatic plants, and animals.

In the realm of ecosystem-based aquaculture, aquatic ecosystems comprise various elements. These elements encompass energy and material inputs, which are partly influenced by human activities, as well as efforts to reduce biodiversity to enhance production, the introduction of artificially stocked dominant organisms, and the potential adverse effects on the surrounding environment from wastewater discharge (Dong, 2023). Aquaculture ecosystems can be categorized based on their energy sources, metabolic traits, nutrient budgets, and ecological constraints. To comprehend the alterations induced by aquaculture systems in aquatic environments, mathematical modeling techniques such as hydrodynamic modeling, nutrient-mass balance modeling, bioenergetic modeling, and trophic and ecological modeling are employed (Moura and Silva, 2019). The Ecosystem Approach to Aquaculture underscores the imperative of preventing harm to ecosystem services, enhancing human well-being, and contextualizing aquaculture within broader sectors and policies. Ecological aquaculture systems integrate various components like settling ponds, ecological treatment ponds, wetland ponds, and aquaculture ponds to achieve cyclic purification and conserve water resources.

19.2.1.1 WATER QUALITY PARAMETERS

Water quality parameters comprise a range of physical, chemical, and biological characteristics employed to assess the state of water (Kim et al., 2010). These parameters encompass various factors, including temperature, pH, electrical conductivity (EC), hardness, chlorides, alkalinity, dissolved oxygen (DO), biochemical oxygen demand (BOD_5), chemical oxygen demand (COD), phosphate, and sulfate (Zangmo et al., 2021). Understanding water quality parameters is vital for gauging the integrity and safety of water sources and discerning the repercussions of pollution on aquatic ecosystems. Various methodologies have been devised to gauge and forecast water quality parameters, such as the modified Gauss-Newton method for parameter estimation and the online sequential extreme learning machine for prediction. These techniques aim to refine the precision and efficacy of water quality evaluation and prognostication, which are pivotal for adept water resource management. Moreover, the advent of water quality parameter spectrum analysis systems has facilitated swift and efficient monitoring of water quality using spectral data.

Water quality parameters relevant to ecosystem-based aquaculture include salinity, pH, dissolved oxygen, temperature, ammonia nitrogen (NH_3-N), chemical oxygen demand (COD), active phosphate (PO_4^{2-}), nitrate nitrogen (NO_3-N), nitrite nitrogen (NO_2-N), inorganic nitrogen, total nitrogen, total phosphorus, suspended solids (SS), and chlorophyll a (chl-a). Precise prediction of these parameters holds paramount importance for promptly identifying problematic areas and implementing effective water quality management strategies within aquaculture industries (Zhang et al., 2023). Various models and methodologies, including random forest regression, deep neural networks, multivariate linear regression, partial least squares, and discrete hidden Markov models, have been deployed to forecast and anticipate these water quality parameters (Li et al., 2022). These models have demonstrated promising outcomes in accurately predicting water quality parameters, empowering decision-makers to undertake preemptive measures and enhance water quality management.

19.2.1.2 BIOLOGICAL DIVERSITY

Aquaculture holds a pivotal role in biodiversity conservation and has the capacity to augment ecosystem functions and services. The correlation between biodiversity and function has been examined in integrated multi-trophic aquaculture (IMTA), revealing that species richness enhances the efficacy of macroalgal bioremediation of wastewater (Knoop et al., 2022). Notably, the diversity of aquaculture species surpasses that of farmed terrestrial species, with aquaculture endeavors playing a role in the preservation of endangered species and the safeguarding of habitats. Nonetheless, the intensification of aquaculture practices has resulted in a decrease in the genetic diversity of cultivated aquatic organisms (Ye et al., 2022). Traditional rice-fish coculture systems have been identified as effective in maintaining robust levels of genetic diversity within carp populations (Nie, 2022). Aquaculture activities have evolved beyond mere food production, encompassing endeavors such as conservation aquaculture, restorative aquaculture, and regenerative aquaculture (Reed, 2022). Collectively, aquaculture harbors the potential to foster biodiversity conservation and sustainable fish production.

Biological diversity stands as a cornerstone within ecosystem-based aquaculture. Aquaculture ecosystems often exhibit diminished biodiversity and are characterized by the introduction of artificially stocked dominant organisms. The Ecosystem Approach to Aquaculture underscores the imperative of safeguarding ecosystem services, enhancing human well-being, and considering intersectoral collaboration and policy integration (Tett, 2017). The microbial diversity present in shrimp culture ponds plays a pivotal role in sustaining ecosystem functionality and health (Hou et al., 2022). Preserving biodiversity is paramount for ensuring the long-term sustainability of natural resource utilization, with Nigeria's aquatic germplasm resources presenting opportunities for genetic enhancement and economic prosperity (Reed, 2022). Aquaculture ecosystems equipped with biological sand filter pools and control systems hold promise for enhancing water treatment and quality, integrating multiple processes and facilitating widespread application and adoption (Wu, 2019).

19.2.1.3 INTERACTIONS AMONG ORGANISMS

Interactions among organisms within ecosystem-based aquaculture are pivotal for the functionality and sustainability of aquatic environments. Biotic interactions, such as those between microbes and other organisms, significantly influence the dynamics of aquaculture systems (Bakker et al., 2022). Microbial populations, encompassing bacteria, fungi, protozoa, and microalgae, are ubiquitous in aquaculture settings and can exert both positive and negative effects on aquatic organisms (Midhun and Arun, 2023).

Leveraging beneficial interactions with microbes presents opportunities to enhance the productivity of fish and shellfish in aquaculture settings (Onianwah et al., 2018). Integrated multi-trophic aquaculture (IMTA) practices involve cultivating species from diverse trophic levels in close proximity, facilitating the recapture of excess nutrients and fostering synergistic relationships among species (Chopin et al., 2013). A comprehensive understanding of microbial community dynamics within aquaculture systems is imperative for preserving the health and performance of fish stocks (Yajima et al., 2023). In essence, the intricate interactions among organisms within ecosystem-based aquaculture systems are multifaceted and carry implications for the productivity and sustainability of aquatic ecosystems.

19.3 DYNAMICS OF AQUATIC ECOSYSTEMS

19.3.1 NUTRIENT CYCLING

Nutrient cycling serves as a cornerstone in the functioning of aquatic ecosystems, facilitating the movement of chemical elements within and among ecosystems and profoundly influencing community structure and ecosystem function. Animals, such as mollusks, play a pivotal role in regulating nutrient cycling in freshwater ecosystems, including the phosphorus (P) cycle in the Laurentian Great Lakes.

Consumer-driven nutrient dynamics (CND) play an additional role in nutrient cycling, where elements like body size, feeding habits, and abiotic features of the ecosystem shape how consumers affect nutrient dynamics (Vanni, 2021). Furthermore, nutrient cycling can represent a significant portion of the overall nutrient availability within food webs, supporting species survival in nutrient-deficient ecosystems while potentially destabilizing ecosystems with excessive nutrient inputs (Atkinson et al., 2017).

Comprehending and managing nutrient cycles in aquatic ecosystems, including aquaponics systems, is imperative for ensuring adequate nutrient availability for plant uptake (Quévreux et al., 2018). This understanding underscores the significance of maintaining balanced nutrient cycling for the health and sustainability of aquatic environments.

19.3.2 TROPHIC RELATIONSHIPS

Trophic relationships serve as fundamental pillars for comprehending the dynamics of aquatic ecosystems, encompassing the transfer of energy and matter among various species and trophic levels. Aquatic trophic networks exhibit dynamic and adaptable characteristics, influenced by both natural phenomena and human activities. To comprehensively assess and forecast the dynamics of trophic networks, a possibilistic model has been devised (Gaucherel et al., 2023). The feeding relationships within aquatic food webs are intricate, involving processes such as trophic positions, grazing and detritus chains, and trophic transfer efficiency, all of which play pivotal roles in energy flow (40). Predation on parasites by predators holds significant implications for community structure and disease risk, underscoring the necessity of studying host-pathogen dynamics within a broader community context (Buck et al., 2015). Contaminant and trophic dynamics within tropical stream ecosystems are shaped by land-use characteristics and trophic relationships, with the lipid content of consumers emerging as a superior predictor of contaminant concentration compared to trophic level. The comprehension of trophic relationships has evolved beyond solely maximizing catches to prioritizing the sustainability of the ecosystem itself (Wolanski et al., 2011). This evolution underscores the importance of holistic ecosystem management approaches in maintaining the health and functionality of aquatic ecosystems.

19.3.3 SUCCESSION AND STABILITY

Aquatic vegetation in fluvial systems frequently exhibits spatial patterns characterized by patches of plants. The stability and development of these patterns are influenced by several factors, such as flow dynamics, vegetation dynamics, and the presence of propagules. The growth rate and migration rate of these vegetation patches depend on factors such as wavenumber and pertinent flow and vegetation parameters (Calvani et al., 2022). Measuring the stabilizing influences in aquatic ecosystems requires evaluating the harmony between local and regional populations of species. Research has shown that stabilization achieved through asynchronous dynamics across different spatial locations is more important than compensatory dynamics resulting from fluctuations in biomass (Hodapp et al., 2023). To evaluate the resilience and stability of ecosystems, a simplified ecohydrological model incorporating plant population interactions and disturbances can be employed. Different plant phenotypes and their characteristics play crucial roles in the ecosystem's recovery following disturbances (Peña-Alzate and Cañón Barriga, 2017). The complexity of an ecosystem can significantly influence its stability, with constrained interaction structures aiding in the prevention of extinctions and rendering the ecosystem more robust (Pettersson et al., 2020).

19.4 PRINCIPLES OF ECOSYSTEM-BASED AQUACULTURE

19.4.1 INTEGRATION WITH NATURAL SYSTEMS

Integration with natural systems is a fundamental tenet of ecosystem-based aquaculture. However, aquaculture ecosystems often exhibit characteristics such as human manipulation of energy and material inputs, reduction of biodiversity, and the artificial stocking of dominant organisms. Regrettably, in fishery resource assessments and stock prediction models, the physical-chemical environment and the dynamics of fished populations are frequently overlooked. The line between aquaculture and capture fisheries is increasingly blurring, necessitating a reduction in reliance on natural ecosystem services within aquaculture systems. Embracing nature's patterns and strategies can offer sustainable solutions to aquaculture challenges, ranging from sex control in captive stocks to the operation of integrated multi-trophic aquaculture systems (Mustafa and Estim, 2018). Examining aquaculture systems through the lens of nature-based solutions is essential for addressing pressing issues such as climate change, biodiversity loss, and the preservation of endangered marine ecosystems (Le Gouvello et al., 2022). This holistic approach can pave the way for more environmentally friendly and resilient aquaculture practices.

19.4.2 BIODIVERSITY CONSERVATION

Aquaculture ecosystems often exhibit characteristics such as the reduction of biodiversity and the artificial stocking of dominant organisms. To mitigate adverse environmental impacts, strategies like integrated multitrophic aquaculture (IMTA) and recirculating aquaculture systems (RAS) have been developed. The concept of the ecosystem approach to aquaculture underscores the importance of aligning goals to minimize reliance on ecosystem services while maximizing productivity (Ming, 2022). This approach mandates that aquaculture operations should not harm ecosystem services and should be developed within the broader context of other sectors and policies (Tett, 2017). Similarly, the emergence of the ecosystem approach to fisheries (EAF) has introduced objectives for maintaining ecosystem health alongside those for fisheries. However, trade-offs and differences in interpretations still persist (Jennings et al., 2014). Overall, the principles of ecosystem-based aquaculture strive to strike a balance between the needs of human populations, ecosystem health, and economic viability while minimizing negative impacts on biodiversity and ecosystem services.

19.4.3 MINIMIZATION OF ENVIRONMENTAL IMPACTS

Aquaculture holds the potential to mitigate environmental impacts through the adoption of restorative practices and principles aimed at supporting the recovery of degraded ecosystems (Alleway et al., 2023). Aquaculture ecosystems typically exhibit characteristics such as controlled energy and material input, reduced biodiversity, artificially stocked organisms, and the potential for negative environmental impacts. Significant research efforts have been dedicated to effectively treating aquaculture effluent, leading to the implementation of technologies such as recirculating aquaculture systems (RAS) and various biological, physical, and chemical treatments (Ahmad et al., 2022). The Ecosystem Approach to Aquaculture emphasizes the imperative for aquaculture operations to refrain from harming ecosystem services, enhance human well-being, and be developed within the broader context of other sectors and policies (Tett, 2017). Sustainable aquaculture practices are geared toward preserving the natural resource base and minimizing anthropogenic impacts. Techniques such as polyculture, green technologies (including RAS, AWE, and Bioflocs), and site-selection employing GIS analysis are among the strategies employed (Olaganathan and Tang, 2017).

19.4.4 SUSTAINABLE RESOURCE USE

Sustainable resource utilization serves as a fundamental principle in ecosystem-based aquaculture, offering the potential to aid in the recovery of degraded ecosystems while addressing the growing demands for food and livelihoods (Alleway et al., 2023). However, aquaculture ecosystems possess characteristics that deviate from those of natural aquatic ecosystems, including reduced biodiversity and the artificial stocking of dominant organisms. To ensure sustainable resource utilization, aquaculture practices must refrain from harming ecosystem services, enhance human well-being, and be integrated within the broader context of other sectors and policies (Jayanthi et al., 2020). It is imperative to monitor and manage the environmental impacts of aquaculture diligently to safeguard profitability, fulfill environmental obligations, and maintain societal acceptance (Tett, 2017). Certification and sustainable seafood ranking programs play crucial roles in upholding ecosystem services and advocating for an ecosystem approach to aquaculture (EAA) (Rector et al., 2021). Nevertheless, there is a pressing need to enhance the representation of ecosystem services, resilience capacity, and cost-benefit analysis within sustainability scheme criteria to fully operationalize the EAA.

19.4.5 ADAPTIVE MANAGEMENT AND MONITORING

Recognizing the inherent uncertainty and complexity of socio-ecological systems, adaptive management and monitoring stand as crucial principles in ecosystem-based aquaculture, offering a structured method for learning and adapting management strategies as needed. This approach involves continuous engagement with stakeholders, incorporating socio-economic considerations into decision-making, and establishing a comprehensive monitoring framework (Negus et al., 2020). Despite the pivotal role monitoring plays in evaluating and guiding environmental management, it frequently encounters challenges such as limited diagnostic capabilities, insufficient long-term resources, and ambiguous objectives (Edmondson and Fanning, 2022). To address these shortcomings, a monitoring framework encompassing causative conceptual models, adaptive monitoring strategies, and management concepts can be implemented. This framework prioritizes monitoring based on ecological risk assessments, validates threats, refines conceptual understandings, and informs efficient management actions. By utilizing updated data and information, repeated risk assessments enable the identification of successful management practices and changes in threat priorities, facilitating adaptive monitoring and management approaches.

19.5 ECOSYSTEM-BASED AQUACULTURE TECHNIQUES

19.5.1 POLYCULTURE AND INTEGRATED MULTI-TROPHIC AQUACULTURE (IMTA)

Polyculture and integrated multi-trophic aquaculture (IMTA) are ecosystem-based aquaculture techniques aimed at mitigating negative environmental impacts and enhancing sustainability (Loayza-Aguilar et al., 2023). IMTA, in particular, involves cultivating species from various trophic levels within a single system, offering broader ecosystem benefits such as nutrient remediation and carbon sequestration (Papageorgiou et al., 2023). The co-cultivation of organisms in IMTA systems can effectively mitigate adverse effects on the marine environment, addressing concerns in both the water column and sediment (Martínez-Porchas et al., 2023). Moreover, IMTA holds promise in contributing to the mitigation of climate change-related processes, such as ocean acidification, by utilizing carbon removal mechanisms facilitated by bivalves (Hossain et al., 2022).

Beyond environmental benefits, IMTA also holds the potential to bolster social and economic sustainability in aquaculture. This includes diversifying products, increasing production rates, and enhancing competitiveness within the industry. However, despite its potential, the widespread adoption of IMTA faces challenges stemming from governance issues and inadequate institutional development within the aquaculture sector. Overcoming these hurdles is crucial for realizing the full potential of IMTA in promoting sustainable aquaculture practices.

19.5.2 USE OF NATURAL FILTERS AND BIOREMEDIATION

Ecosystem-based aquaculture methods, such as natural filters and bioremediation, are gaining attention as effective strategies to alleviate the negative impacts of aquaculture effluents on the environment. Natural filters, like polychaete-assisted sand filters (PASFs), have shown effectiveness in trapping particulate organic matter (POM) and transforming nutrients into valuable biomass, especially within integrated multi-trophic aquaculture (IMTA) setups (Rangaswamy and Singh, 2022). Bioremediation, leveraging live microbes, has proven effective in degrading complex pollutants and enhancing water quality within aquaculture systems (Breil et al., 2022; Martínez-Córdova et al., 2022).

Notably, microbes, including bacteria and microalgae, are harnessed for bioremediation in aquaculture effluents (Elchelwar et al., 2020). These techniques offer sustainable and cost-effective solutions for treating aquaculture wastes while bolstering the overall health of aquatic ecosystems (Jerónimo et al., 2020). Furthermore, the deployment of bioremediation agents, such as heterotrophic nitrifying and denitrifying microbes, holds promise for nitrogen compound removal, thereby contributing to the enhancement of pond ecosystem health. In summary, natural filters and bioremediation stand as effective strategies for curbing the environmental impact of aquaculture practices, offering considerable potential for sustainable and eco-friendly aquaculture management.

19.5.3 HABITAT ENHANCEMENT AND RESTORATION

Habitat enhancement and restoration techniques represent valuable approaches within ecosystem-based aquaculture, aiming to stabilize shorelines, improve fish habitat, and rejuvenate freshwater and estuarine ecosystems. Living shorelines (LS), for instance, have been shown to enhance critical coastal habitats for fish and mobile crustaceans, while the creation of shallow water habitats consistently boosts fish abundance, particularly for juvenile fish (Ferreira de Barros et al., 2023).

However, concerns have been raised regarding the use of plastic structures for habitat enhancement in freshwater systems, owing to limited evidence of ecological benefit and potential pollution from plastic degradation or leaching (Cooke et al., 2023). Aquaculture activities have the potential to contribute significantly to conservation and restoration objectives by facilitating species recovery, habitat restoration, bioremediation, and coastal defense measures (Radinger et al., 2023). Additionally, combining oyster reef restoration initiatives with floating oyster aquaculture technology can lead to the development of innovative ecosystems that improve water quality, reduce erosion, and rehabilitate ecosystem services at higher trophic levels (Overton et al., 2023).

19.5.4 LOW-IMPACT INFRASTRUCTURE AND DESIGN

Low-impact infrastructure and design play pivotal roles in ecosystem-based aquaculture techniques, aiming to enhance aquaculture system performance, improve environmental management, and align with sustainable development goals (Zhao et al., 2022). Aquaculture ecosystems, akin to their natural counterparts, are subject to human intervention, often resulting in reduced biodiversity as production maximization takes precedence. Traditional aquaculture practices, such as paddy-cum-fish culture, boast millennia-long histories and have demonstrated eco-friendly attributes, proving beneficial to rural communities. Furthermore, the establishment of oyster aquaculture in Alabama underscores the intricate interplay between infrastructure, written structures, and the natural environment, emphasizing the imperative of integrating ecological considerations into aquaculture practices. The integration of low-impact infrastructure and design principles into aquaculture systems holds significant potential for fostering sustainable practices and safeguarding ecosystem integrity. These approaches are essential for achieving long-term environmental and socio-economic sustainability within aquaculture operations.

19.5.5 SELECTIVE BREEDING AND GENETIC IMPROVEMENT

Selective breeding and genetic improvement represent crucial techniques within ecosystem-based aquaculture, aiming to enhance the genetic traits of farmed aquatic species, thereby improving production efficiency and resilience to environmental stressors (Sonesson et al., 2023). The application of genetic and genomic technologies, such as genome-wide association studies (GWAS) and genomic selection (GS), has transformed the process of identifying candidate genes and single nucleotide polymorphisms (SNPs) linked to desirable traits, such as heat tolerance and disease resistance.

Nonetheless, various challenges remain, such as the lack of comprehensive genetic information for many farmed species and the need for enhanced business models and financial mechanisms to encourage the widespread adoption of selective breeding in aquaculture (Houston et al., 2022). Despite these obstacles, the utilization of selective breeding and genetic enhancement presents significant potential for improving aquaculture sustainability and bolstering global food security. By harnessing the potential of these techniques, aquaculture can advance toward more resilient and productive systems, thereby contributing to the long-term viability of aquaculture operations and the global food supply.

19.6 CHALLENGES AND FUTURE DIRECTIONS

19.6.1 REGULATORY FRAMEWORKS AND POLICY SUPPORT

Regulatory frameworks and policy support are indispensable for fostering the sustainable development of ecosystem-based aquaculture and mitigating its environmental impact (Gutsulyak et al., 2023). Aquaculture ecosystems possess distinct characteristics, such as human control over energy and material input, biodiversity reduction, and potential environmental repercussions. The European Union's adoption of ecosystem-based fisheries management (EBFM) exemplifies a concerted effort to balance food security with marine ecosystem health, employing regulatory measures and sustainable fisheries partnership agreements (SFPAs) (Carvalho-Souza et al., 2021).

As aquaculture continues to expand, various approaches emerge, including commercial aquaculture, conservation aquaculture, restorative aquaculture, and regenerative aquaculture, necessitating clear definitions and a comprehensive understanding for effective regulation and management (Mizuta and Froehlich, 2022). The DPSIRM framework, when extended to include measures, presents a valuable tool for analyzing and managing fisheries resources, bridging the gap between scientific insights and policy formulation in the assessment and management process (Kang and Zhang, 2023). Robust regulatory frameworks and policy support are vital pillars for ensuring the sustainability and responsible growth of ecosystem-based aquaculture, promoting harmony between industry development and environmental conservation efforts.

19.6.2 TECHNOLOGICAL AND ECONOMIC CONSTRAINTS

Technological and economic constraints within ecosystem-based aquaculture encompass various factors, including insufficient knowledge regarding recommended fish farming practices, limited livelihood assets, and ineffective legislation and enforcement mechanisms (Mat Zain and Lee, 2022). Moreover, the integration of emergy synthesis into aquaculture systems encounters hurdles stemming from methodological disparities in unit emergy values for water and feed, challenges in accurately quantifying and categorizing water resources, and insufficient consideration of ecosystem services and disservices (David et al., 2021).

The growth and expansion of marine aquaculture face obstacles such as conflicts with other maritime activities, bureaucratic administrative procedures, and the imperative for enhanced management practices and technological advancements. Meanwhile, traditional aquaculture practices confront issues associated with water quality degradation, pollution, disease outbreaks, and the potential transmission of diseases to wild fish populations (Galparsoro et al., 2020).

To address these constraints effectively, measures such as the provision of extension and advisory services, widespread adoption of aquaculture technology, standardization and simplification of regulatory frameworks, enhanced governance structures, and active stakeholder engagement are imperative (Leow and Tan, 2020). Furthermore, the implementation of circular economies and integrated multi-trophic aquaculture holds promise for fostering sustainable production practices and environmental preservation. These approaches are essential for overcoming technological and economic barriers and advancing the sustainability of ecosystem-based aquaculture.

19.6.3 SOCIAL ACCEPTANCE AND STAKEHOLDER ENGAGEMENT

Stakeholder engagement stands as a pivotal element for the success of ecosystem-based aquaculture endeavors. Understanding the perceptions, expectations, and roles of diverse stakeholders within community-based projects is essential for achieving sustainable development outcomes. Effective stakeholder engagement necessitates a nuanced understanding of local socio-hydrologic conditions and the adaptation of engagement strategies to accommodate these variations (Smyth et al., 2021).

Purposeful engagement of stakeholders should be prioritized, with a focus on fostering meaningful and productive involvement through participatory processes (Galparsoro et al., 2020). Stakeholder participation offers valuable insights into both the environmental and human dimensions of marine management, thereby facilitating the implementation of ecosystem-based marine spatial planning initiatives (Baporikar, 2021). Furthermore, stakeholder engagement serves to garner broader support for aquatic ecosystem management endeavors, thereby ensuring the sustainable provision of ecosystem services (Flotemersch et al., 2019).

In order to advance the incorporation of ecologically and socially beneficial data into extensive water monitoring initiatives, surveys should take into account a range of factors, including local knowledge, proximity to the resource, demographic profiles, and reasons for participation. By integrating these aspects, stakeholders can be engaged more effectively, thus promoting the progress of ecosystem-based aquaculture initiatives.

19.6.4 RESEARCH NEEDS AND KNOWLEDGE GAPS

Research needs and knowledge gaps in ecosystem-based aquaculture encompass various facets critical for understanding and advancing sustainable practices in the field. These include:

1. **Understanding Ecological Requirements:** Research is needed to elucidate the ecological needs and requirements for successful restoration and sustainable aquaculture practices. This involves identifying crucial research areas, such as investigating the impact of certification and sustainable seafood ranking programs on the preservation of ecosystem services and the promotion of an Ecosystem Approach to Aquaculture (EAA).
2. **Integration of Restoration Practices:** There is a pressing need to explore the feasibility of integrating restoration efforts within aquaculture practices to engender positive environmental changes and optimize space utilization.
3. **Settlement Stage Dynamics:** Knowledge gaps persist regarding the settlement stage in aquaculture, encompassing aspects such as the timing, location, and materials for the placement of settlement plates. Addressing these gaps is crucial for both restoration success and the feasibility of aquaculture ventures.
4. **Resilience Capacity and Cost-Benefit Analysis:** Additional research is needed to evaluate the resilience of ecosystem services, identify methods for enhancing these services, and recognize attributes conducive to cost-benefit analysis of aquaculture practices. Understanding these dynamics is crucial for informing decision-making processes and advocating for sustainable aquaculture practices.

19.6.5 OPPORTUNITIES FOR SCALING UP AND MAINSTREAMING

Opportunities for scaling up and mainstreaming ecosystem-based aquaculture (EBA) are abundant, driven by several key factors:

1. **Ecosystem Approach to Aquaculture (EAA):** The adoption of EAA principles has heightened awareness of holistic and participatory approaches in aquaculture, steering the sector toward greater sustainability.
2. **Spatial Planning and Blue Growth Initiatives:** Efforts to implement the Ecosystem Approach to Aquaculture (EAA) emphasize spatial planning, which closely aligns with broader initiatives such as 'blue growth.' This convergence presents significant opportunities for the future expansion of ecosystem-based aquaculture (Kumar, 2022).
3. **Investment in Extractive Aquaculture:** Extractive aquaculture practices, including bivalve and seaweed farming, provide valuable ecosystem services such as nitrogen removal and habitat provision, which carry economic value. Investing in these practices can further promote sustainable aquaculture development (Brugere et al., 2019).
4. **Technological Advancements:** Technology plays a pivotal role in enhancing the sustainability of aquaculture farming. Innovative solutions addressing waste management issues and promoting ocean health preservation contribute to the mainstreaming of ecosystem-based aquaculture (Barrett et al., 2022).
5. **Eco-Tourism Opportunities:** Harnessing eco-tourism opportunities can act as a catalyst for the adoption of sustainable aquaculture technologies. Moreover, eco-tourism initiatives contribute to raising public awareness and education regarding the significance of sustainable practices within the aquaculture sector (Leow and Tan, 2020).

19.7 CONCLUSION

In conclusion, ecosystem-based aquaculture presents a promising approach to sustainable aquaculture practices, emphasizing the integration of ecological principles to enhance both production and environmental conservation. By mimicking natural ecosystems and utilizing techniques such as polyculture, habitat enhancement, and nutrient recycling, ecosystem-based aquaculture strives to minimize negative environmental impacts while promoting the health and resilience of aquatic ecosystems. This holistic approach not only benefits the environment but also supports the long-term viability and profitability of aquaculture operations. Embracing ecosystem-based principles can lead to thriving aquatic environments, resilient communities, and a more sustainable seafood industry for future generations.

KEYWORDS

- **adaptive monitoring**
- **aquaculture**
- **biodiversity**
- **bioremediation**
- **environmental conservation**
- **habitat**
- **restoration**
- **sustainable seafood industry**

REFERENCES

Ahmad, A. L., Chin, J. Y., Harun, M. H. Z. M., & Low, S. C. (2022). Environmental impacts and imperative technologies towards sustainable treatment of aquaculture wastewater: A review. *Journal of Water Process Engineering*, *46*, 102553.

Alleway, H. K., Waters, T. J., Brummett, R., Cai, J., Cao, L., Cayten, M. R., Costa-Pierce, B. A., Dong, Y. W., Brandstrup Hansen, S. C., Liu, S., & Liu, Q. (2023). Global principles for restorative aquaculture to foster aquaculture practices that benefit the environment. *Conservation Science and Practice*, *5*(8), e12982.

Anthony, R. (2022). Fish production and biodiversity conservation: An interplay for life sustenance. *Sustainable Development and Biodiversity*.

Atkinson, C. L., Capps, K. A., Rugenski, A. T., & Vanni, M. J. (2017). Consumer-driven nutrient dynamics in freshwater ecosystems: From individuals to ecosystems. *Biological Reviews*, *92*(4), 2003–2023.

Bakker, A. M., de Vries, J., Verdonschot, R. C., & Verdonschot, P. F. (2022). *Biotische Interacties in Aquatische Ecosystemen* (No. 2022-12). Stichting Toegepast Onderzoek Waterbeheer (STOWA).

Baporikar, N. (2021). Outlook on aquaculture community-based projects for socio-economic development: A case study. In *Behavioral-Based Interventions for Improving Public Policies* (pp. 155–175). IGI Global.

Barrett, L. T., Theuerkauf, S. J., Rose, J. M., Alleway, H. K., Bricker, S. B., Parker, M., Petrolia, D. R., & Jones, R. C. (2022). Sustainable growth of non-fed aquaculture can generate valuable ecosystem benefits. *Ecosystem Services*, *53*, 101396.

Beodeul, K., Chang, I., & Zhang, K. (2023). Stepwise analysis of the ecosystem-based fisheries assessment and management approach linked to the DPSIR framework. *Journal of Fisheries and Marine Sciences Education.*

Breil, P., Pons, M. N., Armani, G., Amer, R., Pienaar, H., Oberholster, P., & Namour, P. (2022). Natural-based solutions for bioremediation in water environment. In *Sustainable Solutions for Environmental Pollution: Air, Water and Soil Reclamation* (pp. 1–93).

Brugère, C., Aguilar-Manjarrez, J., Beveridge, M. C., & Soto, D. (2019). The ecosystem approach to aquaculture 10 years on–A critical review and consideration of its future role in blue growth. *Reviews in Aquaculture*, *11*(3), 493–514.

Buck, J. C., Scholz, K. I., Rohr, J. R., & Blaustein, A. R. (2015). Trophic dynamics in an aquatic community: Interactions among primary producers, grazers, and a pathogenic fungus. *Oecologia*, *178*, 239–248.

Calvani, G., Carbonari, C., & Solari, L. (2022). Stability analysis of submerged vegetation patterns in rivers. *Water Resources Research*, *58*(8).

Chopin, T., MacDonald, B., Robinson, S., Cross, S., Pearce, C., Knowler, D., Noce, A., Reid, G., Cooper, A., Speare, D., & Burridge, L. (2013). The Canadian Integrated Multi-Trophic Aquaculture Network (CIMTAN)—A network for a new ERA of ecosystem responsible aquaculture. *Fisheries*, *38*(7), 297–308.

Cooke, S. J., Piczak, M. L., Vermaire, J. C., & Kirkwood, A. E. (2023). On the troubling use of plastic 'habitat' structures for fish in freshwater ecosystems–Or–When restoration is just littering. *Facets*, *8*(1), 1–19.

David, L. H., Pinho, S. M., Agostinho, F., Kimpara, J. M., Keesman, K. J., & Garcia, F. (2021). Emergy synthesis for aquaculture: A review on its constraints and potentials. *Reviews in Aquaculture*, *13*(2), 1119–1138.

de Barros, M. S., Rodriguez, A. R., Bland, A., & Baker, R. (2023). Assessing habitat enhancement by living shoreline restoration: Exploring potential caveats of nekton community metrics. *Restoration Ecology*, *31*(5), e13935.

de Carvalho-Souza, G. F., Torres, M. Á., Farias, C., Acosta, J. J., Tornero, J., Sobrino, I., Ramos, F., & Llope, M. (2021). International politics must be considered together with climate and fisheries regulation as a driver of marine ecosystems. *Global Environmental Change*, *69*, 102288.

Dong, S. L., Gao, Q. F., & Li, L. (2023). Aquaculture ecosystem. In S. L. Dong, X. L. Tian, Q. F. Gao, & Y. W. Dong (Eds.), *Aquaculture Ecology*. Springer, Singapore.

Dong, Y. W. (2023). Aquaculture mapping in the context of climate change. In *Aquaculture Ecology* (pp. 473–490). Springer Nature Singapore.

Edmondson, E., & Fanning, L. (2022). Implementing adaptive management within a fisheries management context: A systematic literature review revealing gaps, challenges, and ways forward. *Sustainability*, *14*(12), 7249.

Elchelwar, V. R., Vasava, R. J., Deepak, A. P., Gora Shiva Prasad, C. S., Yusufzai, S. I., Shrivastava, V., & Prabhakar, P. (2020). Bioremediation: A tool for sustainable development of aquaculture. *Journal of Entomology and Zoology Studies*, *8*(2), 1331–1333.

Eze, E., Kirby, S., Attridge, J., & Ajmal, T. (2023). Aquaculture 4.0: Hybrid neural network multivariate water quality parameters forecasting model. *Scientific Reports*, *13*(1), 16129.

FAO. (2010). *Aquaculture Development. 4. Ecosystem Approach to Aquaculture.* FAO Technical Guidelines for Responsible Fisheries No. 5, Suppl. 4. Rome: FAO. Available from: http://www.fao.org/docrep/013/i1750e/i1750e00.htm (accessed on 10th July 2025).

Flotemersch, J. E., Shattuck, S. M., Aho, K. B., Cox, C. E., & Cairns, M. R. (2019). Factors influencing social demands of aquatic ecosystems. *Ecology and Society: A Journal of Integrative Science for Resilience and Sustainability*, *24*(4), 1.

Galparsoro, I., Murillas, A., Pinarbasi, K., Sequeira, A. M., Stelzenmüller, V., Borja, Á., O'Hagan, A. M., Boyd, A., Bricker, S., Garmendia, J. M., & Gimpel, A. (2020). Global stakeholder vision for ecosystem-based marine aquaculture expansion from coastal to offshore areas. *Reviews in Aquaculture*, *12*(4), 2061–2079.

Gaucherel, C., Fayolle, S., Savelli, R., Philippine, O., Pommereau, F., & Dupuy, C. (2023). Diagnosis of planktonic trophic network dynamics with sharp qualitative changes. *bioRxiv*, pp.2023-06.

Gutsulyak, S. A., Magzanova, D. K., & Sokolskaya, E. (2023). Regulatory and legal framework in the field of sustainable development and provision of aquaculture at the present stage. *Rybnoe hozâjstvo*. https://www.fishindustry.ru (accessed on 11th July 2025).

Hodapp, D., Armonies, W., Dannheim, J., Downing, J. A., Filstrup, C. T., & Hillebrand, H. (2023). Individual species and site dynamics are the main drivers of spatial scaling of stability in aquatic communities. *Frontiers in Ecology and Evolution*, *11*, 864534.

Hossain, A., Senff, P., & Glaser, M. (2022). Lessons for coastal applications of IMTA as a way towards sustainable development: A review. *Applied Sciences*, *12*(23), 11920.

Hou, D., Zhou, R., Wei, D., Zeng, S., Weng, S., Yan, Q., He, J., & Huang, Z. (2022). Abundant and rare microbial communities respectively contribute to an aquaculture pond ecosystem. *Frontiers in Marine Science*, *9*, 856126.

Houston, R. D., Kriaridou, C., & Robledo, D. (2022). Animal board invited review: Widespread adoption of genetic technologies is key to sustainable expansion of global aquaculture. *Animal*, *16*(10), 100642.

Jayanthi, M., Duraisamy, M., Thirumurthy, S., Samynathan, M., Kabiraj, S., Manimaran, K., & Muralidhar, M. (2020). Ecosystem characteristics and environmental regulations based geospatial planning for sustainable aquaculture development. *Land Degradation & Development*, *31*(16), 2430–2445.

Jennings, S., Smith, A. D., Fulton, E. A., & Smith, D. C. (2014). The ecosystem approach to fisheries: Management at the dynamic interface between biodiversity conservation and sustainable use. *Annals of the New York Academy of Sciences*, *1322*(1), 48–60.

Jerónimo, D., Lillebø, A. I., Santos, A., Cremades, J., & Calado, R. (2020). Performance of polychaete assisted sand filters under contrasting nutrient loads in an integrated multi-trophic aquaculture (IMTA) system. *Scientific Reports*, *10*(1), 20871.

Kim, T., & Sheng, Y. P. (2010). Estimation of water quality model parameters. *KSCE Journal of Civil Engineering*. https://doi.org/10.1007/s12205-010-0000-0 (accessed on 11th July 2025).

Knoop, J., Barrento, S., Lewis, R., Walter, B., & Griffin, J. N. (2022). Incorporating concepts of biodiversity into modern aquaculture: Macroalgal species richness enhances bioremediation efficiency in a lumpfish hatchery. *Algae*, *37*(3), 213–226.

Kumar, A. (2022). *Ecosystem-Based Adaptation: Approaches to Sustainable Management of Aquatic Resources*. Elsevier.

Le Gouvello, R., Brugere, C., & Simard, F. (2022). *Aquaculture and Nature-Based Solutions. Identifying Synergies Between Sustainable Development of Coastal Communities, Aquaculture, And Marine and Coastal Conservation*. Gland: IUCN.

Leow, B. T., & Tan, H. K. (2020). Technology-driven sustainable aquaculture for eco-tourism. In *WCFS2019: Proceedings of the World Conference on Floating Solutions* (pp. 209–218). Springer Singapore.

Li, D., Sun, Y., Sun, J., Wang, X., & Zhang, X. (2022). An advanced approach for the precise prediction of water quality using a discrete hidden Markov model. *Journal of Hydrology*, *609*, 127659.

Loayza-Aguilar, R. E., Huamancondor-Paz, Y. P., Saldaña-Rojas, G. B., & Olivos-Ramirez, G. E. (2023). Integrated multi-trophic aquaculture (IMTA): Strategic model for sustainable mariculture in Samanco Bay, Peru. *Frontiers in Marine Science*, *10*, 1151810.

Martínez-Córdova, L. R., Robles-Porchas, G. R., Vargas-Albores, F., Porchas-Cornejo, M. A., & Martínez-Porchas, M. (2022). Microbial bioremediation of aquaculture effluents. In *Microbial Biodegradation and Bioremediation* (pp. 409–417). Elsevier.

Martínez-Porchas, M., Vargas-Albores, F., Garibay-Valdez, E., & Martínez-Córdova, L. R. (2023). Editorial: Advances and perspectives in integrated multi-trophic aquaculture. *Frontiers in Marine Science*, *10*.

Midhun, S. J., & Arun, D. (2023). History and development of microbial technology in aquaculture. In *Recent Advances in Aquaculture Microbial Technology* (pp. 1–13). Academic Press.

Mizuta, D. D., Froehlich, H. E., & Wilson, J. R. (2023). The changing role and definitions of aquaculture for environmental purposes. *Reviews in Aquaculture*, *15*(1), 130–141.

Moura, R. S. T. D., & Henry-Silva, G. G. (2019). Food web and ecological models used to assess aquatic ecosystems submitted to aquaculture activities. *Ciência Rural*, *49*, e20180050.

Mustafa, S., & Estim, A. (2018). Adopt and adapt nature's design principles to create sustainable aquaculture systems. *Current Science*, *115*(6), 1195–1197.

Negus, P., Blessing, J., Clifford, S., & Marshall, J. (2020). Adaptive monitoring using causative conceptual models: Assessment of ecological integrity of aquatic ecosystems. *Australasian Journal of Environmental Management*, *27*(2), 224–240.

Nie, P. (2023). More and diverse contributions from aquaculture. *Reviews in Aquaculture*, *15*(1), 2–2.

Nur, S. M. Z., & Lai, K. L. (2022). Sustainable aquaculture technologies: Adoption readiness, perceived behaviour and determinant barriers among the fish farmers in Penang, Malaysia. *Global Conference on Business and Social Sciences*, *2*(53).

Olaganathan, R., & Kar Mun, A. T. (2017). Assessing the efficiency of different sustainable farming practices in reducing the environmental impacts caused by aquaculture. *International Journal of Advanced Biotechnology and Research*, *8*, 305–314.

Onianwah, I. F., Stanley, H. O., & Oyakhire, M. O. (2018). Microorganisms in sustainable aquaculture development. *Global Advanced Research Journal of Microbiology*, *7*(8), 127–131.

Overton, K., Dempster, T., Swearer, S. E., Morris, R. L., & Barrett, L. T. (2024). Achieving conservation and restoration outcomes through ecologically beneficial aquaculture. *Conservation Biology*, *38*(1), e14065.

Papageorgiou, N., Dimitriou, P. D., Chatzivasileiou, D., Tsapakis, M., & Karakassis, I. (2023). Can IMTA provide added ecosystem value services in the fish farms of Greece? *Frontiers in Marine Science*, *9*.

Peña-Alzate, S., & Cañón Barriga, J. E. (2017). Approaching the concepts of ecosystems resilience and stability through spatiotemporal system dynamics and agent-based modelling. *Revita Faulted de Ingeniería Universidad de Antioquia*, *84*, 84–96.

Pettersson, S., Savage, V. M., & Jacobi, M. N. (2020). Stability of ecosystems enhanced by species-interaction constraints. *Physical Review E*, *102*(6), 062405.

Phub, Z., Singh, D., Raza, Y., & Narayan, R. (2021). A review on water quality parameter. *Journal of Emerging Technologies and Innovative Research*.

Pingbing, M. (2022). *A Primer of Ecological Aquaculture*. https://doi.org/10.1093/oso/9780198850229.001.0001 (accessed on 11th July 2025)

Quévreux, P., Barot, S., & Thébault, É. (2018). Impact of nutrient cycling on food web stability. *bioRxiv*, 276592. https://doi.org/10.1101/276592 (accessed on 11th July 2025)

Radinger, J., Matern, S., Klefoth, T., Wolter, C., Feldhege, F., Monk, C. T., & Arlinghaus, R. (2023). Ecosystem-based management outperforms species-focused stocking for enhancing fish populations. *Science*, *379*(6635), 946–951.

Rangaswamy, B., & Singh, I. B. (2022). PCR-based fingerprinting techniques to transpire the microbial populations diversity of biofilm-based nitrifying packed bed bioreactor in the bioremediation of aquaculture wastewater. *Green Technology, Resilience, and Sustainability*, *2*(1), 5.

Rector, M. E., Filgueira, R., & Grant, J. (2021). Ecosystem services in salmon aquaculture sustainability schemes. *Ecosystem Services*, *52*, 101379.

Rusdi, R., & Kodiran, T. (2022). Sustainable management of whiteleg shrimp (*Litopenaeus vannamei*) aquaculture in Barru Regency, South Sulawesi Province based on EAA (Ecosystem Approach to Aquaculture). In *IOP Conference Series: Earth and Environmental Science* (*Vol. 1119*, No. 1, p. 012074). IOP Publishing. https://doi.org/10.1088/1755-1315/1119/1/012074 (accessed on 11th July 2025).

Smyth, R. L., Fatima, U., Segarra, M., Borre, L., Zilio, M. I., Reid, B., Pincetl, S., Astorga, A., Cisneros, M. A. H., Conde, D., & Harmon, T. (2021). Engaging stakeholders across a socio-environmentally diverse network of water research sites in North and South America. *Environmental Development*, *38*, 100582.

Sonesson, A. K., Hallerman, E., Humphries, F., Hilsdorf, A. W. S., Leskien, D., Rosendal, K., Bartley, D., Hu, X., Garcia Gomez, R., & Mair, G. C. (2023). Sustainable management and improvement of genetic resources for aquaculture. *Journal of the World Aquaculture Society*, *54*(2), 364–396.

Vanni, M. J. (2021). Invasive mussels regulate nutrient cycling in the largest freshwater ecosystem on Earth. *Proceedings of the National Academy of Sciences*, *118*(8), e2100275118.

Yajima, D., Fujita, H., Hayashi, I., Shima, G., Suzuki, K., & Toju, H. (2023). Core species and interactions prominent in fish-associated microbiome dynamics. *Microbiome*, *11*(1), 53.

Ye, Y., Ren, W., Zhang, S., Zhao, L., Tang, J., Hu, L., & Chen, X. (2022). Genetic diversity of fish in aquaculture and of common carp (*Cyprinus carpio*) in traditional rice–fish coculture. *Agriculture*, *12*(7), 997.

Zhang, Y., Jing, W., Deng, Y., Zhou, W., Yang, J., Li, Y., Cai, Y., Hu, Y., Peng, X., Lan, W., & Peng, M. (2023). Water quality parameters retrieval of coastal mariculture ponds based on UAV multispectral remote sensing. *Frontiers in Environmental Science*, *11*, 1079397.

Zhao, Y., Xue, B., Bi, C., Ren, X., & Liu, Y. (2023). Influence mechanisms of macro-infrastructure on micro-environments in the recirculating aquaculture system and biofloc technology system. *Reviews in Aquaculture*, *15*(3), 991–1009.

Zhou, X., Fang, J., Shi, S., Yang, L., Xie, B., & Huang, L. (2022). Ecosystem services of ecosystem approach to mariculture: Providing an unprecedented opportunity for the reform of China's sustainable aquaculture. *Frontiers in Marine Science*, *9*, 909231.

CHAPTER 20

Nutrition and Feeding of Striped Catfish, *Pangasianodon hypophthalmus*

JEBARSON SOLOMON[1] and TINCY VARGHESE[2*]

[1]*ICAR–Central Institute of Fisheries Education, Mumbai, Maharashtra, India*

[2]*ICAR – Central Institute of Fisheries Technology, Kochi, Kerala, India*

*Corresponding author

ABSTRACT

The striped catfish, *Pangasius* (*Pangasianodon hypophthalmus*), is a well-renowned fish species in aquaculture. Despite its low market value, it is a popular choice among farmers due to its fast growth rate, robust resilience, and year-round production. As feed plays a crucial role in the growth of fish and the expenditure of farming, formulating a complete and cost-effective diet tailored specifically for *Pangasius* helps enhance its growth, flesh quality, and farming yield. A well-balanced diet, with a gross energy content of approximately 4025 kcal, crude protein ranging from 35–40%, and lipids of 10.1%, along with essential amino acids, fatty acids, and other minor nutrients, is adequate for optimal growth of *Pangasius*. The existing nutrient requirement studies for *Pangasius* are limited, and the available data covers only major nutrients and a few minor nutrients. This chapter consolidates the available nutrient requirement data for *Pangasius*, offers practical feed formulation, and discusses recommended feeding practices for successful farming. The presented references and feed formulations, aligning with the known nutrient requirements of *Pangasius*, serve as a valuable guide for aquaculture professionals to improve feeding management during cultivation.

Sustainable Aquaculture: Recent Advances and Techniques. Dhanaji W. Patil, Sandip S. Markad (Eds.)

DOI: 10.1201/9781779641113-20

20.1 PRODUCTION SCENARIO OF FARMED PANGASIANODON HYPOPHTHALMUS

The striped catfish, *Pangasius* (*Pangasianodon hypophthalmus*), is an influential fish species, contributing approximately 5.1% of the total inland finfish culture production in 2020 (SOFIA, 2022). It is highly valued in many parts of Asia, including Vietnam, Singapore, Indonesia, Taiwan, Bangladesh, the Philippines, Malaysia, Nepal, and India, as an economically important aquaculture species due to its rapid growth, high productivity, and year-round production. *Pangasius* is known for its resilience and adaptability to various environmental conditions, such as temperature, salinity, turbidity, and dissolved oxygen levels (Chondar, 1999). It is suitable for both monoculture and polyculture, often being farmed with carp in open water bodies and cages. Worldwide, there is a steady growth in *Pangasius* production (Figure 20.1). Vietnam, the fish's native country, leads in both production and export, producing about 1.2 million tons with an export turnover of approximately $1.8 billion (Thuy et al., 2014). Bangladesh produces around 0.3 million tons annually (Hossain and Rahman, 2014). FAO and the World Bank consider this fish species significant and evolving in aquaculture in the Western hemisphere (McGee, 2014). Thailand, for instance, initiated cage culture of *Pangasius*, contributing to 5.4% of freshwater aquaculture production, with an annual net production of 0.25 million tons (Pongpat and Tongpool, 2013). In Malaysia, it is widely cultured in freshwater and brackish water cages, with an annual production of around 0.6 million tons (Idris et al., 2013). *Pangasius* is a crucial player in aquaculture trade, exported to over 130 countries (Phan et al., 2009). The United States, Russia, European Union, and Southeast Asia are major markets for this fish. Vietnam is the leading exporter, and the United States is the primary importer. This fish is primarily exported as white fillets and has become a significant whitefish product in the global market in the past decade.

20.2 FEEDING BEHAVIOR OF PANGASIANODON HYPOPHTHALMUS

In their natural habitat, *Pangasius* is omnivorous, feeding on a variety of food items including small fish, crustaceans, insects, and plant matter (Fishbase). In aquaculture settings, they are usually fed a diet consisting

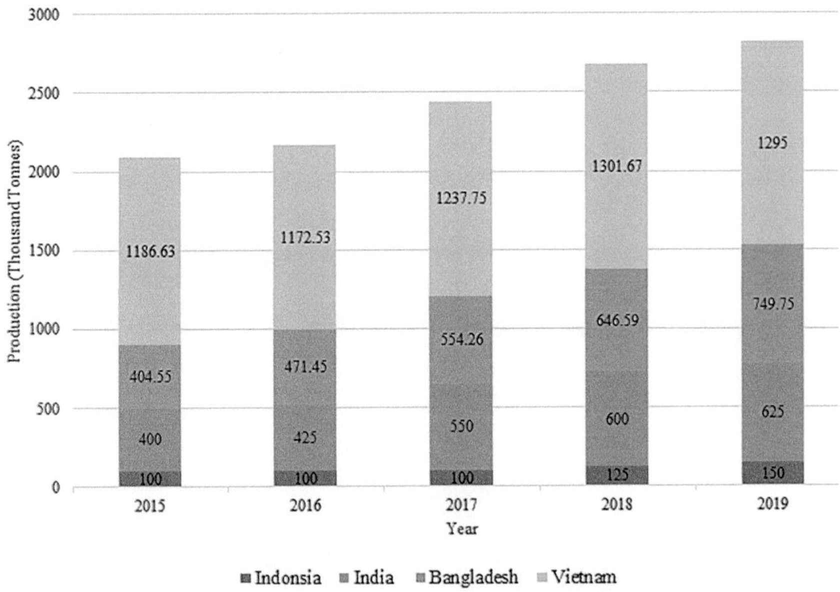

FIGURE 20.1 Production trends of *Pangasianodon hypophthalmus* in leading countries.
Source: Adapted from Platas-Rosado et al. (2021).

of commercial pellets or flakes designed for omnivorous fish. They are bottom-dwelling fish, often found near the substrate of rivers and lakes. Their downward-facing mouth is adapted for scavenging food from the bottom (FAO). They are more active during the night and dawn, which is when they tend to feed (Mukai et al., 2010). In captivity, they should be fed during these times for optimal feeding response.

20.3 ENERGY REQUIREMENTS OF PANGASIANODON HYPOPHTHALMUS

The energy content of the diet plays a crucial role in the growth and overall health of fish. The dietary energy supply should meet the energy requirements for maintaining vital cellular functions and productive purposes such as growth and reproduction (Kamalam et al., 2020) (Figure 20.2).

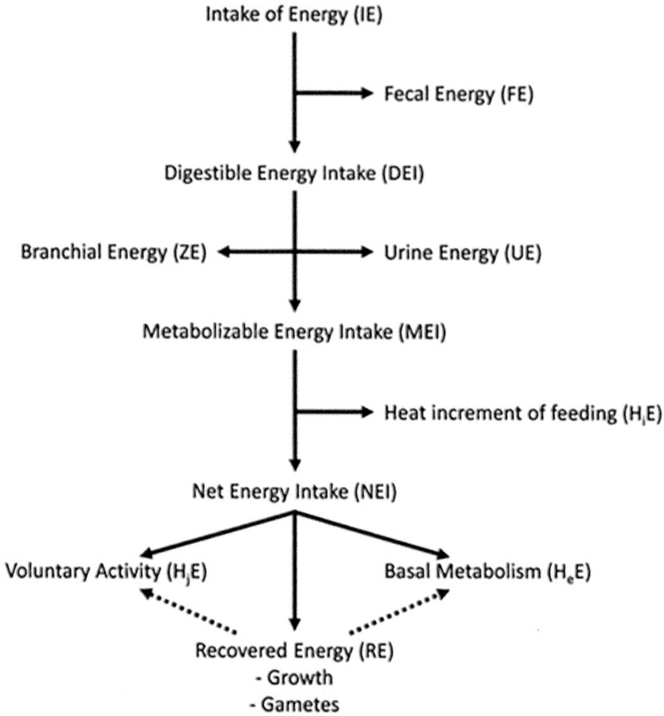

FIGURE 20.2 Energy budget: Schematic overview of the flow of energy through fish and shrimp.

Source: Adapted from National Research Council (NRC), (2011).

Fish derive energy from the breakdown and oxidation of proteins, fats, and carbohydrates from both dietary sources and body stores. The energy requirement of fish depends on the physiological state of development and environmental conditions (Xuemei, 2012). For optimal growth, the dietary energy requirement of *Pangasius* juveniles was estimated to be 4025 kcal per kg of diet at ambient temperature (24.5°C) (Ranjan et al., 2018). Furthermore, raising the environmental temperature to 32°C increases the energy requirement to 4100 kcal per kg of diet (Ranjan et al., 2018). However, the energy requirement of *Pangasius* during the grow-out stage and the detailed information on partitioned energy for this species are yet to be known. Further research is required to gain a better understanding of the energy partitioning of *Pangasius* and how it can be optimized to support its growth and overall health.

20.4 NUTRIENT REQUIREMENTS OF PANGASIANODON HYPOPHTHALMUS

Feeds formulated for *Pangasius* should provide all essential nutrients, including energy-yielding macronutrients (protein, lipid, carbohydrates, etc.), essential amino acids, essential fatty acids, as well as non-energy-yielding micronutrients such as vitamins and minerals. Currently, the knowledge of dietary requirements for essential nutrients in *Pangasius* is limited, and comprehensive studies are only available for its fingerling stage. Since the requirements for most essential nutrients in *Pangasius* are unknown, it is recommended to incorporate available data on requirements from related species, such as channel catfish, to formulate a complete diet with balanced nutrients for *Pangasius* (Sivaramakrishnan et al., 2017) (Table 20.1).

20.4.1 PROTEIN AND AMINO ACID REQUIREMENTS

Dietary protein is the most expensive and crucial nutrient for the growth and survival of fish (Lovell, 2013). Insufficient protein in the diet leads to a decrease or complete cessation of growth, while excessive protein results in the production of more ammonia and negatively impacts fish growth and voluntary feed intake (Yadata et al., 2020). Therefore, it is highly advised to supplement cultured fish with the optimal protein level in their diet. For *Pangasius* juveniles, the dietary protein requirement was found to be in the range of 35–40%. Hung et al. (2004) reported that *Pangasius* fingerlings require 35% dietary protein for optimal growth performance. A similar result was also reported by Jayant et al. (2018), concluding that *Pangasius* juveniles require 37.1% protein in their diet. Bano et al. (2023), in their study, found a slightly higher protein requirement of about 40% in the diet of *Pangasius* juveniles. Considering the market value of *Pangasius* and the economic success of farming, this fish species is often not fed up to its protein requirement by farmers. Insufficient protein intake hinders growth and reduces the feed conversion ratio (FCR). This problem can be overcome through two strategic feeding approaches: first, protein sparing with other energy-yielding nutrients like lipids and carbohydrates; and second, maintaining the optimal ratio of indispensable amino acids (IAAs) and dispensable amino acids (DAAs) in the diet (Kamalam et al., 2020). Unfortunately, quantitative data on the recommended dietary levels of

TABLE 20.1 Nutrient Requirements of *Pangasius* and Reference Data for Unknown Nutrients from Channel Catfish

Nutrient	Requirement	References
	Energy-Yielding Nutrient	
Protein (%)	**35**	Hung et al. (2004)
	37.1	Jayant et al. (2018)
	40	Bano et al. (2023)
	Essential Amino Acid (%)	
Lysine	**1.2**	Rachmawati and Nurhayati (2022)
Arginine	*1*	Robinson et al. (1981)
Histidine	*0.4*	Wilson et al. (1980)
Isoleucine	*0.6*	Wilson et al. (1980)
Leucine	*0.8*	Wilson et al. (1980)
Methionine	*0.6*	Harding et al. (1977)
Phenylalanine	*0.5*	Robinson et al. (1980)
Threonine	*0.5*	Wilson et al. (1978)
Tryptophan	*0.1*	Wilson et al. (1978)
Valine	*0.7*	Wilson et al. (1980)
Lipid (%)	**10.1**	Sivaramakrishnan et al. (2017)
	Essential Fatty Acids (%)	
n–3, HUFA (EPA:DHA, 1:2)	**1**	Kumar et al. (2014)
Carbohydrate (%)	NR	
	Non-Energy Yielding Nutrient	
	Vitamins	
	Water-Soluble Vitamins (mg/kg)	
Vitamin C	**46–76**	Daniel et al. (2018)
Vitamin B		
Thiamin	*1*	Murai and Andrews (1978a)
Riboflavin	*6–9*	Murai and Andrews (1978b); Serrini et al. (1996)
Pantothenic acid	*10–15*	Murai and Andrews (1979); Wilson et al. (1983)
Pyridoxine	*3*	Andrews and Murai (1979)
Niacin	*7.4–14*	Andrews and Murai (1978); Ng et al. (1997)
Biotin	R	
Folate	*1*	Duncan et al. (1993)
Cobalamin	R	
Choline	*400*	Wilson and Poe (1988)
Myoinositol	*NR*	Burtle and Lovell (1989)

TABLE 20.1 *(Continued)*

Nutrient	Requirement	References
	Fat-Soluble Vitamins	
Vitamin A (mg/kg)	*0.3–0.6*	Dupree, (1970)
Vitamin D (μg/kg)	*12.5–25*	Andrews et al. (1980); Lovell and Li (1978)
Vitamin E (mg/kg)	*25–50*	Murai and Andrews (1974); Wilson et al. (1984)
Vitamin K	R	
	Minerals (mg/kg)	
	Macrominerals	
Calcium	*1.5*	Andrews et al. (1973)
Phosphorus	*0.8*	Andrews et al. (1973)
Magnesium	*0.04*	Gatlin et al. (1982)
Sodium	*0.06*	NRC (2011)
Potassium	*0.26*	Wilson and Naggar (1992)
	Microminerals	
Zinc	**32**	Singh et al. (2017)
Copper	**1-2**	El-Sharawy et al. (2022)
Chromium	**2.82**	Akter et al. (2021)
Iodine	*1.1*	NRC (2011)
Manganese	*2.4*	Gatlin and Wilson (1984a)
Selenium	*0.25*	Gatlin and Wilson (1984b)
Iron	*30*	Gatlin and Wilson (1986)

Note: Values in bold – requirement data of *Pangasius*; Values in italics – reference data from channel catfish; NR – not required under practical conditions; R – required in diet but quantity not yet determined; HUFA – highly unsaturated fatty acid; EPA – eicosapentaenoic acid; DHA – docosahexaenoic acid.

many (9 out of 10) IAAs for *Pangasius* are currently unavailable. The only known requirement data of IAAs for *Pangasius* juveniles is lysine. Lysine is a critical amino acid that should be supplemented in the diet because of its deficiency in plant-based ingredients (cereals), and it is essential for various physiological functions, including growth and protein synthesis (Cai et al., 2018). The lysine requirement of *Pangasius* juveniles was found to be 1.2% (Rachmawati and Nurhayati, 2022). Apart from IAAs, there are conditionally essential amino acids (CEAA), which the body can synthesise but not in quantities sufficient to meet the physiological maintenance of the fish (e.g., taurine) (Li et al., 2009). Several freshwater fish species require taurine supplementation to achieve optimal growth performance, feed consumption, digestion, assimilation, immunity, and other physiological functions (Adeshina and Abdel-Tawwab, 2020). The dietary taurine requirement for *Pangasius* fry was found to be 1.5–1.63% for optimal growth (Peter et al., 2022).

20.4.2 LIPID AND FATTY ACID REQUIREMENTS

The exact lipid requirements for fish species cannot be precisely determined, as they depend on the levels of other energy-yielding nutrients in the diet, like protein and carbohydrates (Stickney and Hardy, 1989). Protein being costly, it is ideal to design a diet where protein meets only growth needs, and lipids and carbohydrates meet energy requirements. This strategy reduces feed costs and nitrogenous waste (Prabu et al., 2017). Sparing the protein by including lipid at high levels increases fat deposition, harming the flesh quality—a common issue in *Pangasius* (Phan et al., 2021). Hence, optimum lipid levels should be included in the diet. Liu et al. (2011) reported that dietary levels of 45% protein and 9% lipid are adequate for the best growth performance of *Pangasius*. Later, Sivaramakrishnan et al. (2017) determined the optimal dietary lipid requirement for *Pangasius* to be around 10.1%. The protein-sparing effect of lipid was also evident in *Pangasius*. Phumee et al. (2009) showed that raising lipid levels from 6% to 12% allows for a reduction of the protein requirement by up to 10%, from 40% to 30%, without compromising growth. Therefore, the recommended lipid incorporation in the *Pangasius* diet is 9–12%. This range depends on the lipid source and the presence of essential fatty acids (EFA). As a freshwater fish, *Pangasius* has the enzymatic mechanism to convert polyunsaturated fatty acids (PUFAs)—linolenic and linoleic acid—into highly unsaturated fatty acids (HUFAs) such as eicosapentaenoic acid (EPA), docosahexaenoic acid (DHA), and arachidonic acid. Hence, they require linolenic acid (18:3n-3) and linoleic acid (18:2n-6) in their diet (Sargent et al., 1999). So far, there is no specific EFA requirement study for *Pangasius*. Only Kumar et al. (2014) reported that supplementing the diet containing a protein-energy ratio of 133 mg/kg with 1% HUFAs (EPA:DHA, 1:2) enhances the growth and flesh quality of *Pangasius*. EFA (18:3n-3, 18:2n-6) requirements of various freshwater and diadromous fish species are in the range of 1–2%, and warmwater species require higher levels of 18:2n-6 than 18:3n-3 (National Research

Council [NRC], 2011). Therefore, it is recommended to provide a diet with a lipid level of 9–12%, including C18 PUFAs (18:3n-3 and 18:2n-6) at 1–2%, with a high level of 18:2n-6, to achieve optimal growth performance in *Pangasius*.

20.4.3 CARBOHYDRATES

Carbohydrates are widely used as a cost-effective source of dietary energy in animal feed (Wilson, 1994). However, excessive quantities of carbohydrates in fish feed can negatively impact growth and lead to metabolic issues due to the limited ability of fish to digest and metabolize carbohydrates. Fish do not have an intrinsic dietary requirement for carbohydrates; they can still survive and grow without them because of their ability to efficiently synthesize glucose from non-carbohydrate precursors (National Research Council [NRC], 2011). Despite this, carbohydrates are present in practical feeds, as they cannot be avoided when using plant ingredients. Feeding excessive carbohydrates to fish has been shown to reduce their growth rate due to poor feed utilization (Hemre et al., 2002). The ability of fish to utilize dietary carbohydrates varies depending on their species and habitat. Freshwater and warm-water fish have a higher capacity to utilize carbohydrates in their diet compared to cold-water or marine fish (Wilson, 1994). To reduce the cost of feed without affecting the growth of cultured fish, it is crucial to include the optimum level of carbohydrates in their diet. According to research, it has been recommended that up to 40% of carbohydrates can be included in the diet of *Pangasius* fingerlings without any negative impact on their growth or physiological health (Bhumarkar et al., 2017). Moreover, *Pangasius* fingerlings were found to have good starch utilization ability; they showed high growth performance at 10% and 20% starch in the diet (Hung et al., 2003). This indicates the greater potential of *Pangasius* to utilize carbohydrates in comparison to other cultured fish species.

20.4.4 VITAMIN REQUIREMENTS

Vitamins are essential organic compounds required in small quantities in the diet for various bodily functions, including metabolism, immune system, growth, and development (Godswill et al., 2020). Among water-soluble vitamins, the B complex vitamins (thiamine, riboflavin, niacin, pantothenic acid, pyridoxine, biotin, folate, cobalamin) are necessary in relatively small amounts and primarily serve as coenzymes. Three other water-soluble vitamins (choline, inositol, and vitamin C) are required in larger quantities and have various other roles beyond coenzymes (NRC, 2011). Fat-soluble vitamins mainly function independently of enzymes, with only a few having coenzyme roles (such as vitamin K) (NRC, 2011). The interplay between nutrients influences the dietary requirements for various vitamins; for example, pyridoxine requirements vary with dietary protein levels (Akhtar and Ciji, 2020). If conditions are met, some dietary essential vitamins are synthesized from other nutrients (e.g., choline synthesis, which occurs if enough methyl donors like methionine are present in the diet) (Case et al., 1976). In warm-water fishes, some water-soluble vitamins can be sourced from microorganisms in the gastrointestinal tract (Limsuwan and Lovell, 1981). Excessive supplementation of vitamins also results in adverse effects (hypervitaminosis) (Elango et al., 2015). Therefore, formulating a diet with proper vitamin supplementation is essential. In practical situations, vitamins are fortified slightly higher than required to compensate for possible losses during feed manufacturing, storage, and leaching (NRC, 2011). In the case of *Pangasius*, specific requirement data for many vitamins are limited. Al-Noor et al. (2014) reported that *Pangasius* requires vitamin E (alpha-tocopherol acetate) between 90 and 300 mg/kg. The vitamin C (L-ascorbyl 2-polyphosphate) requirement for *Pangasius* was 46 to 76 mg/kg (Daniel et al., 2018). Under stressful conditions (heavy metal [arsenic] and thermal stress), the riboflavin requirement for *Pangasius* was found to be 10 mg/kg (Kumar, 2021). However, details regarding the requirements of other vitamins for *Pangasius* are currently unknown.

20.4.5 MINERAL REQUIREMENTS

In an aquaculture setting, fish acquire essential minerals from both the water they inhabit and through feed (Lall, 2022). The possibility of leaching of minerals from feed to the surrounding water is high; hence, determining the source of the minerals entering the fish body and participating in metabolic processes is challenging. This complexity makes it difficult for researchers to conduct studies on the mineral requirements of fish, leading to limited information in this area. Furthermore, several factors complicate the assessment of mineral needs in fish. These include interactions between minerals themselves and with other nutrients, the choice of ingredients in the diet as mineral content varies with ingredients, and the form in which minerals are supplied (organic, inorganic, chelated with ligands, or as nanoparticles) (Bell and Cowey, 1989; Kamunde and MacPhail, 2011; Antony et al., 2016). The major minerals include sodium, chlorine, calcium, magnesium, potassium, and phosphorus and are required in higher concentrations in fish diets. These minerals play crucial roles in forming skeletal structures, hard tissues (such as fin rays, scales, teeth, and exoskeleton), electron transfer, regulation of acid-base equilibrium, production of membrane potentials, and osmoregulation. Trace minerals are needed in smaller quantities, which include chromium, copper, iodine, iron, manganese, selenium, and zinc (NRC, 2011). For warm-water omnivorous fish species like *Pangasius*, diets are designed with more plant ingredients to reduce cost, limiting the availability of various minerals (Daniel, 2018). Therefore, formulating an optimized diet that meets all mineral requirements is essential. Only for a few minerals are the requirement data available for *Pangasius*. Akter et al. (2021) determined that *Pangasius* requires 2.82 mg/kg of chromium in its diet. The dietary zinc requirement was found to be 32 mg/kg (Singh et al., 2017). In another study, when zinc was supplemented in the form of a zinc amino acid complex, the requirement was found to be 50 mg/kg, and when augmented as a proportion of ZnSO4:ZnAA, the requirement was increased to 100 mg/kg (Jintasataporn et al., 2015). Selenium supplementation in the form of nanoparticles showed a requirement range of 1.0 to 1.11 mg/kg (El-Sharawy et al., 2021). Growth performance was enhanced in *Pangasius* when supplemented with copper in the range of 1–2 mg/kg (El-Sharawy et al., 2022). Careful consideration is necessary to formulate an optimal diet for *Pangasius*, considering the essentiality of minerals for growth and well-being.

20.5 PRACTICAL FEED FORMULATION FOR PANGASIANODON HYPOPHTHALMUS

Feed formulation begins by determining the levels of essential nutrients to be present in the feed, based on the requirements of the target fish species. Ingredients are then selected considering their availability, cost, and nutrient content. The quantity of each ingredient used is then determined, considering its nutrient content and the fish's ability to utilize it. Regarding the essential nutrient level, the protein requirement for *Pangasius* is typically between 35–40% (Bano et al., 2023; Hung et al., 2004; Jayant et al., 2018). However, because of its low market value, a protein level of less than the requirement is often provided in the diet, ranging from 26–30% (Seafish). To further reduce the feed cost, plant ingredients are used over 90%, and fish meal is included only around 5% (Seafish). When reducing fishmeal, it is important to supplement the diet with limiting amino acids, vitamins, and minerals based on the specific nutrient content of other ingredients and the requirements of *Pangasius* (Daniel, 2018). In terms of lipid sources, research indicates that a diet with vegetable oils, particularly crude palm oil and soybean oil, leads to better growth in *Pangasius* compared to a fish oil diet (Asdari et al., 2011). Therefore, it is recommended to replace fish oil with crude palm oil and soybean oil to reduce the feed price. Carbohydrate sources like rice bran, wheat bran, and corn starch are incorporated into feed formulations to enhance stability, floatability, and pellet binding (Kamalam et al., 2017; Patil et al., 2019). This careful balance of nutrients and ingredients aims to optimize the economic efficiency of *Pangasius* farming (Table 20.2).

TABLE 20.2 Generalized Feed Formulations for *Pangasianodon hypophthalmus*

Ingredient	Inclusion (%)
Fish meal	5
Plant protein (soybean meal and/or other oilseed meals, e.g., canola, sunflower, etc.)	44
Wheat flour	15
Rice bran	20
Corn flour	5
Vegetable oils (crude palm oil and/or soybean oil)	8
Others (e.g., vitamins, minerals, antioxidants, inorganics, etc.)	3
Total	100

20.6 FEEDING PRACTICES

The purpose of developing a nutritionally complete diet will only be satisfied if it is delivered to the cultured fish according to its feeding behavior and needs. Several key factors must be considered to enhance the delivery of the formulated feed, including feed size, feeding level (ration size), feeding frequency, and feeding time (Kamalam et al., 2020). The size of the feed pellet should be determined based on the fish's mouth size, ensuring it is easy for them to grab and swallow. A mismatch between feed size and mouth size leads to unnecessary wastage of feed (Lovell, 2003). The feeding level and frequency also depend on the fish's size and the water temperature in which they are raised (Lall and Tibbetts, 2009). Smaller fish and those cultured in cooler temperatures require less feed than larger fish and those in warmer temperatures. Due to their limited body reserves, small fish require more frequent feeding than larger fish (Craig et al., 2009). The next crucial factor in effective feed delivery is feeding time, which is determined mainly by the behavioral nature of the fish. Socially active fish tend to feed more during the daytime, whereas shy fish prefer feeding during the night or in low-light conditions. In the case of *Pangasius*, which is known for its shyness, studies have shown higher growth and survival rates when cultured in dark or dim light conditions compared to daytime (Mukai et al., 2010). Therefore, feeding *Pangasius* more during nighttime is recommended for overall high production and farming success (Table 20.3).

TABLE 20.3 Recommended Feeding Level and Frequencies for Culturing *Pangasius* During the Grow-Out Phase

Culture Months and Fish Sizes	Feeding Rate (% bW)	Feeding Frequency (time/day)
Month 1st and 2nd (<100 g)	5–6	3
Month 3rd to 5th (>100 to 800 g)	2–3	2
Month 6th and 7th (800 g to harvest)	1.5–2	1

Source: Nguyen (2013).

20.7 CONCLUSION AND FUTURE TRENDS

In conclusion, aquaculturists can optimize the feed cost and feed efficiency of *Pangasius* by carefully selecting ingredients and formulating a feed that meets the energy, protein, lipid, carbohydrate, vitamin, and mineral requirements. Although there is a lack of comprehensive data on nutrient requirements, especially during the crucial grow-out stage, leveraging available knowledge from the fingerling stage and related species can provide valuable insights that should be considered in feed formulation. Recognizing the existing data gaps, further research into the specific nutrient requirements of *Pangasius*, particularly during the grow-out stage, is essential. This research would not only contribute to a more accurate understanding of the species' nutritional needs but also facilitate the development of cost-effective and customized feeds to enhance production outcomes.

KEYWORDS

- **carbohydrate**
- **feed efficiency**
- **feeding behavior**
- **feeding practices**
- **lipid**
- **nutrient requirements**
- **nutrition**
- ***Pangasianodon hypophthalmus***
- **Pangasius**
- **protein**

REFERENCES

Adeshina, I., & Abdel-Tawwab, M. (2020). Dietary taurine incorporation to high plant protein-based diets improved growth, biochemical, immunity, and antioxidants biomarkers of African catfish, *Clarias gariepinus* (B.). *Fish Physiology and Biochemistry*, *46*(4), 1323–1335.

Akhtar, M. S., & Ciji, A. (2020). Pyridoxine and its biological functions in fish: Current knowledge and perspectives in aquaculture. *Reviews in Fisheries Science & Aquaculture*, *29*(2), 260–278.

Akter, S., Jahan, N., Rohani, M. F., Akter, Y., & Shahjahan, M. (2021). Chromium supplementation in diet enhances growth and feed utilization of striped catfish (*Pangasianodon hypophthalmus*). *Biological Trace Element Research*, *199*(12), 4811–4819.

Al-Noor, S. M., Hossain, M. D., & Islam, M. A. (2014). The study of fillet proximate composition, growth performance and survival rate of striped catfish (*Pangasius hypophthalmus*) fed with diets containing different amounts of alpha-tocopherol (Vitamin-E). *Journal of Bio-Science, 20*, 67–74.

Andrews, J. W., & Murai, T. (1978). Dietary niacin requirements for channel catfish. *The Journal of Nutrition, 108*(9), 1508–1511.

Andrews, J. W., & Murai, T. (1979). Pyridoxine requirements of channel catfish. *The Journal of Nutrition, 109*(4), 533–537.

Andrews, J. W., Murai, T., & Campbell, C. (1973). Effects of dietary calcium and phosphorus on growth, food conversion, bone ash and hematocrit levels of catfish. *The Journal of Nutrition, 103*(5), 766–771.

Andrews, J. W., Murai, T., & Page, J. W. (1980). Effects of dietary cholecalciferol and ergocalciferol on catfish. *Aquaculture, 19*(1), 49–54.

Antony Jesu Prabhu, P., Schrama, J. W., & Kaushik, S. J. (2016). Mineral requirements of fish: A systematic review. *Reviews in Aquaculture, 8*(2), 172–219.

Asdari, R., Aliyu-Paiko, M., Hashim, R., & Ramachandran, S. (2011). Effects of different dietary lipid sources in the diet for *Pangasius hypophthalmus* (Sauvage, 1878) juvenile on growth performance, nutrient utilization, body indices and muscle and liver fatty acid composition. *Aquaculture Nutrition, 17*(1), 44–53.

Bano, S., Khan, N., Arslan, M., Fatima, M., Khalique, A., & Wan, A. H. L. (2023). Impact of various dietary protein levels on the growth performance, nutrient profile, and digestive enzymes activities to provide an effective diet for striped catfish (*Pangasius hypophthalmus*). *Turkish Journal of Fisheries and Aquatic Sciences, 23*(7).

Bell, J. G., & Cowey, C. B. (1989). Digestibility and bioavailability of dietary selenium from fishmeal, selenite, selenomethionine and selenocystine in Atlantic salmon (*Salmo salar*). *Aquaculture, 81*(1), 61–68.

Bhumarkar, R., Sahoo, S., Kumar, S., Aklakur, M., & Sahu, N. P. (2017). Feeding of different level of digestible carbohydrate to *Pangasianodon hypophthalmus* fingerlings: Effect on growth, amylase activity and gene expression. *Cellulose, 28*, 8.

Burtle, G. J., & Lovell, R. T. (1989). Lack of response of channel catfish (*Ictalurus punctatus*) to dietary myo-inositol. *Canadian Journal of Fisheries and Aquatic Sciences, 46*(2), 218–222.

Cai, W. C., Liu, W. B., Jiang, G. Z., Wang, K. Z., Sun, C. X., & Li, X. F. (2018). Lysine supplement benefits the growth performance, protein synthesis, and muscle development of *Megalobrama amblycephala* fed diets with fish meal replaced by rice protein concentrate. *Fish Physiology and Biochemistry, 44*(4), 1159–1174.

Case, G. L., Mitchell, A. D., Harper, A. E., & Benevenga, N. J. (1976). Significance of choline synthesis in the oxidation of the methionine methyl group in rats. *The Journal of Nutrition, 106*(6), 735–746.

Chondar, S. L. (1999). *Biology of Finfish and Shellfish.* SCSC Publishers (India), West Bengal.

Craig, S., Helfrich, L. A., Kuhn, D., & Schwarz, M. (2009). *Understanding Fish Nutrition, Feeds, and Feeding.* Virginia Cooperative Extension. https://www.pubs.ext.vt.edu (accessed on 11th July 2025).

Daniel, N. (2018). A review on replacing fish meal in aqua feeds using plant protein sources. *International Journal of Fisheries and Aquatic Studies, 6*(2), 164–179.

Daniel, N., Muralidhar, A. P., Srivastava, P. P., Jain, K. K., Pani Prasad, K., Manish, J., & Sivaramakrishnan, T. (2018). Dietary ascorbic acid requirement for growth of striped catfish, *Pangasianodon hypophthalmus* (Sauvage, 1878) juveniles. *Aquaculture Nutrition, 24*(1), 616–624.

Duncan, P. L., Lovell, R. T., Butterworth, C., Freeberg, L. E., & Tamura, T. (1993). Dietary folate requirement determined for channel catfish, *Ictalurus punctatus*. *The Journal of Nutrition, 123*(11), 1888–1897.

Dupree, H. K. (1970). Dietary requirement of vitamin A acetate and beta carotene. *Progress in Sport Fishery Research*, 148–150.

Elango, G., Venkataraman, D. D., Rao, S. V., & Kiran, V. R. (2015). Hypervitaminosis. *International Journal of Biomedical Research, 6*(3), 151–154.

El-Sharawy, M. E., Hamouda, M., Soliman, A. A., Amer, A. A., El-Zayat, A. M., Sewilam, H., Younis, E. M., Abdel-Warith, A. W. A., & Dawood, M. A. O. (2021). Selenium nanoparticles are required for the optimum growth behavior, antioxidative capacity, and liver wellbeing of striped catfish (*Pangasianodon hypophthalmus*). *Saudi Journal of Biological Sciences, 28*(12), 7241–7247.

El-Sharawy, M. E., Mahmoud, A. A., Soliman, A. A., Amer, A. A., Mahmoud, M. A., Alkafafy, M., & Dawood, M. A. O. (2022). Studying the influence of copper on the growth behavior, antioxidative status, and histology of the intestine and liver of striped catfish (*Pangasianodon hypophthalmus*). *Biological Trace Element Research, 200*(3), 1331–1338.

FAO. (2022). *The State of World Fisheries and Aquaculture 2022: Towards Blue Transformation*. Rome, FAO. https://doi.org/10.4060/cc0461en (accessed on 11th July 2025).

FAO. (n.d.). *Pangasianodon hypophthalmus* (Sauvage, 1878) [Pangasiidae]. https://www.fao.org (accessed on 11th July 2025).

FishBase. (n.d.). *Pangasianodon hypophthalmus*, striped catfish: Fisheries, aquaculture, aquarium. https://www.fishbase.se/summary/Pangasianodon-hypophthalmus (accessed on 11th July 2025).

Gatlin, D. M., & Wilson, R. P. (1984a). Studies on the manganese requirement of fingerling channel catfish. *Aquaculture, 41*(2), 85–92.

Gatlin, D. M., & Wilson, R. P. (1984b). Dietary selenium requirement of fingerling channel catfish. *The Journal of Nutrition, 114*(3), 627–633.

Gatlin, D. M., & Wilson, R. P. (1986). Characterization of iron deficiency and the dietary iron requirement of fingerling channel catfish. *Aquaculture, 52*(3), 191–198.

Gatlin, D. M., Robinson, E. H., Poe, W. E., & Wilson, R. P. (1982). Magnesium requirement of fingerling channel catfish and signs of magnesium deficiency. *The Journal of Nutrition, 112*(6), 1182–1187.

Godswill, A. G., Somtochukwu, I. V., Ikechukwu, A. O., & Kate, E. C. (2020). Health benefits of micronutrients (vitamins and minerals) and their associated deficiency diseases: A systematic review. *International Journal of Food Sciences, 3*(1), 1–32.

Harding, D. E., Allen, O. W., & Wilson, R. P. (1977). Sulfur amino acid requirement of channel catfish: L-methionine and L-cystine. *The Journal of Nutrition, 107*(11), 2031–2035.

Hemre, G. I., Mommsen, T. P., & Krogdahl, Å. (2002). Carbohydrates in fish nutrition: Effects on growth, glucose metabolism and hepatic enzymes. *Aquaculture Nutrition, 8*(3), 175–194.

Hossain, M. R. A., & Rahman, B. S. (2014). Thai pangas (*Pangasius sutchi*) in Bangladesh—A review. *International Journal of Business and Social Science Research, 1*(2), 98–106.

Hung, L. T., Lazard, J., Mariojouls, C., & Moreau, Y. (2003). Comparison of starch utilization in fingerlings of two Asian catfishes from the Mekong River (*Pangasius bocourti* Sauvage, 1880, *Pangasius hypophthalmus* Sauvage, 1878). *Aquaculture Nutrition, 9*(4), 215–222.

Hung, L. T., Suhenda, N., Slembrouck, J., Lazard, J., & Moreau, Y. (2004). Comparison of dietary protein and energy utilization in three Asian catfishes (*Pangasius bocourti, P. hypophthalmus* and *P. djambal*). *Aquaculture Nutrition, 10*(5), 317–326.

Idris, K., Azman, A., D'Silva, J. L., Man, N., & Shaffril, H. A. M. (2013). Benefits and problems in caged rearing of *Lates calcarifer, Oreochromis niloticus* and *Pangasius sutchi*: The case of entrepreneurs in Pekan, Pahang. *Life Science Journal, 10*(4).

Jayant, M., Muralidhar, A. P., Sahu, N. P., Jain, K. K., Pal, A. K., & Srivastava, P. P. (2018). Protein requirement of juvenile striped catfish, *Pangasianodon hypophthalmus. Aquaculture International, 26*(1), 375–389.

Jintasataporn, O., Ward, T., & Kattakdad, S. (2015). The efficacy of organic zinc amino acid complex (AvailaZn®) on growth performance and immunity of Pangasius catfish (*Pangasianodon hypophthalmus*). *Aquacultura Indonesiana, 15*(2).

Kamalam, B. S., Medale, F., & Panserat, S. (2017). Utilisation of dietary carbohydrates in farmed fishes: New insights on influencing factors, biological limitations and future strategies. *Aquaculture, 467*, 3–27.

Kamalam, B. S., Rajesh, M., & Kaushik, S. (2020). Nutrition and feeding of rainbow trout (*Oncorhynchus mykiss*). In *Fish Nutrition and Its Relevance to Human Health* (pp. 299–332).

Kamunde, C., & MacPhail, R. (2011). Metal–metal interactions of dietary cadmium, copper and zinc in rainbow trout, *Oncorhynchus mykiss. Ecotoxicology and Environmental Safety, 74*(4), 658–667.

Kumar, N. (2021). Dietary riboflavin enhances immunity and anti-oxidative status against arsenic and high temperature in *Pangasianodon hypophthalmus. Aquaculture, 533*, 736209.

Kumar, S., Munilkumar, S., Gupta, S. K., Jain, K. K., Pal, A. K., G. M., S., & Prakasha, C. (2014). Protein/energy ratio and HUFA content in the diet of *Pangasianodon hypophthalmus* (Sauvage, 1878) fingerlings: Effect on growth and flesh quality. *Israeli Journal of Aquaculture–Bamidgeh, 65*.

Lall, S. P. (2022). Chapter 6–The minerals. In R. W. Hardy & S. J. Kaushik (Eds.), *Fish Nutrition* (Fourth Edition).

Lall, S. P., & Tibbetts, S. M. (2009). Nutrition, feeding, and behavior of fish. *Veterinary Clinics of North America: Exotic Animal Practice, 12*(2), 361–372.

Li, P., Mai, K., Trushenski, J., & Wu, G. (2009). New developments in fish amino acid nutrition: Towards functional and environmentally oriented aquafeeds. *Amino Acids, 37*(1), 43–53.

Limsuwan, T., & Lovell, R. T. (1981). Intestinal synthesis and absorption of vitamin B-12 in channel catfish. *The Journal of Nutrition, 111*(12), 2125–2132.

Liu, X. Y., Wang, Y., & Ji, W. X. (2011). Growth, feed utilization and body composition of Asian catfish (*Pangasius hypophthalmus*) fed at different dietary protein and lipid levels. *Aquaculture Nutrition, 17*(5), 578–584.

Lovell, R. T. (2003). 13–Diet and fish husbandry. In J. E. Halver & R. W. Hardy (Eds.), *Fish Nutrition* (Third Edition).

Lovell, R. T., & Li, Y. P. (1978). Essentiality of vitamin D in diets of channel catfish (*Ictalurus punctatus*). *Transactions of the American Fisheries Society, 107*(6), 809–811.

Lovell, T. (2013). *Nutrition and Feeding of Fish*. Springer.

McGee, M. V. (2014). Pangasius culture in western hemisphere: Strong market support could aid expansion in region. *Global Aquaculture Advocate*, November/December, 73–75.

Mukai, Y., Tuzan, A. D., Lim, L. S., & Yahaya, S. (2010). Feeding behavior under dark conditions in larvae of sutchi catfish *Pangasianodon hypophthalmus*. *Fisheries Science, 76*(3), 457–461.

Murai, T., & Andrews, J. W. (1974). Interactions of dietary α-tocopherol, oxidized menhaden oil and ethoxyquin on channel catfish (*Ictalurus punctatus*). *The Journal of Nutrition, 104*(11), 1416–1431.

Murai, T., & Andrews, J. W. (1978a). Thiamin requirement of channel catfish fingerlings. *The Journal of Nutrition, 108*(1), 176–180.

Murai, T., & Andrews, J. W. (1978b). Riboflavin requirement of channel catfish fingerlings. *The Journal of Nutrition, 108*(9), 1512–1517.

Murai, T., & Andrews, J. W. (1979). Pantothenic acid requirements of channel catfish fingerlings. *The Journal of Nutrition, 109*(7), 1140–1142.

National Research Council. (2011). *Nutrient Requirements of Fish and Shrimp*. National Academies Press.

Ng, W. K., Serrini, G., Zhang, Z., & Wilson, R. P. (1997). Niacin requirement and inability of tryptophan to act as a precursor of NAD^+ in channel catfish, *Ictalurus punctatus*. *Aquaculture, 152*(1), 273–285.

Nguyen, T. P. (2013). On-farm feed management practices for striped catfish (*Pangasianodon hypophthalmus*) in Mekong River Delta, Viet Nam. In *On-farm feeding and feed management in aquaculture* (pp. 241–267).

Patil, D. W., Gaikwad, J. M., & Markad, S. S. (2019). Delineation of effect of α-amylase supplementation in diet on growth and survival of common carp (*Cyprinus carpio*) larvae. *International Journal of Research and Analytical Reviews, 6*(1), 1089–1096.

Peter, N., Pradhan, C., Dileep, N., Musharraf, M., & Thazhakot Vasunambisan, S. (2022). Dietary taurine improved growth performance, nutrient utilization, and antioxidant enzyme activities in pangasius (*Pangasianodon hypophthalmus*). *Journal of the World Aquaculture Society, 53*(1), 106–121.

Phan, L. T. T., Kals, J., Masagounder, K., Mas-Muñoz, J., La, N. T. H., & Schrama, J. W. (2021). Effect of dietary carbohydrate and fat supplementation on the yield and chemical composition of fillet and the location of fat deposition in striped catfish (*Pangasius hypophthalmus*), African catfish (*Clarias gariepinus*) and snakehead (*Channa striata*). *Aquaculture Reports, 21*, 100806.

Phan, L. T., Bui, T. M., Nguyen, T. T. T., Gooley, G. J., Ingram, B. A., Nguyen, H. V., Nguyen, P. T., & De Silva, S. S. (2009). Current status of farming practices of striped catfish, *Pangasianodon hypophthalmus* in the Mekong Delta, Vietnam. *Aquaculture, 296*(3), 227–236.

Phumee, P., Hashim, R., Aliyu-Paiko, M., & Shu-Chien, A. C. (2009). Effects of dietary protein and lipid content on growth performance and biological indices of iridescent shark (*Pangasius hypophthalmus*, Sauvage 1878) fry. *Aquaculture Research*, *40*(4), 456–463.

Platas-Rosado, D. E., González-Reynoso, L., Hernández-Arzaba, J. C., & Torres-Tadeo, C. M. (2021). Economic impact and feasibility of striped catfish farming (*Pangasius hypophthalmus*) in Mexico. *Agro Productividad*, *14*(4).

Pongpat, P., & Tongpool, R. (2013). Life cycle assessment of fish culture in Thailand: Case study of Nile tilapia and striped catfish. *International Journal of Environmental Science and Development*, 608–612.

Prabu, E., Felix, S., Felix, N., Ahilan, B., & Ruby, P. (2017). An overview on significance of fish nutrition in aquaculture industry. *International Journal of Fisheries and Aquatic Studies*, *5*(6), 349–355.

Rachmawati, D., & Nurhayati, D. (2022). Effect of dietary lysine on the growth performance of *Pangasius hypophthalmus*. *Depik*, *11*(2), 111–116.

Ranjan, A., Jain, K. K., Srivastava, P. P., & Muralidhar, P. A. (2018). Dietary energy requirement of *Pangasianodon hypophthalmus* (Sauvage, 1878) juveniles reared at two temperatures. *Turkish Journal of Fisheries and Aquatic Sciences*, *18*(1).

Robinson, E. H., Wilson, R. P., & Poe, W. E. (1980). Total aromatic amino acid requirement, phenylalanine requirement and tyrosine replacement value for fingerling channel catfish. *The Journal of Nutrition*, *110*(9), 1805–1812.

Robinson, E. H., Wilson, R. P., & Poe, W. E. (1981). Arginine requirement and apparent absence of a lysine-arginine antagonist in fingerling channel catfish. *The Journal of Nutrition*, *111*(1), 46–52.

Sargent, J., Bell, G., McEvoy, L., Tocher, D., & Estevez, A. (1999). Recent developments in the essential fatty acid nutrition of fish. *Aquaculture*, *177*(1), 191–199.

Seafish. (n.d.). *Pangasius Feed*. https://www.seafish.org/responsible-sourcing/aquaculture-farming-seafood/species-farmed-in-aquaculture/aquaculture-profiles/pangasius/feed/ (accessed on 11th July 2025).

Serrini, G., Zhang, Z., & Wilson, R. P. (1996). Dietary riboflavin requirement of fingerling channel catfish (*Ictalurus punctatus*). *Aquaculture*, *139*(3), 285–290.

Singh, D. K., Kumar, M., Ranjan, A., Udit, U. K., Vimal, B., Nayak, B. B., & Gupta, S. (2017). Growth and nutrient utilization of *Pangasianodon hypophthalmus* (Sauvage, 1878) fed with graded level of zinc. *International Journal of Current Microbiology and Applied Sciences*, *6*(6), 671–683.

Sivaramakrishnan, T., Sahu, N. P., Jain, K. K., Muralidhar, A. P., Saravanan, K., Ferosekhan, S., Praveenraj, J., & Artheeswaran, N. (2017). Optimum dietary lipid requirement of *Pangasianodon hypophthalmus* juveniles in relation to growth, fatty acid profile, body indices and digestive enzyme activity. *Aquaculture International*, *25*(2), 941–954.

Stickney, R. R., & Hardy, R. W. (1989). Lipid requirements of some warmwater species. *Aquaculture*, *79*(1), 145–156.

Thuy, C. X., Lam, T. B., Toan, H. T., & McCommick, K. (2014). Comparison of the chemical components and functional properties of fish protein isolate from *Pangasius hypophthalmus* byproducts to other protein isolates. *Asian Journal of Food and Agro-Industry*, *7*(1), 47–56.

Wilson, R. P. (1994). Utilization of dietary carbohydrate by fish. *Aquaculture*, *124*(1), 67–80.

Wilson, R. P., & Naggar, G. E. (1992). Potassium requirement of fingerling channel catfish, *Ictalurus punctatus*. *Aquaculture*, *108*(1), 169–175.

Wilson, R. P., & Poe, W. E. (1988). Choline nutrition of fingerling channel catfish. *Aquaculture*, *68*(1), 65–71.

Wilson, R. P., Allen, O. W., Robinson, E. H., & Poe, W. E. (1978). Tryptophan and threonine requirements of fingerling channel catfish. *The Journal of Nutrition*, *108*(10), 1595–1599.

Wilson, R. P., Bowser, P. R., & Poe, W. E. (1983). Dietary pantothenic acid requirement of fingerling channel catfish. *The Journal of Nutrition*, *113*(10), 2124–2128.

Wilson, R. P., Bowser, P. R., & Poe, W. E. (1984). Dietary vitamin E requirement of fingerling channel catfish. *The Journal of Nutrition*, *114*(11), 2053–2058.

Wilson, R. P., Poe, W. E., & Robinson, E. H. (1980). Leucine, isoleucine, valine and histidine requirements of fingerling channel catfish. *The Journal of Nutrition*, *110*(4), 627–633.

Xuemei, Z. (2012). Research advances on several factors affecting energy budget of fish. *Chinese Fishery Quality and Standards*.

Yadata, G. W., Ji, K., Liang, H., Ren, M., Ge, X., & Yang, Q. (2020). Effects of dietary protein levels with various stocking density on growth performance, whole body composition, plasma parameters, nitrogen emission and gene expression related to TOR signaling of juvenile blunt snout bream (*Megalobrama amblycephala*). *Aquaculture*, *519*, 734730.

CHAPTER 21

Investigating the Impact of Vegetable Oils on Immune Modulation in Fin Fish: Unraveling the Relationship Between Dietary Lipids and Immunological Responses

CHRISTIAN LARBI AYISI,[1,*] GYAMFUA AFRIYIE,[2] and GERTRUDE DZIFA MENSAH[3]

[1]*Department of Water Resources and Aquaculture Management, School of Sustainable Development, University of Environment and Sustainable Development, Somanya, Eastern Region, Ghana*

[2]*Department of Water Resources and Sustainable Development, School of Sustainable Development, University of Environment and Sustainable Development, Somanya, Eastern Region, Ghana*

[3]*CSIR College of Science and Technology, Cantonments, Accra, Ghana*

*Corresponding author

ABSTRACT

Although aquaculture is faced with numerous challenges, it has grown to be a multibillion-dollar industry. With varying degrees of success, some researchers have studied the use of vegetable oils as an alternative source of lipid in fish feeds to increase output and optimize profits. The relationship between enzymes, genes, and signaling factors that control the immune system of fin fish is reviewed in this study. By concentrating on the innate and adaptive immune systems, this study sheds light on the immune system

Sustainable Aquaculture: Recent Advances and Techniques. Dhanaji W. Patil, Sandip S. Markad (Eds.)

DOI: 10.1201/9781779641113-21

of fish. Furthermore, the roles and functions of signaling factors, enzymes, and genes involved in fish immune system modulation are discussed. The study concludes that different vegetable oils have different effects on immune parameters. While there has been much research on the impact of dietary lipids on fish health, this specific study which focus on immune responses provides new insights. This study makes a substantial contribution to our knowledge of fish nutrition and health management and offers insightful information about the possible application of vegetable oils to regulate immune function, which can improve aquaculture methods and help create diets specifically designed for fish that will give them the best possible health and performance.

21.1 INTRODUCTION

The demand for fishery products such as shellfish and fish, which continues to increase, must be met with an increase in production. One of the main factors determining the production and profitability of fish farming businesses is the high cost of feed. In an effort to reduce the price of fish feed, numerous studies have been conducted, including analyses of the effects of various lipid sources on growth, feed consumption, and immune responses (Aminikhoei et al. 2013; Ayisi et al. 2018; Kim et al. 2012; Wang et al. 2016; Yong et al. 2015). The main goal of these studies was to find alternative oils to fish oil (FO), which has been a standard lipid source in fish feed formulations.

Among these alternatives, vegetable oils (VO) have proven to be a suitable option because, at a lower cost, larger quantities can be produced. Additionally, they are highly concentrated in C18 fatty acids, primarily in the forms of α-linolenic acid (ALA; 18:3n-3), oleic acid (OA; 18:1n-9), and linoleic acid (LA; 18:2n-6). However, vegetable oils lack long-chain polyunsaturated fatty acids (LC-PUFAs), which are necessary for fish growth and metabolism. These deficiencies include arachidonic acid (ARA; 20:4n-6), docosahexaenoic acid (DHA; 22:6n-3), and eicosapentaenoic acid (EPA; 20:5n-3) (Rosenlund et al. 2001; Torstensen et al. 2007).

Although vegetable oils can serve as a suitable alternative to replace fish oil, the application of vegetable oils in fish feed in large or excessive quantities may affect growth and immune response (Brandsen et al. 2003). It has been shown that sudden changes in dietary fatty acid levels caused by the addition of such vegetable oils impact non-specific immune responses, such as respiratory burst activity (Mourente et al. 2005), as well as the specific immune response (Fracalossi and Lovell 1994), and disease resistance in fish (Apraku et al. 2017). Balfry et al. (2001) and Brandsen et al. (2003) reported a significant difference in fish resistance to pathogens when the dietary n-6/n-3 ratio was manipulated.

There appears to be very little information available regarding the role of these dietary fatty acids as modulators of the expression of genes involved in immune responses, and vice versa, despite numerous studies on the effects of substituting dietary fish oil with vegetable oil in fish (Montero et al. 2010). This is likely because much of the research conducted has focused on different fish species and different sources of dietary vegetable oil.

Therefore, this paper reviews the genes, enzymes, and transcription factors that play essential roles in the immune response of fish, especially the genes responsible for modulating immunity-related genes. This paper also reviews studies on the effect of replacing fish oil with vegetable oil on these genes, enzymes, and transcription factors, as well as on bacterial and viral diseases. The ultimate aim of this review is to provide future research on the impact of switching from fish oil to vegetable oil on the immune response and immunity-related genes in cultured fish with a solid and well-founded foundation.

21.2 IMMUNE SYSTEM OF FISH

Fish immune systems share many characteristics, such as soluble components, receptors, pathways, and cells, with those of other vertebrates (Esteban et al. 2024). The biological defenses employed by the immune system to keep pathogens out of living organisms are called defenses against invasion. The immune system of fish has developed to distinguish between substances that are non-self (such as infections and foreign bodies) and self (such as host tissue) (Dezfuli et al. 2016). In fish, the immune system comprises both the adaptive (specific) and innate (non-specific) immune systems.

21.2.1 INNATE IMMUNE SYSTEM

The first line of defense when an infection occurs, whether caused by bacteria, fungi, or viruses, is the response generated by the innate immune system (Yang et al. 2016). This system comprises a large group of cells, antimicrobial proteins, body fluids, glycoproteins, peptides, and peptides in tissues (Robertsen 1999). Fish's innate immune system responds quickly to infections but does not provide long-lasting defense (Turvey and Broide 2010). Broadly specific receptors encoded with germline information mediate the recognition process in innate immunity (Dezfuli et al. 2016). In innate immunity, monocytes and macrophages essentially control cellular processes, such as cell adhesion to target cells or tissues, through the functional activation of cell adhesion molecules. The innate immune system in fish is composed of humoral, physical, and cellular components (Kordon et al. 2018) (Figure 21.1).

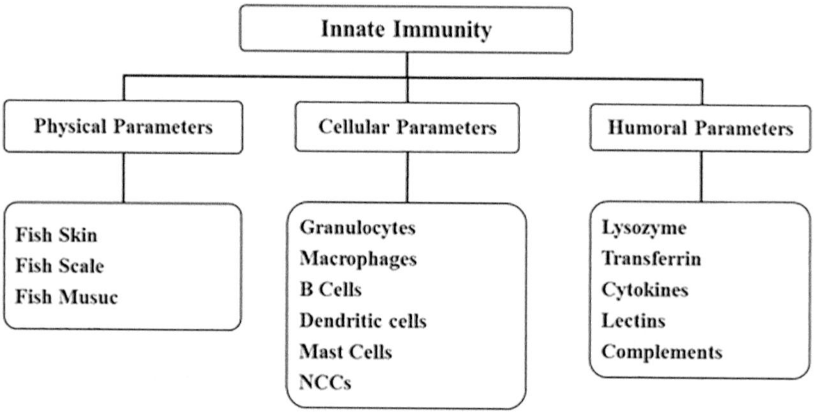

FIGURE 21.1 The elements of the fish innate immune system.
Source: Modified from Kordon et al. (2018).

Physical barriers, such as the gills, the epithelial layer of the gastro-intestinal tract, and the scales and mucus on the skin, act as the first lines of defense in the fish's innate immune system (Magnadottir 2006). The skin plays a crucial role in early defense against pathogen invasion (Smith et al. 2019). Skin-associated lymphoid tissue (SALT) has been identified in fish skin. SALT consists of various cell types, including granulocytes, macrophages, secretory cells (such as goblet cells), lymphocytes (such as B and T cells), and Langerhans-like cells (Esteban 2012; Xu et al. 2013). Fish mucus contains lectins, pentraxins, lysozymes, complement proteins, antibacterial peptides, and immunoglobulin M (IgM), all of which are essential for blocking pathogen entry (Saurabh and Sahoo 2008).

The cellular component of the fish's innate immune system consists of several cell types, including neutrophils, dendritic cells, NK cells, granulocytes such as mast cells and eosinophilic granule cells, and monocytes/macrophages (Sahoo et al. 2021). Macrophages, produced and released into the extracellular fluids after a pathogen infects a host, are responsible for mediating humoral responses (Smith et al. 2019). Humoral immune parameters include a range of soluble compounds that act as prophylactics by blocking the pathogen's vital enzymes and preventing microbial proliferation (Mokhta et al. 2023).

21.2.2 ADAPTIVE/ACQUIRED IMMUNE SYSTEM

The immune system that has the capacity to effectively detect specific pathogens and generate immunological memories is termed as either adaptive or specific immunity (Harikrishnan et al., 2011). A range of soluble chemicals known as humoral immune parameters act as prophylactics by blocking the pathogen's vital enzymes and stopping the microbe from proliferating (Uribe et al., 2011). The primary distinct elements of the adaptive immune system include recombination-activating gene, immunoglobulins, major histocompatibility complex, and T cell receptors (Brazeau and Friedman, 2015; Flajnik and Kasahara, 2010; Kordon et al., 2022). In order to protect against recurrent infections and mount a prompt and effective defense against the same pathogens, the adaptive immune system produces memory cells that offer long-lasting specific immunity (Tangye and Tarlinton, 2009). Secombes and Wang (2012) state that the two primary lymphocyte groups involved in the response are T cells, which mediate immune responses, and B cells, which mediate antibody (humoral) responses.

21.3 GENES, TRANSCRIPTION FACTORS, AND ENZYMES RESPONSIBLE FOR IMMUNE RESPONSES IN FISH

Several genes, transcription factors, and enzymes play different roles in the functioning of the fish immune system. In this section, each of these genes, transcription factors, and enzymes are briefly described along with highlights of their operational mechanisms.

21.3.1 TOLL-LIKE RECEPTOR (TLR)

To ensure a favorable response to foreign invaders, the significant family of pattern recognition receptors (PRRs) known as toll-like receptors (TLRs) is closely regulated by complex processes involving multiple molecules (Zhou et al., 2018). TLRs are a family of conserved PRRs. They are crucial in identifying pathogen-associated molecular patterns (PAMPs), such as lipopolysaccharide (LPS), lipoprotein, lipoteichoic acid, nucleic acid, and flagellin (Palti, 2011). TLRs are essential for host defense in the innate immune system against microbial pathogens (Gao et al., 2013). The function of TLRs is activated mainly when identifying PAMPs, which include numerous types such as peptidoglycan, lipoprotein, lipopolysaccharide, dsRNA, and flagellin (Purcell et al., 2006). TLR domains immediately trigger intracellular signals after ectodomains have identified the PAMP. This is achieved through cytoplasmic myeloid differentiation primary, which subsequently recruits TNF receptor-associated factor 6 (TRAF6) and interleukin-1 receptor-associated kinases. TRAF6 then activates downstream protein kinases and transcription factors (Palti, 2011).

21.3.2 MYELOID DIFFERENTIATION FACTOR 88

An adapter protein called myeloid differentiation factor 88 (MyD88) can activate nuclear factor-kappa B (NF-κB) by receiving signals from TLRs and the interleukin-1 receptor (Yao et al., 2009). It is indispensable for the Toll-like receptor (TLR) family's signal transduction (Gong et al., 2023). The two parts of MyD88 are the N-terminal death domain and the C-terminal Toll/interleukin (IL)-IR homology domain (Tang et al., 2022). Following PAMP stimulation of TLR, MyD88 binds to TLR via the TIR domain. This causes IL-1-receptor-associated kinase 4 (IRAK4) to be recruited, enabling MyD88 to interact with TLR via its DD (Gong et al., 2023). When IRAK is phosphorylated, proinflammatory cytokines like TNF-α, IL-1, and IL-12 are produced, and nuclear factor kappa B (NF-κB) and transcription factors activating protein-1 (AP-1) are activated (Gong et al., 2023). MyD88 is crucial for the innate immune system's initial responses (Zuo et al., 2012). Invertebrate and vertebrate animals' TLR-mediated inflammatory responses depend critically on MyD88 (Tang et al., 2023). MyD88 has been identified and described in a number of fish species, including Japanese flounder (*Paralichthys olivaceus*) (Takano et al., 2006), zebrafish (*Danio rerio*) (Jault et al., 2004), Atlantic salmon (*Salmo salar*) (Skjæveland et al., 2009), yellow catfish (*Pelteobagrus fulvidraco*) (Yu et al., 2018), and spotted knifejaw (*Oplegnathus punctatus*) (Liu et al., 2019).

21.3.3 INTERFERON-RELATED FACTOR (IRF)

A class of multifunctional polypeptides known as cytokines, interferons (IFNs) have the ability to both directly prevent viral replication and stimulate immune cells. Type I IFNs—which include IFN-α and IFN-β—and type II IFN (IFN-γ) are the two major classes of interferons (Krishnan et al., 2019). IFNs are described as having non-specific antiviral activity via cellular metabolic pathways that include RNA and protein synthesis (Anonymous, 1980). As transcriptional mediators of bacterial, viral, and IFN-induced signaling pathways, interferon-related factors (IRFs) are crucial for immune response, apoptosis, cell growth regulation, and antiviral defense (Tamura et al., 2008).

21.3.4 TUMOR NECROSIS FACTOR (TNF)

Tumor necrosis factor (TNF) is a multifunctional cytokine that is critical in the survival, differentiation, proliferation, and death of cells. During the differentiation of terminal T cells, TNF plays an important role by regulating the switch from "antigen mode" to "inflammation mode." One of the most significant adipokines released from adipose tissue is TNF. It begins as a 26 kDa transmembrane protein that is cleaved into a 17 kDa soluble molecule by metalloproteinase (Kriegler et al., 1988). TNF-α, a type II transmembrane glycoprotein, binds to both TNFR-1 and TNFR-2 receptors. TNF-α has been found in several fish species, including turbot (*Scophthalmus maximus*) (Liu et al., 2016), common carp (*Cyprinus carpio*) (Zhao et al., 2012), zebrafish (*Danio rerio*) (Lei et al., 2016), and mandarin fish (*Siniperca chuatsi*) (Xiao et al., 2007). According to Zou et al. (2003), channel catfish (*Ictalurus punctatus*) possess a TNF-α-like gene that codes for a propeptide consisting of 230 amino acids and a mature peptide of 162 amino acids.

21.3.5 NUCLEAR FACTOR KAPPA B

Nuclear factor kappa B (NF-κB) is a family of transcription factors essential for immunity, inflammation, apoptosis, and cell division (Oeckinghaus and Ghosh, 2009). The NF-κB pathway regulates numerous genes involved in inflammatory and immune responses (Qiu et al., 2020). According to Iwasaki and Medzhitov (2010), NF-κB may serve as the principal regulator of both innate and adaptive immune responses. NF-κB signaling is essential for a wide range of immunological transcriptional processes, including the development of secondary lymphoid organs and the activation and proliferation of adaptive immune cells (Alcamo et al., 2002; Kaileh and Sen, 2012; Cildir et al., 2016; Dorrington and Fraser, 2019).

21.3.6 CYTOKINES

Cytokines have pleiotropic functions, often overlapping or even contradicting one another, which makes their classification complex. They are produced by epithelial cells, lymphocytes, mast cells, macrophages, and granulocytes. The major categories of cytokines include chemokines, colony-stimulating factors, interferons (IFNs), interleukins (ILs), and tumor necrosis factors (TNFs). Upon stimulation by bacterial, parasitic, or viral pathogens, cytokines are secreted by immune-related cells (Ferreira et al., 2018). Cytokines play a key role in immune system regulation. They are transported to infection sites by immune cells and induce inflammatory responses to control both resident and newly arrived phagocytes (Wang and Secombes, 2013). The cytokine repertoire in fish appears to be comparable to that of mammals. Several cytokine suppressors and homologues, including IL-6, IFN, TNF-α, and IL-1β, have been cloned in different fish species (Reyes-Cerpa et al., 2012).

21.3.7 INTERLEUKIN (IL)

Interleukins (ILs) are a class of cytokines initially believed to be expressed exclusively by leukocytes, but they are now known to be produced by a wide range of body cells. ILs bind to high-affinity receptors on cell surfaces to initiate a range of cellular responses. Their significant roles include the activation, proliferation, maturation, differentiation, migration, and adhesion of immune cells. ILs also possess both pro-inflammatory and anti-inflammatory properties. They are primarily responsible for regulating growth, differentiation, and immune response during inflammation (Vaillant and Qurie, 2021). ILs facilitate communication between different cell populations (Moreau and Meurens, 2017). Based on functional similarities, receptor chains, and sequence homology, ILs are grouped into several families: IL-1, IL-10, IL-12, and IL-17 (Akdis et al., 2011).

21.3.8 ARGINASE

Arginase is a ubiquitous ureohydrolase enzyme found in a wide range of living things, including bacteria, yeast, plants, and animals (Dzik, 2014). Arginase exists in two forms, namely cytosolic and mitochondrial forms. Arginase I, a cytosolic form that is highly expressed in the liver, is primarily believed to be involved in ureagenesis. It has been suggested that Arginase II, a mitochondrial form, is more extensively expressed and plays a role in the inflammatory process, the production of polyamines, and the amino acids ornithine, proline, and glutamate (Ferreira et al., 2004). It catalyzes the Krebs–Henseleit urea cycle's last stage, which converts *L*-arginine to *L*-ornithine and urea and eliminates toxic ammonia (Dzik, 2014).

21.3.9 LYSOZYME

The bacteriolytic enzyme lysozyme is extensively dispersed throughout the body and is a component of most animals' nonspecific defense mechanisms (Uribe et al., 2011). Lysozyme is an enzymatic protein found in animals, plants, bacteria, and viruses. By cleaving the peptidoglycan component of bacterial cell walls, it acts as an antimicrobial agent and causes cell death (Oliver and Wells, 2015). In innate immunity, the antimicrobial peptide lysozyme is very essential (Klüter et al., 2014). The primary lysozyme sources are neutrophils, macrophages, and monocytes (Klüter et al., 2014). Nonetheless, research has found this enzyme in the granules of the intestinal eosinophilic granular cells (Sveinbjornsson et al., 2006).

21.3.10 LECTINS

All living things contain lectins, which are proteins that bind carbohydrates and belong to the class of pattern recognition receptors (PRRs). They are thought to play key roles in the innate immune response by mediating pathogen recognition in the fish immune system. Additionally, fish lectins are crucial to fertilization, embryogenesis, and morphogenesis (Elumalai et al., 2019). Because lectins recognize pathogen-associated molecular patterns such as galactoses, mannoses, lipopolysaccharides, β-glucans, α-mannans, lipoteichoic acids, and peptidoglycans, they can identify carbohydrates and their conjugates through non-reversible covalent binding and the formation of cell aggregations (Yang et al., 2016).

21.3.11 IMMUNOGLOBULINS (IG)

Immunoglobulins (Ig) are composed of two light chains and two heavy chains. They are heterodimeric proteins with variable and constant regions. Whereas the constant region determines effector functions, the variable region is responsible for binding antigens (Hikima et al., 2011). Fish are among the major groups of jawed vertebrates that possess an adaptive immune system based in part on Ig and T-cell receptors. Immunoglobulins play a role in mediating cellular toxicity, neutralizing infectious pathogens, and initiating the complement pathway in enhancing the fish immune system (Ye et al., 2013).

21.3.12 SUPEROXIDE DISMUTASE (SOD)

Superoxide dismutase is a crucial antioxidant defense found in almost all living cells that come into contact with oxygen. Superoxide (O_2^-) radicals are dismutated into two less damaging molecules—hydrogen peroxide (H_2O_2) or molecular oxygen (O_2)—by SOD. Oxygen metabolism produces SOD, which could cause several types of cell damage if left unregulated (Desai and Bhilave, 2018). SOD is an important indicator used to evaluate oxidative stress in fish. This is due to the fact that SOD plays a crucial part in fish defense mechanisms. It prevents cells from being damaged by the free radical process and has the ability to defend fish against the toxic effects of reactive oxygen species (ROS) (Pedrajas et al., 1995). SOD is therefore considered to be the first enzyme that cleanses ROS (Chien et al., 2003).

21.3.13 PEROXIREDOXINS (PRXS)

A group of antioxidant proteins called peroxiredoxins shields cells from oxidative damage brought on by various peroxide forms (Zheng et al., 2010). Peroxiredoxins are also known as natural killer enhancing factor, thiol-specific oxidant, torin alkylhydroperoxide reductase, thioredoxin peroxidase, and calpromotin (Arockiaraj et al., 2012). Importantly, they reduce toxic peroxide compounds to products that are non-toxic; hence, they are considered critical in cellular oxidative stress (Priyathilaka et al., 2016). PRXs are characterized by their ability to regulate intracellular levels of ROS through detoxification of H_2O_2. They are also involved in cell differentiation (Rabilloud et al., 1995), immune responses (Rabilloud et al., 1995), and proliferation (Prosperi et al., 1993).

21.3.14 LABORATORY OF GENETICS AND PHYSIOLOGY 2 (LGP2)

A type of receptor similar to retinoic acid-inducible gene I-like receptors (RLRs), the laboratory of genetics and physiology 2 (LGP2) is a member of the RLR family. Due to the absence of caspase recruitment domains, LGP2 is the most enigmatic RLR (Malur et al., 2012); therefore, it doesn't interact with Promor Stimulator 1 of interferon-β. It plays an essential role in viral RNA recognition in the cytoplasm and regulation during innate immune responses. LGP2 is recognized as a positive or negative regulator of signal transduction in RLR pathways (Sato et al., 2010). In fish, LGP2 activates interferon signaling by binding to poly(I:C) (Chang et al., 2011).

21.3.15 SIGNAL TRANSDUCER AND ACTIVATOR OF TRANSCRIPTION (STAT)

Cell proliferation, immune regulation, transformation, and inflammation are important roles played by signal transducer and activator of transcription (STAT), making it a critical transcription factor (Huang et al., 2014). STAT mediates functional responses such as cytokine signaling (Lee et al., 2015). In addition, STAT has been reported to regulate autophagy and apoptosis. Through phosphorylation at serine 727 and tyrosine 701, STAT proteins undergo activation (Bourke et al., 2013). STAT is also involved in tumorigenesis. The Janus kinase (JAK) family proteins are connected to the STAT pathway upstream, and STAT can integrate inputs from various signaling pathways (Loh et al., 2019).

21.3.16 MYXOVIRUS RESISTANCE (MX) GENE

Fish *Mx* gene induction is caused by type-I interferon, the primary cytokine that mediates the innate immune antiviral function. The specific role of the *Mx* gene in preventing the replication of viruses as well as its assistance to type-I interferon is not fully known (Das et al., 2019). A recent study in Chinese giant salamander showed that *Mx* may play an important antiviral role in the immune responses against iridovirus (Liu et al., 2020).

The majority of *Mx* activity has been focused on RNA viruses, including *Bunya*, *Orthomyxo*, *Rhabdo*, and *Paramyxo* viruses (Meier et al., 1990; Zurcher et al., 1992). Additionally, it has been demonstrated that *Mx*'s intrinsic GTPase activity is essential for its antiviral activity against specific viruses (Johannes et al., 1997).

21.3.17 NITRIC OXIDE SYNTHASE (NOS)

Natural signaling molecules such as nitric oxide are produced by vertebrates, including fish, and have a variety of physiological uses in their organ systems (Peter et al., 2022). Nitric oxide synthase (NOS) is known to be expressed in the central nervous system of fishes. Peter and Gayathry (2021) reported that temperature elevation (Jorgensen et al., 2014), hyperammonia (Choudhury and Saha, 2012a), hyperosmotic stress (Gerber et al., 2018), desiccation stress (Choudhury and Saha, 2012b; Garofalo et al., 2015), and hypoxia/anoxia (Hansen et al., 2016; Imbrogno et al., 2014; Jensen et al., 2015; Sandvik et al., 2012) are among the numerous environmental stressors that fish respond to via the NOS/NO system. The three isoforms of NOS that are expressed by fish are endothelial, neuronal, and inducible (Eddy, 2005). While neuronal and endothelial NOS are constitutive enzymes regulated by intracellular Ca^{2+}/calmodulin, inducible NOS is induced at the level of gene transcription and is Ca^{2+}-independent in target cells (Althaus, 2012).

21.3.18 ALPHA-2-MACROGLOBULIN (A2M)

Within the vertebrate innate immune system, alpha-2-macroglobulin (A2M) is a broad-spectrum protease-binding protein that has been evolutionarily conserved. As a result of trapping and inhibiting all classes of parasitic protease and microbes, A2M protects the host against invading pathogens (Mohindra et al., 2019). Mohindra et al. (2019) revealed a wide variety of tissue-specific A2M transcripts in hilsa herring *Tenualosa ilisha*.

Alpha-2-macroglobulin, a non-specific protease inhibitor, plays a significant role in the host's defense mechanism by inhibiting both exogenous and endogenous proteases. A2M contains both an alpha and a beta chain in carp (*Cyprinus carpio*). The bioactive sections are found in the alpha chain. The only A2M isoform with significantly elevated transcription during infection was A2M3, indicating that this specific isoform is of utmost biological significance (Onara et al., 2008).

21.3.19 RESPIRATORY BURST

Respiratory burst, also known as oxidative burst, is the rapid release of hydrogen peroxide and superoxide radical (reactive oxygen species) from various types of cells. Reactive oxygen species (ROS) are produced and used more frequently when phagocytic cell membranes are stimulated, which in turn activates membranes linked to NADP oxidase. It has been reported that the killing activity of hemocytes or bacterial pathogens by phagocytes—which essentially indicates the mechanism of cellular immunity in fish—correlates with respiratory activity (Srivastava and Pandey, 2015).

An extremely effective non-specific cellular defense mechanism in phagocytic cells, such as neutrophils and monocytes/macrophages, is respiratory burst, a potent oxygen-dependent killing mechanism. Since this type of non-specific immunity is the most important defense against infections, it may be vital in certain fish species (Haugland et al., 2012).

21.3.20 COMPLEMENT

One major role of the complement system is to serve as a main link between both adaptive and innate immune responses in fish (Hawlisch et al., 2006). A system of serum proteins called complement is essential to many defense mechanisms. Freshwater and marine fishes exhibit both the classical and alternative complement pathways of activation (Holland et al., 2002). Fish serum is known for the alternative complement pathway activity, which does not depend on antibodies (Boshra et al., 2006).

One function of complement is to kill pathogens through activation of phagocytes and opsonization. Fish complement systems generally do not work against virulent gram-negative or gram-positive bacteria, despite having bactericidal activity against non-virulent gram-negative bacteria (Boshra et al., 2006). Activation of complement either by antibody-antigen (Ag-Ig) complexes or microorganisms results in complement-mediated killing. Due to Ag-Ig complex activation, Srivastava and Pandey (2015) reported that complement is a crucial effector mechanism for the adaptive immune response.

21.3.21 GLUTATHIONE PEROXIDASE (GPX)

Glutathione peroxidases (GPx) are essential for the detoxification of lipids and hydrogen peroxide. They are a component of the first line of defense against hydrogen peroxide, peroxides, and the superoxide anion. GPx performs this function with the concomitant oxidation of glutathione (Almar et al., 1998). As one of the antioxidases in the glutathione antioxidative system, it keeps the levels of reactive oxygen species (ROS) in cells in check and guards against cellular damage.

As the primary enzymatic defense mechanism of infectious parasites, GPx protects organisms from exposure to toxic ROS in the host. Therefore, the parasite's ability to fend off oxidative damage is critical to its survival, suggesting that GPx may be a useful antigen (Zelck et al., 2004).

21.3.23 CATALASE (CAT)

All oxygen-breathing organisms possess the antioxidant enzyme catalase, which is a crucial component of their defense system (Zamocky et al., 2008). The enzyme has the potential to remove reactive oxygen species, preventing oxidative damage to cells and tissues by decomposing hydrogen peroxide (H_2O_2) into water and molecular oxygen (Sampaio et al., 2008).

Upregulation or downregulation of catalase can influence diverse biological processes such as cell differentiation, proliferation, apoptosis, and migration (Mates et al., 1999), as H_2O_2 is involved in sensing and signaling of these biological events (Veal et al., 2007). Catalase has been shown to protect freshwater fish *Oreochromis niloticus* from oxidative stress caused by different heavy metals by increasing or decreasing CAT activities in various fish organs. Thus, CAT is considered a bioindicator that is highly sensitive to the antioxidant defense system (Atli et al., 2006).

21.4 RELATIONSHIP BETWEEN DIETARY VEGETABLE OILS ON THE EXPRESSION OF TRANSCRIPTION OR SIGNALING FACTORS AND GENES RELATED TO IMMUNE FUNCTION

Over the past few decades, aquaculture nutritionists have focused on strategies to find suitable alternatives to dietary fish oil. One of the suggested alternatives that has proven to be ideal is vegetable oil. Vegetable oils, however, contain less *n*-3 long-chain polyunsaturated fatty acids (EPA and DHA) (Hixson, 2014).

Cui et al. (2020) studied the regulation of inflammatory gene expression in marine fish *Larimichthys crocea* diets by substituting four vegetable oils (VO)—soybean, palm, sesame, and olive—for dietary fish oil (FO). The results showed that the FO group had lower levels of mRNA expression of inflammatory genes such as interleukin (IL)-1β, IL-6, tumor necrosis factor (TNF) α, and interferon (IFN) γ than the four VO groups. The authors concluded that one of the reasons VO causes inflammation in marine fish may be that it reduces the expression of inflammatory genes induced by LPS through the TLR signaling pathway. This could be due to substituting VO, which lack arachidonic acid (ARA), for dietary FO.

Kutluyer et al. (2017) showed that various vegetable oils such as safflower seed oil (SSO), soybean oil (SBO), linseed oil (LO), and the blends of SBO+LO and SSO+LO, when used as substitutes for cod liver oil (CO) in rainbow trout, resulted in differential expression of immune system-related genes. SSO upregulated the expression of *gpx* in liver and muscle, *cat* in liver, and *gst* and *sod* in muscle. SBO upregulated the expression of *gr* in liver. The blend of SBO+LO upregulated the expression of *cat* in muscle, while the blend of SSO+LO upregulated the expression of *gst* in liver.

Mu et al. (2020) examined the expression of inflammatory responses in large yellow croakers (*Larimichthys crocea*) fed different lipid sources. Rapeseed oil was used to replace fish oil at 50% and 100%. When rapeseed oil was substituted, the expression of *arginase I* and *interleukin 10* decreased. However, the expression of *tnfα*, *interleukin 1β*, *toll-like receptor 22*, and *myeloid differentiation factor 88* was upregulated in the liver. Additionally, *glutathione peroxidase*, *superoxide dismutase 1 and 2*, *nuclear factor erythroid 2-related factor 2*, and *glutathione* were downregulated, implying that increased rapeseed oil in *L. crocea* could induce inflammation.

Similarly, Tan et al. (2016) evaluated the effects of substituting fish oil with dietary vegetable oils on immune-related genes in *L. crocea.* The results showed an upregulation in *il1β* and *tnfα* (pro-inflammatory cytokines), a downregulation of *arginase I* and *il10* (anti-inflammatory cytokine) gene expression, and an increase in the expression of *ikkα/β*, *iκbα* phosphorylation, and nuclear *p65* protein.

In a study by Montero et al. (2010), the effects of switching from anchovy oil to linseed or soybean oil were investigated in gilthead sea bream (*Sparus aurata*), focusing on the expression of two pro-inflammatory cytokines (*interleukin 1β* and *tumor necrosis factor-α*) in the head, kidney, and intestine following exposure to the bacterial pathogen *Photobacterium damselae* subsp. *piscicida*. Before infection, groups fed linseed and soybean oil showed higher levels of *il1β* and *tnfα* expression. However, after infection, there was a further increase in *il1β* and *tnfα* levels in the fish oil-fed groups.

In hybrid grouper, replacement of fish oil with alternative dietary oils led to downregulation of *nrf2* expression but stimulated the expression of *il1β*, *tnfα*, *tlr22*, and *myd88* (Yan et al., 2020).

This review highlights that different vegetable oils produce varying effects on the expression of immune-related genes. The observed differential expression can be explained by the fact that vegetable oils modulate immune parameters differently depending on the fish species, VO source, and substitution level (Montero et al., 2010). According to Geay et al. (2015), these variations may stem from differences in the fish species' capacities for endogenous long-chain polyunsaturated fatty acid (LC-PUFA) biosynthesis, which could compensate for LC-PUFA deficiencies in VO-based diets.

Based on the various studies discussed above, replacement of FO with VO in the feed of fish species influences the biological processes both positively and negatively. Consequently, it is suggested to use both VO and FO simultaneously in suitable quantities to ensure healthy fish production and substantial profit returns. Yan et al. (2020) claim that pro-inflammatory factors are expressed in tissues and that dietary VO substitution alters the histological morphology of hepatocyte cell tissue. This was discovered following the formulation of seven experimental diets based on fish meal that were isonitrogenous and isolipidic. For eight weeks, 40 fish in triplicate groups (initial mean body weight ± standard error = 15.09 ± 0.01 g) were fed diets containing 5% fish oil (FO), which served as the controls, along with equal amounts of various vegetable oils (VOs), including corn oil (CO), sunflower oil (SO), tea oil (TO), olive oil (OO), rice oil (RO), and mixed oil (MO). The results showed that VO enhanced the activity of non-specific immune enzymes in addition to downregulating *nrf2* expression and promoting the expression of pro-inflammatory factors (*il1β*, *tnfα*, *tlr22*, and *myd88*) in the kidney.

21.5 CONCLUSION

In conclusion, aquaculture is a dynamic industry with significant economic implications, but it is also fraught with difficulties, especially when it comes to obtaining high-quality fish feed and the constant threat of disease. Researchers looking to maximize production efficiency and financial returns are focusing on the search for alternative lipid sources, particularly vegetable oils. This review explores the complex interactions between fin fish immune systems and vegetable oils, illuminating the diverse functions of genes, enzymes, and signaling factors in the regulation of innate and adaptive immunity. Varieties of vegetable oils have different effects on immune system responses in fish. This could be attributed to the differences in vegetable oils, fish species, age, and size of the fish under consideration. The ability of fish to defend itself against invading pathogens is essential in aquaculture. Therefore, the impact of VO on the immunity of fish must be considered and further studied. Understanding the subtle effects of vegetable oils on fish immunity is crucial for making informed decisions as aquaculture moves toward more sustainable practices. This knowledge lays the groundwork for future research and strategic developments aimed at achieving the best possible fish health and production results.

KEYWORDS

- **aquaculture**
- **dynamic industry**
- **fin fish**
- **gene expression**
- **genetic modulation**
- **immune system**
- **sustainable practices**
- **vegetable oils**

REFERENCES

Akdis, M., Burgler, S., Crameri, R., Eiwegger, T., Fujita, H., Gomez, E., Klunker, S., Meyer, N., O'Mahony, L., Palomares, O., Rhyner, C., Quaked, N., Schaffartzik, A., Van De Veen, W., Zeller, S., Zimmermann, M., & Akdis, C. A. (2011). Interleukins, from 1 to 37, and interferon-γ: Receptors, functions, and roles in diseases. *Journal of Allergy and Clinical Immunology, 127*(3), 1–70.

Alcamo, E., Hacohen, N., Schulte, L. C., Rennert, P. D., Hynes, R. O., & Baltimore, D. (2002). Requirement for the NF-kappaB family member RelA in the development of secondary lymphoid organs. *Journal of Experimental Medicine, 195*, 233–244. https://doi.org/10.1084/jem.20011885 (accessed on 11th July 2025).

Almar, M., Otero, L., Santos, C., & Gallego, J. G. (1998). Liver glutathione content and glutathione-dependent enzymes of two species of freshwater fish as bioindicators of chemical pollution. *Journal of Environmental Science and Health, B33*(6), 769–783.

Althaus, M. (2012). Gasotransmitters: Novel regulators of epithelial Na^+ transport? *Frontiers in Physiology, 3*, 83. https://doi.org/10.3389/fphys.2012.00083 (accessed on 11th July 2025).

Aminikhoei, Z., Choi, J., & Lee, S. M. (2013). Effects of different dietary lipid sources on growth performance, fatty acid composition, and antioxidant enzyme activity of juvenile rockfish, *Sebastes schlegeli. Journal of the World Aquaculture Society, 44*(5), 716–725.

Apraku, A., Liu, L., Leng, X., Rupia, E. J., & Ayisi, C. L. (2017). Evaluation of blended virgin coconut oil and fish oil on growth performance and resistance to *Streptococcus iniae* challenge of Nile tilapia (*Oreochromis niloticus*). *Egyptian Journal of Basic and Applied Sciences, 4*(3), 175–184.

Arockiaraj, J., Easwvaran, S., Vanaraja, P., Singh, A., Othman, R. Y., & Bhassu, S. (2012). Immunological role of thiol-dependent peroxiredoxin gene in *Macrobrachium rosenbergii. Fish & Shellfish Immunology, 33*, 121–129.

Atli, G., Alptekin, Ö., Tükel, S., & Canli, M. (2006). Response of catalase activity to Ag^{+}, Cd^{2+}, Cr^{6+}, Cu^{2+} and Zn^{2+} in five tissues of freshwater fish *Oreochromis niloticus*. *Comparative Biochemistry and Physiology Part C: Toxicology & Pharmacology*, *143*(2), 218–224. https://doi.org/10.1016/j.cbpc.2006.02.003 (accessed on 11th July 2025).

Ayisi, C. L., Zhao, J., & Wu, J. W. (2018). Replacement of fish oil with palm oil: Effects on growth performance, innate immune response, antioxidant capacity and disease resistance in Nile tilapia (*Oreochromis niloticus*). *PLOS ONE*, *13*(4), e0196100. https://doi.org/10.1371/journal.pone.0196100 (accessed on 11th July 2025).

Balfry, S. K., & Higgs, D. A. (2001). Influence of dietary lipid composition on the immune system and disease resistance of finfish. In C. Lim & C. D. Webster (Eds.), *Nutrition and Fish Health* (pp. 213–234). Haworth Press.

Boshra, H., Li, J., & Sunyer, J. O. (2006). Recent advances on the complement system of teleost fish. *Fish & Shellfish Immunology*, *20*(2), 239–262.

Bourke, L. T., Knight, R. A., Latchman, D. S., Stephanou, A., & McCormick, J. J. (2013). Signal transducer and activator of transcription-1 localizes to the mitochondria and modulates mitophagy. *JAK-STAT*, *2*(4). https://doi.org/10.4161/jkst.25666 (accessed on 11th July 2025).

Brandsen, M. P., Carter, C. G., & Nichols, P. D. (2003). Replacement of fish oil with sunflower oil in feeds for Atlantic salmon (*Salmo salar* L.): Effect on growth performance, tissue fatty acid composition and disease resistance. *Comparative Biochemistry and Physiology Part B*, *135*, 611–625.

Brazeau, M. D., & Friedman, M. (2015). The origin and early phylogenetic history of jawed vertebrates. *Nature*, *520*(7548), 490–497.

Chang, M., Collet, B., Nie, P., Lester, K., Campbell, S., Secombes, C. J., & Zou, J. (2011). Expression and functional characterization of the RIG-I-like receptors MDA5 and LGP2 in rainbow trout (*Oncorhynchus mykiss*). *Journal of Virology*, *85*, 8403–8412.

Chien, Y. H., Pan, C. H., & Hunter, B. (2003). The resistance to physical stresses by *Penaeus monodon* juveniles fed diets supplemented with astaxanthin. *Aquaculture*, *216*, 177–191.

Cildir, G., Low, K. C., & Tergaonkar, V. (2016). Noncanonical NF-κB signaling in health and disease. *Trends in Molecular Medicine*, *22*, 414–429. https://doi.org/10.1016/j.molmed.2016.03.002 (accessed on 11th July 2025).

Cioni, C., Angiulli, E., & Toni, M. (2019). Nitric oxide and the neuroendocrine control of the osmotic stress response in teleosts. *International Journal of Molecular Sciences*, *20*(3), 489. https://doi.org/10.3390/ijms20030489 (accessed on 11th July 2025).

Cui, K., Li, X., Chen, Q., Li, Q., Gao, S., Tan, P., Mai, K., & Ai, Q. (2020). Effect of replacement of dietary fish oil with four vegetable oils on prostaglandin E2 synthetic pathway and expression of inflammatory genes in marine fish *Larimichthys crocea*. *Fish & Shellfish Immunology*, *107*, 529–536. https://doi.org/10.1016/j.fsi.2020.09.038 (accessed on 11th July 2025).

Das, B. K., Roy, P., Rout, A. K., Sahoo, D. R., Panda, S. P., Pattanaik, S., Dehury, B., Behera, K., & Mishra, S. S. (2019). Molecular cloning, GTP recognition mechanism and tissue-specific expression profiling of myxovirus resistance (Mx) protein in *Labeo rohita* (Hamilton) after Poly I:C induction. *Scientific Reports*, *9*, 3956.

Desai, T. H., & Bhilave, M. P. (2018). Toxicological studies of methanol on superoxide dismutase (SOD) activity of freshwater fish *Cirrhinus mrigala*. *International Journal of Fisheries and Aquatic Studies*, *6*(5), 20–22.

Dezfuli, B. S., Bosi, G., DePasquale, J. A., Manera, M., & Giari, L. (2016). Fish innate immunity against intestinal helminths. *Fish & Shellfish Immunology*, *50*, 274–287. https://doi.org/10.1016/j.fsi.2016.02.002 (accessed on 11th July 2025).

Dorrington, M. G., & Fraser, I. D. C. (2019). NF-κB signaling in macrophages: Dynamics, crosstalk, and signal integration. *Frontiers in Immunology*, *10*, 705. https://doi.org/10.3389/fimmu.2019.00705 (accessed on 11th July 2025).

Dzik, J. M. (2014). Evolutionary roots of arginase expression and regulation. *Frontiers in Immunology*, *5*(544), 1–11.

Eddy, F. B. (2005). Role of nitric oxide in larval and juvenile fish. *Comparative Biochemistry and Physiology Part A: Molecular & Integrative Physiology*, *142*(2), 221–230. https://doi.org/10.1016/j.cbpb.2005.05.038 (accessed on 11th July 2025).

Elumalai, P., Rubeena, A. S., Arockiaraj, J., Wongpanya, R., Cammarata, M., Ringø, E., & Vaseeharan, B. (2019). The role of lectins in finfish: A review. *Reviews in Fisheries Science and Aquaculture*, *27*(2), 152–169.

Esteban, M. Á. (2012). An overview of the immunological defenses in fish skin. *ISRN Immunology*, *2012*, 853470. https://doi.org/10.5402/2012/853470 (accessed on 11th July 2025).

Esteban, M. A. (2024). A review of soluble factors and receptors involved in fish skin immunity: The tip of the iceberg. *Fish & Shellfish Immunology*, *145*, 109311. https://doi.org/10.1016/j.fsi.2023.109311 (accessed on 11th July 2025).

Ferreira, V. L., Helena, H. L., Bonetti, B. A. F., Leonart, L. P., & Pontarolo, R. (2018). Cytokines and interferons: Types and functions. In *IntechOpen*. https://doi.org/10.5772/intechopen.74550 (accessed on 11th July 2025).

Flajnik, M. F., & Kasahara, M. (2010). Origin and evolution of the adaptive immune system: Genetic events and selective pressures. *Nature Reviews Genetics*, *11*(1), 47–59. https://doi.org/10.1038/nrg2703 (accessed on 11th July 2025).

Fracalossi, D. M., & Lovell, R. T. (1994). Dietary lipid sources influence responses of channel catfish (*Ictalurus punctatus*) to challenge with the pathogen *Edwardsiella ictaluri*. *Aquaculture*, *119*, 287–298.

Gao, H., Wu, L., Sun, J. S., Geng, X. Y., & Pan, B. P. (2013). Molecular characterization and expression analysis of Toll-like receptor 21 cDNA from *Paralichthys olivaceus*. *Fish & Shellfish Immunology*, *35*(4), 1138–1145.

Geay, F., Mellery, J., Tinti, E., Douxfils, J., Larondelle, Y., Mandiki, S. N. M., & Kestemont, P. (2015). Effects of dietary linseed oil on innate immune system of Eurasian perch and disease resistance after exposure to *Aeromonas salmonicida achromogenes*. *Fish & Shellfish Immunology*, *47*(2), 782–796.

Gong, X., Li, M., Zhang, L., Huang, S., & Wang, G. (2023). Identification and functional analysis of myeloid differentiation factor 88 (MyD88) in early development of *Haliotis diversicolor*. *Fish & Shellfish Immunology*, *142*, 109085. https://doi.org/10.1016/j.fsi.2023.109085 (accessed on 11th July 2025).

Harikrishnan, R., Balasundaram, C., & Heo, M. S. (2011). Impact of plant products on innate and adaptive immune system of cultured finfish and shellfish. *Aquaculture*, *317*, 1–15.

Haugland, G. T., Jakobsen, R. A., Vestvik, N., Ulven, K., Stokka, L., & Wergeland, H. I. (2012). Phagocytosis and respiratory burst activity in lumpsucker (*Cyclopterus lumpus* L.) leucocytes analysed by flow cytometry. *PLOS ONE, 7*(10), e47909.

Hawlisch, H., & Kohl, J. (2006). Complement and Toll-like receptors: Key regulators of adaptive immune responses. *Molecular Immunology, 43*, 13–21.

Hikima, J. I., Jung, T. S., & Aoki, T. (2011). Immunoglobulin genes and their transcriptional control in teleosts. *Developmental & Comparative Immunology, 35*(9), 924–936.

Hixson, S. M. (2014). Fish nutrition and current issues in aquaculture: The balance in providing safe and nutritious seafood, in an environmentally sustainable manner. *Journal of Aquaculture Research & Development, 5*, 234.

Holland, J. W., Pottinger, T. G., & Secombes, C. J. (2002). Recombinant interleukin-1 beta activates the hypothalamic–pituitary–interrenal axis in rainbow trout, *Oncorhynchus mykiss*. *Journal of Endocrinology, 175*, 261–267.

Huang, X., Huang, Y., Yang, Y., Wei, S., & Qin, Q. (2014). Involvement of fish signal transducer and activator of transcription 3 (STAT3) in SGIV replication and virus induced paraptosis. *Fish & Shellfish Immunology, 41*(2), 308–316. https://doi.org/10.1016/j.fsi.2014.09.011 (accessed on 11th July 2025).

Iwasaki, A., & Medzhitov, R. (2010). Regulation of adaptive immunity by the innate immune system. *Science, 327*, 291–295.

Jault, C., Pichon, L., & Chluba, J. (2004). Toll-like receptor gene family and TIR-domain adapters in *Danio rerio*. *Molecular Immunology, 40*, 759–771.

Johannes, L., Kambadur, R., Lee-Hellmich, H., Hodgkinson, C. A., Arnheiter, H., & Meier, E. (1997). Antiviral determinants of rat Mx GTPases map to the carboxyterminal half. *Journal of Virology, 71*(12), 9792–9795.

Kaileh, M., & Sen, R. (2012). NF-κB function in B lymphocytes. *Immunological Reviews, 246*, 254–271. https://doi.org/10.1111/j.1600-065X.2012.01106.x (accessed on 11th July 2025).

Kim, D. K., Kim, K. D., Seo, J. Y., & Lee, S. M. (2012). Effects of dietary lipid source and level on growth performance, blood parameters and flesh quality of sub-adult olive flounder (*Paralichthys olivaceus*). *Asian-Australasian Journal of Animal Sciences, 25*(6), 869–879. https://doi.org/10.5713/ajas.2011.11470 (accessed on 11th July 2025).

Klüter, T., Fitschen-Oestern, S., Lippross, S., Weuster, M., Mentlein, R., Steubesand, N., Neunaber, C., Hildebrand, F., Pufe, T., Tohidnezhad, M., Beyer, A., Seekamp, A., & Varoga, D. (2014). The antimicrobial peptide lysozyme is induced after multiple trauma. *Mediators of Inflammation, 2014*, Article 303106. https://doi.org/10.1155/2014/303106 (accessed on 11th July 2025).

Kordon, A. O., Karsi, A., & Pinchuk, L. (2018). Innate immune responses in fish: Antigen presenting cells and professional phagocytes. *Turkish Journal of Fisheries and Aquatic Sciences, 18*, 1123–1139.

Kordon, A. O., Pinchuk, L., & Karsi, A. (2022). Adaptive immune system in fish. *Turkish Journal of Fisheries and Aquatic Sciences, 22*(4), TRJFAS20235. http://doi.org/10.4194/TRJFAS20235 (accessed on 11th July 2025).

Kriegler, M., Perez, C., DeFay, K., Albert, I., & Lu, S. D. (1988). A novel form of TNF/cachectin is a cell surface cytotoxic transmembrane protein: Ramifications for the complex physiology of TNF. *Cell, 53*, 45–53.

Krishnan, R., Kurcheti, P. P., Mushtaq, Z., Jeena, K., & Vismai, N. T. (2019). Interferon regulatory factors, IRF3 and IRF7 in Asian seabass, *Lates calcarifer*: Characterization, ontogeny and transcriptional modulation upon challenge with nervous necrosis virus. *Fish & Shellfish Immunology*, *89*, 468–476. https://doi.org/10.1016/j.fsi.2019.03.073 (accessed on 11th July 2025).

Kutluyer, F., Sirkecioglu, A. N., Aksakal, E., Aksakal, F. İ., Tunç, A., & Günaydın, E. (2017). Effect of dietary fish oil replacement with plant oils on growth performance and gene expression in juvenile rainbow trout (*Oncorhynchus mykiss*). *Annals of Animal Science*, *17*(4), 1135–1153.

Lei, X. D., Sun, Y., Cai, S. J., Fang, Y. W., Cui, J. L., & Li, Y. H. (2016). Role of tumor necrosis factor-alpha in zebrafish retinal neurogenesis and myelination. *International Journal of Ophthalmology*, *9*(6), 831–837. https://doi.org/10.18240/ijo.2016.06.07 (accessed on 11th July 2025).

Liu, D., Mai, K., Zhang, Y., Xu, W., & Ai, Q. (2016). Tumor necrosis factor-α inhibits hepatic lipid deposition through GSK-3β/β-catenin signaling in juvenile turbot (*Scophthalmus maximus* L.). *General and Comparative Endocrinology*, *228*, 1–8. https://doi.org/10.1016/j.ygcen.2015.12.027 (accessed on 11th July 2025).

Liu, X., Li, X., Du, X., Sun, M., Wang, X., Li, W., Zhai, J., Liu, J., Yu, H., & Zhang, Q. (2019). Spotted knifejaw (*Oplegnathus punctatus*) MyD88: Intracellular localization, signal transduction function and immune responses to bacterial infection. *Fish & Shellfish Immunology*, *89*, 719–726.

Liu, Y., Li, Y., Zhou, Y., Jiang, N., Fan, Y., & Zeng, L. (2020). Characterization, expression pattern and antiviral activities of Mx gene in Chinese giant salamander, *Andrias davidianus*. *International Journal of Molecular Sciences*, *21*, 2246. https://doi.org/10.3390/ijms21062246 (accessed on 11th July 2025).

Loh, C. Y., Arya, A., Naema, A. F., Wong, W. F., Sethi, G., & Looi, C. Y. (2019). Signal transducer and activator of transcription (STATs) proteins in cancer and inflammation: Functions and therapeutic implication. *Frontiers in Oncology*, *9*, Article 48.

Magnadottir, B. (2006). Innate immunity of fish (overview). *Fish & Shellfish Immunology*, *20*, 137–151. https://doi.org/10.1016/j.fsi.2004.09.006 (accessed on 11th July 2025).

Malur, M., Gale, M., & Krug, R. M. (2012). LGP2 downregulates interferon production during infection with seasonal human influenza A viruses that activate interferon regulatory factor 3. *Journal of Virology*, *86*, 10733–10738.

Mates, J. M., Perez-Gomez, C., & De Castro, I. N. (1999). Antioxidant enzymes and human diseases. *Clinical Biochemistry*, *32*, 595–603.

Meier, E., Kunz, G., Haller, O., & Arnheiter, H. (1990). Activity of rat Mx proteins against a rhabdovirus. *Journal of Virology*, *64*(12), 6263–6269.

Mohindra, V., Dangi, T., Chowdhury, L. M., & Jena, J. K. (2019). Tissue specific alpha-2-macroglobulin (A2M) splice isoform diversity in Hilsa shad, *Tenualosa ilisha* (Hamilton, 1822). *PLOS ONE*, *14*(7), e0216144.

Mokhtar, D. M., Zaccone, G., Alesci, A., Kuciel, M., Hussein, M. T., & Sayed, R. K. A. (2023). Main components of fish immunity: An overview of the fish immune system. *Fishes*, *8*, 93. https://doi.org/10.3390/fishes8020093 (accessed on 11th July 2025).

Montero, D., Mathlouthi, F., Tort, L., Afonso, J. M., Torrecillas, S., Fernández Vaquero, A., Negrin, D., & Izquierdo, M. S. (2010). Replacement of dietary fish oil by vegetable oils affects humoral immunity and expression of pro-inflammatory cytokines genes in gilthead sea bream *Sparus aurata*. *Fish & Shellfish Immunology*, *29*, 1073–1081.

Mourente, G., Good, J. E., & Bell, J. G. (2005). Partial substitution of fish oil with rapeseed, linseed and olive oils in diets for European sea bass (*Dicentrarchus labrax* L.): Effects on flesh fatty acid composition, plasma prostaglandins E2 and F2α, immune function and effectiveness of a fish oil finishing diet. *Aquaculture Nutrition*, *11*, 25–40.

Mu, H., Wei, C., Xua, W., Gao, W., Zhang, W., & Mai, K. (2020). Effects of replacement of dietary fish oil by rapeseed oil on growth performance, anti-oxidative capacity and inflammatory response in large yellow croaker *Larimichthys crocea*. *Aquaculture Reports*, *16*, 100251.

Oeckinghaus, A., & Ghosh, S. (2009). The NF-kappaB family of transcription factors and its regulation. *Cold Spring Harbor Perspectives in Biology*, *1*, a000034.

Oliver, W. T., & Wells, J. E. (2013). Lysozyme as an alternative to antibiotics improves growth performance and small intestinal morphology in nursery pigs. *Journal of Animal Science*, *91*, 3129–3136. https://doi.org/10.2527/jas.2012-5782 (accessed on 11th July 2025).

Onara, D. F., Forlenza, M., Gonzalez, S. F., Rakus, K. Ł., Pilarczyk, A., Irnazarow, I., & Wiegertjes, G. F. (2008). Differential transcription of multiple forms of alpha-2-macroglobulin in carp (*Cyprinus carpio*) infected with parasites. *Developmental & Comparative Immunology*, *32*(4), 339–347. https://doi.org/10.1016/j.dci.2007.06.007 (accessed on 11th July 2025).

Palti, Y. (2011). Toll-like receptors in bony fish: From genomics to function. *Developmental and Comparative Immunology*, *35*, 1263–1272.

Pedrajas, J. R., Peinado, J., & López-Barea, J. (1995). Oxidative stress in fish exposed to model xenobiotics. Oxidatively modified forms of Cu, Zn-superoxide dismutase as potential biomarkers. *Chemico-Biological Interactions*, *98*, 267–282.

Penglee, R., Gao, L., Huang, Y., Liu, L., Nimitkul, S., & Bao, B. (2021). The role of nitric oxide and neuronal nitric oxide synthase in zebrafish (*Danio rerio*) shoaling. *Aquaculture and Fisheries*, *6*(6), 565–573. https://doi.org/10.1016/j.aaf.2020.08.007 (accessed on 11th July 2025).

Peter, M. C. S., & Gayathry, R. (2021). Nitric oxide synthase (NOS) inhibitor L-NAME activates inducible NOS/NO system and drives multidimensional regulation of Na^+/K^+ ATPase in ionocyte epithelia of immersion-stressed air-breathing fish (*Anabas testudineus* Bloch). *Journal of Experimental Zoology*, *335*, 396–416. https://doi.org/10.1002/jez.2454 (accessed on 11th July 2025).

Peter, M. C. S., & Gayathry, R. (2022). Inducible nitric oxide synthase/nitric oxide system as a biomarker for stress and ease response in fish: Implication on Na^+ homeostasis during hypoxia. *Frontiers in Physiology*, *13*, 821300. https://doi.org/10.3389/fphys.2022.821300 (accessed on 11th July 2025).

Priyathilaka, T. T., Kim, Y., Udayantha, H. M., Lee, S., Herath, H. M., Lakmal, H. H., Elvitigala, D. A., Umasuthan, N., Godahewa, G. I., & Kang, S. I. (2016). Identification and molecular characterization of peroxiredoxin 6 from Japanese eel (*Anguilla japonica*) revealing its potent antioxidant properties and putative immune relevancy. *Fish & Shellfish Immunology*, *51*, 291–302.

Prosperi, M. T., Ferbus, D., Karczinski, I., & Goubin, G. (1993). A human cDNA corresponding to a gene overexpressed during cell proliferation encodes a product sharing homology with amoebic and bacterial proteins. *Journal of Biological Chemistry, 268*, 11050–11056.

Purcell, M. K., Smith, K. D., Hood, L., Winton, J. R., & Roach, J. C. (2006). Conservation of Toll-like receptor signaling pathways in teleost fish. *Comparative Biochemistry and Physiology–Part D: Genomics and Proteomics, 1*, 77–88.

Qiu, W., Hu, J., Magnuson, J. T., Greer, J., Yang, M., Chen, Q., Fang, M., Zheng, C., & Schlenk, D. (2020). Evidence linking exposure of fish primary macrophages to antibiotics activates the NF-κB pathway. *Environment International, 138*, 105624. https://doi.org/10.1016/j.envint.2020.105624 (accessed on 11th July 2025).

Rabilloud, T., Berthier, R., Vincon, M., Ferbus, D., Goubin, G., & Lawrence, J. J. (1995). Early events in erythroid differentiation: Accumulation of the acidic peroxidoxin (PRP/TSA/NKEF-B). *Biochemical Journal, 312*(3), 699–705.

Reyes-Cerpa, S., Maisey, K., Reyes-López, F., Toro-Ascuy, D., Sandino, A. M., & Imarai, M. (2012). Fish cytokines and immune response. In *New Advances and Contributions to Fish Biology*. https://doi.org/10.5772/53504 (accessed on 11th July 2025).

Robertsen, B. (2006). The interferon system of teleost fish. *Fish & Shellfish Immunology, 20*, 172–191.

Rosenlund, G., Obach, A., Sandberg, M. G., Standal, H., & Tveit, K. (2001). Effect of alternative lipid sources on long-term growth performance and quality of Atlantic salmon (*Salmo salar* L.). *Aquaculture Research, 32*, 323–328.

Sahoo, S., Banu, H., Prakash, A., & Tripathi, G. G. (2021). Immune system of fish: An evolutionary perspective. *IntechOpen*. https://doi.org/10.5772/intechopen.99541 (accessed on 11th July 2025).

Sampaio, F. G., Boijink, C. L., Oba, E. T., Santos, L. R. B. B., Kalinin, A. L., & Rantin, F. T. (2008). Antioxidant defenses and biochemical changes in pacu (*Piaractus mesopotamicus*) in response to single and combined copper and hypoxia exposure. *Comparative Biochemistry and Physiology Part C, 147*, 43–51.

Sato, K., Takami, A., Isozaki, T., Hikida, T., Shimono, A., & Imamura, T. (2010). Mass spectrometric study of secondary organic aerosol formed from the photo-oxidation of aromatic hydrocarbons. *Atmospheric Environment, 44*(8), 1080–1087.

Saurabh, S., & Sahoo, P. K. (2008). Lysozyme: An important defense molecule of fish innate immune system. *Aquaculture Research, 39*, 223–239.

Secombes, C. J., & Wang, T. (2012). The innate and adaptive immune system of fish. In *Infectious Disease in Aquaculture: Prevention and Control* (pp. 3–68). Elsevier.

Skjæveland, I., Iliev, D. B., Strandskog, G., & Jørgensen, J. B. (2009). Identification and characterization of TLR8 and MyD88 homologs in Atlantic salmon (*Salmo salar*). *Developmental & Comparative Immunology, 33*, 1011–1017.

Smith, N. C., Rise, M. L., & Christian, S. L. (2019). A comparison of the innate and adaptive immune systems in cartilaginous fish, ray-finned fish, and lobe-finned fish. *Frontiers in Immunology, 10*, 2292. https://doi.org/10.3389/fimmu.2019.02292 (accessed on 11th July 2025).

Srivastava, P. K., & Pandey, A. K. (2015). Role of immunostimulant in immune responses of fish and shellfish. *Biochemical and Cellular Archives, 15*(1), 47–73.

Sveinbjornsson, B., Olsen, R., & Paulsen, S. (2006). Immunocytochemical localization of lysozyme in intestinal eosinophilic granule cells (EGCs) of Atlantic salmon, *Salmo salar* L. *Journal of Fish Diseases, 19*(5), 349–355.

Takano, T., Kondo, H., Hirono, I., Saito-Taki, T., Endo, M., & Aoki, T. (2006). Identification and characterization of a myeloid differentiation factor 88 (MyD88) cDNA and gene in Japanese flounder, *Paralichthys olivaceus*. *Developmental & Comparative Immunology*, *30*, 807–816.

Tamura, T., Yanai, H., Savitsky, D., & Taniguchi, T. (2008). The IRF family transcription factors in immunity and oncogenesis. *Annual Review of Immunology*, *26*, 535–584.

Tan, P., Dong, X., Mai, K., Xu, W., & Ai, Q. (2016). Vegetable oil induced inflammatory response by altering TLR-NF-κB signalling, macrophage infiltration and polarization in adipose tissue of large yellow croaker (*Larimichthys crocea*). *Fish & Shellfish Immunology*, *59*, 398–405.

Tang, S., Xu, X., Yu, D., Long, M., Xia, H., Lu, Y., & Gan, Z. (2023). Evolutionary and functional conservation of myeloid differentiation factor 88 (MyD88) in amphibian *Xenopus tropicalis*. *Gene*, *865*, 147332. https://doi.org/10.1016/j.gene.2023.147332 (accessed on 11th July 2025).

Tang, X., Yang, M., Liu, J., Zheng, L., Xu, D., Chi, C., Lv, Z., & Liu, H. (2022). Identification, functional characterization and expression pattern of myeloid differentiation factor 88 (MyD88) in *Nibea albiflora*. *Fish & Shellfish Immunology*, *124*, 380–390. https://doi.org/10.1016/j.fsi.2022.04.027 (accessed on 11th July 2025).

Tangye, S. G., & Tarlinton, D. M. (2009). Memory B cells: Effectors of long-lived immune responses. *European Journal of Immunology*, *39*(8), 2065–2075. https://doi.org/10.1002/eji.200939531 (accessed on 11th July 2025).

Torstensen, B. E., Bell, J. G., Sargent, J. R., Rosenlund, G., Henderson, R. J., & Graff, I. E. (2005). Tailoring of Atlantic salmon (*Salmo salar* L.) flesh lipid composition and sensory quality by replacing fish oil with a vegetable oil blend. *Journal of Agricultural and Food Chemistry*, *53*, 10166–10178.

Turvey, S. E., & Broide, D. H. (2010). Innate immunity. *The Journal of Allergy and Clinical Immunology*, *125*(2 Suppl 2), S24–S32. http://dx.doi.org/10.1016/j.jaci.2009.07.016 (accessed on 11th July 2025).

Uribe, C., Folch, H., Enriquez, R., & Moran, G. (2011). Innate and adaptive immunity in teleost fish: A review. *Veterinarni Medicina*, *56*(10), 486–503.

Vaillant, A. A. J., & Qurie, A. (2021). Interleukin. In *StatPearls* [Internet]. Treasure Island (FL): StatPearls Publishing. PMID: 29763015.

Veal, E. A., Day, A. M., & Morgan, B. A. (2007). Hydrogen peroxide sensing and signaling. *Molecular Cell*, *26*, 1–14.

Wang, L., Lu, Q., Luo, S., Zhan, W., Chen, R., Lou, B., & Xu, D. (2016). Effect of dietary lipid on growth performance, body composition, plasma biochemical parameters and liver fatty acids content of juvenile yellow drum *Nibea albiflora*. *Aquaculture Reports*, *4*, 10–16.

Wang, T., & Secombes, C. J. (2013). The cytokine networks of adaptive immunity in fish. *Fish & Shellfish Immunology*, *35*(6), 1703–1718.

Wang, X., Hinson, E. R., & Cresswell, P. (2007). The interferon-inducible protein viperin inhibits influenza virus release by perturbing lipid rafts. *Cell Host & Microbe*, *2*, 96–105.

Xiao, J., Zhou, Z. C., Chen, C., Huo, W. L., Yin, Z. X., Weng, S. P., Chan, S. M., Yu, X. Q., & He, J. G. (2007). Tumor necrosis factor-alpha gene from mandarin fish, *Siniperca chuatsi*: Molecular cloning, cytotoxicity analysis and expression profile. *Molecular*

Immunology, *44*(14), 3615–3622. https://doi.org/10.1016/j.molimm.2007.03.016 (accessed on 11th July 2025).

Xu, Z., Parra, D., Gomez, D., Salinas, I., Zhang, Y. A., von Gersdorff Jorgensen, L., Heinecke, R. D., Buchmann, K., LaPatra, S., & Sunyer, J. O. (2013). Teleost skin, an ancient mucosal surface that elicits gut-like immune responses. *Proceedings of the National Academy of Sciences of the United States of America*, *110*, 13097–13102. https://doi.org/10.1073/pnas.1304319110 (accessed on 11th July 2025).

Yan, X. B., Dong, X. H., Tan, B. P., Zhang, S., Chi, S. Y., Yang, Q. H., Liu, H. Y., & Yang, Y. Z. (2020). Influence of different oil sources on growth, disease resistance, immune response and immune-related gene expression on the hybrid grouper (♀ *Epinephelus fuscoguttatus* × ♂ *E. lanceolatus*), to *Vibrio parahaemolyticus* challenge. *Fish & Shellfish Immunology*, *99*, 310–321. https://doi.org/10.1016/j.fsi.2020.02.025 (accessed on 11th July 2025).

Yang, C. G., Jiang, M., Wu, F., Yu, L., Tian, J., Liu, W., Lu, X., & Wen, H. (2016). Identification of a C-type lectin from tilapia (*Oreochromis niloticus*) and its functional characterization under low-temperature stress. *Fish & Shellfish Immunology*, *58*, 631–640.

Yao, C. L., Kong, P., Wang, Z. Y., Ji, P. F., Cai, M. Y., & Liu, X. D. (2009). Molecular cloning and expression of MyD88 in large yellow croaker, *Pseudosciaena crocea*. *Fish & Shellfish Immunology*, *26*, 249–255.

Ye, J., Kaattari, I. M., Ma, C., & Kaattari, S. (2013). The teleost humoral immune response. *Fish & Shellfish Immunology*, *35*(6), 1719–1728.

Yong, A. S. K., Ooi, S. Y., Shapawi, R., Biswas, K., & Kenji, T. (2015). Effects of dietary lipid increments on growth performance, feed utilization, carcass composition and intraperitoneal fat of marble goby, *Oxyeleotris marmorata*, juveniles. *Turkish Journal of Fisheries and Aquatic Sciences*, *15*, 653–660.

Yu, L., Zhang, L., Yang, H., Gui, G., Liu, Y., & Xiao, Y. (2018). Identification and characterization of the myeloid differentiation factor 88 gene in yellow catfish. *3 Biotech*, *8*, 430.

Zamocky, M., Furtmüller, P. G., & Obinger, C. (2008). Evolution of catalases from bacteria to humans. *Antioxidants & Redox Signaling*, *10*(9), 1527–1548. https://doi.org/10.1089/ars.2008.2046 (accessed on 11th July 2025).

Zelck, U. E., & Von Janowsky, B. (2004). Antioxidant enzymes in intramolluscan *Schistosoma mansoni* and ROS-induced changes in expression. *Parasitology*, *128*(5), 493–500.

Zhao, X., Duan, D., Feng, X., Chen, Y., Sun, Z., Jia, S., He, J., Wang, B., Li, W., Zhang, J., Wang, W., Yang, Z., & Lu, Q. (2012). Molecular cloning and expression analysis of common carp *Cyprinus carpio* tumor necrosis factor-α. *Fisheries Science*, *78*(6), 1229–1236. https://doi.org/10.1007/s12562-012-0563-8 (accessed on 11th July 2025).

Zheng, W. J., Hu, Y. H., Zhang, M., & Sun, L. (2010). Analysis of the expression and antioxidative property of a peroxiredoxin 6 from *Scophthalmus maximus*. *Fish & Shellfish Immunology*, *29*, 305–311.

Zou, J., Secombes, C. J., Long, S., Miller, N., Clem, L. W., & Chinchar, V. G. (2003). Molecular identification and expression analysis of tumor necrosis factor in channel catfish (*Ictalurus punctatus*). *Developmental & Comparative Immunology*, *27*, 845–858.

Zuo, R., Ai, Q., Mai, K., Xu, W., Wang, J., Xu, H., Liufu, Z., & Zhang, Y. (2012). Effects of dietary n-3 highly unsaturated fatty acids on growth, nonspecific immunity, expression of some immune-related genes and disease resistance of large yellow croaker (*Larimichthys crocea*) following natural infestation of parasites (*Cryptocaryon irritans*). *Fish & Shellfish Immunology*, *32*(2), 249–258.
Zurcher, T., Pavlovic, J., & Staeheli, P. (1992). Mouse Mx2 protein inhibits vesicular stomatitis virus but not influenza virus. *Virology*, *187*(2), 796–800.

CHAPTER 22

Comparative Evaluation of *Saccharomyces cerevisiae* and *Bacillus subtilis* 200 on Growth Performance, Proximate Composition, and Feed Utilization of African Bony Tongue (*Heterotis niloticus*)

CHRISTIAN LARBI AYISI

Department of Water Resources and Aquaculture Management, School of Sustainable Development, University of Environment and Sustainable Development, Somanya, Eastern Region, Ghana

*Corresponding author

ABSTRACT

The present study was performed to evaluate the effects of dietary incorporation of *Saccaromyeces cerevisae* (SC) and *Bacillus subtilis* 200 (BS) on the growth performance, proximate composition and feed utilization of African bony tongue (*Heterotis niloticus*). *H. niloticus* was fed four different diets for eight weeks, viz. control diet (Ctrl), diet with 4g BS per kg (BS), diet with 4g SC per kg (SC) and diet containing 2g BS and 2g SC per kg (BS+SC). The outcome revealed significant differences ($P<0.05$) between treatments with respect to final weight (FW), percentage weight gain (WG%), weight gain (WG) and specific growth rate (SGR) whereas, difference in condition factor and feed intake was insignificantly ($P>0.05$). The test species fed

Sustainable Aquaculture: Recent Advances and Techniques. Dhanaji W. Patil, Sandip S. Markad (Eds.)

DOI: 10.1201/9781779641113-22

with BS+SC diet had significantly higher growth performance (FW, WG, WG%, and SGR), while FCR (1.04±0.17) for diet Ctrl was significantly higher ($P<0.05$). Protein efficiency ratio for diet BS+SC (3.39±0.11) was significantly higher ($P<0.05$). Significant difference ($P<0.05$) was found in crude lipid and moisture contents. The study concludes that, dietary incorporation of symbiotics of *Saccharomyces cerevisae* and *Bacillus subtilis* 200 in African bony tongue can improve growth performance and feed utilization without adverse impact on body's proximate composition.

22.1 INTRODUCTION

Intensifying aquaculture production to meet the growing demand for fish globally has resulted in an increase in outbreaks of infections, including viral and bacterial diseases (Gayeed et al., 2021; Naylor et al., 2021), due to increased stress and weakened immunity, making cultured *spp.* susceptible to pathogens (Dawood et al., 2016; Shadrack et al., 2021). The outbreak of diseases results in economic loss, which subsequently hinders the sustainable development and expansion of the aquaculture industry (Cerezuela et al., 2011).

Generally, chemotherapy and antibiotics are used for controlling fish disease outbreaks. Numerous studies have reported the usage of antibiotics during the treatment of various health disorders (Das et al., 2021; Islam et al., 2021; Maas et al., 2021; Shadrack et al., 2021). In cultured species, overuse of antibiotics and chemotherapy may result in the build-up of toxins, drug residues, and bacterial resistance. These might have detrimental effects on both the environment and people (Dash et al., 2015). Finding eco-friendly alternatives with minimal to no side effects is therefore essential (Huynh et al., 2018).

Probiotics improve the growth, health, and absorption of nutrients in cultured fish (Dawood et al., 2015). Probiotics are environmentally friendly feed additives that increase fish production (Akhter et al., 2015; Chowdhury and Roy, 2020; Lazado and Caipang, 2014) by enhancing the action of different digestive enzymes to improve feed utilization, which increases nutrient availability (Dawood and Koshio, 2016; Dawood et al., 2020). Furthermore, by enhancing gut health, dietary probiotics promote the growth of the cultured species (Andani et al., 2012; Han et al., 2015). Again, by changing the make-up of the gut's microbial community, probiotics improve the body's capacity to produce enzymes (lipases, amylases, proteases, etc.) in the digestive system (Kuebutornye et al., 2020). Additionally, probiotics increase the availability of nutrients like short-chain fatty acids (SCFA) and amino acids (Dawood et al., 2019; Maas et al., 2021).

Saccharomyces cerevisiae is one of the probiotics that has been seen as a promising candidate in aquafarming (Abu-Elala et al., 2013; Yang et al., 2020) because it modifies the morphological structure and microbiota of the fish intestine, which affects the health of fish that are cultured (He et al., 2011). *Bacillus* spp. has also been utilized in aquaculture extensively (Chen et al., 2020; Van Doan et al., 2016; Wang et al., 2008) and shown to improve feed intake, alter gut microbiota, activate immune systems, and strengthen resistance to disease (Cao et al., 2011; Das et al., 2013; Ran et al., 2012; Reda and Selim, 2015; Ridha and Azad, 2012; Ringo et al., 2014; Ringo and Song, 2016). Additionally, it has been noted that combining different probiotic strains enhances the health advantages for cultivated species (Giri et al., 2014; Niu et al., 2019; Shadrack et al., 2021).

In sub-Saharan Africa, river basins are home to the African bonytongue (*Heterotis niloticus*), which is harvested by fisheries. As a result of *H. niloticus*'s rapid growth, its value for human consumption, its ability to reach a weight of 7 kg, and its omnivorous diet, it is a promising aquaculture candidate (Goure-Bi et al., 2018; Monentcham et al., 2010).

Several studies have reported the application of probiotics in diets of fish, Park et al. (2016) (*Oncorhynchus mykiss*), Rodigues et al. (2020) (*Rhamdia quelen*), Siddik et al. (2021) (*Lates calcarifer*), Amoah et al. (2021) (*Sillago sihama Forsskal, 1775*), Shadrack et al. (2021) (*Seriola dumerili*), Maas et al. (2021) (*O. niloticus*), Das et al. (2021) (*Heteropneustes fossilis*), and Mohammadian et al. (2022) (*Cyprinio carpio*). The impact of probiotics on *H. niloticus* is currently the subject of less investigation. Thus, the purpose of this research is to assess how *Saccharomyces cerevisiae* and *Bacillus subtilis* 200 affect *H. niloticus* growth performance, feed utilization, and proximate composition.

22.2 MATERIALS AND METHODS

22.2.1 FEED PREPARATION

Commercial *Bacillus cereus* and *Saccharomyces cerevisiae* were procured and used for diet preparation in this study. Table 22.1 shows the basal diet (control diet) with crude protein of 34% used for this study. Three experimental diets, excluding the control, were prepared to contain 4 g/

kg of *Bacillus cereus* (Diet BS), 4 g/kg of *Saccharomyces cerevisiae* (Diet SC), and 2 g/kg each of *Bacillus cereus* and *Saccharomyces cerevisiae* (Diet BS+SC). Following procedures previously described by Xia et al. (2018), the formulated diets were prepared. Briefly, ingredients that had been ground were mixed thoroughly using the hand. The mixture was then combined with water and oil, and the appropriate amount of *Saccharomyces cerevisiae* and *Bacillus cereus* were added to form a soft dough. The diets were prepared fresh every week and stored in plastic bags at 40°C to guarantee the viability of the added probiotics.

TABLE 22.1 Formulation and Calculated Chemical Compositions of the Basal Diet

Ingredients	**Percentage (%)**
Fish meal	20
Soybean meal	30
Wheat bran	15
Maize	13
Salt	2
Vitamins premix	2
Mineral pre-mix	2
Starch	6
Oil	10
Probiotics	0
Proximate Composition	
Protein	40.0
Lipid	9.52
Moisture	10.79
Ash	5.38

22.2.2 EXPERIMENTAL FISH AND REARING CONDITION

Fingerlings of *Heterotis niloticus* were purchased from the Aquaculture Research and Development Centre (ARDEC) of the CSIR-Water Research Institute, Akosombo. After being delivered to the culture facility, fingerlings underwent a two-week period of acclimation. For the duration of the acclimatization period, fingerlings were kept in a concrete tank measuring 2 meters by 2 meters by 1 meter. The culturing tanks were connected to one reservoir, which provided a fresh water supply on a regular basis. Every two days, 30% of the water was replaced. Each culturing tank, with dimensions of 2m × 1m × 1m, was stocked with twenty (*H. niloticus*) fingerlings (initial weight of 9.16 ± 0.36). All tanks were aerated during the rearing period.

22.2.3 SAMPLE COLLECTION

All surviving fish after the feeding regime were counted. Body weight of each fish was measured in grams (g) using a CAS electronic scale, while total length was measured in centimeters (cm) using a measuring board. On a random basis, the fish (whole body) were sampled for the analysis of their respective lipid, ash, moisture, and protein content.

22.2.4 GROWTH AND FEED UTILIZATION

The following metrics were used to evaluate feed utilization and growth performance. Mutalib et al. (2023) have previously used these standard equations.

- The total amount of feed (g) consumed during the 56-day trial is the feed intake (FI).
- $\text{Feed conversion ratio (FCR)} = \frac{\text{Fee intake(g)}}{\text{weight gain (g)}}$
- $\text{Protein efficiency ratio (PER)} = \frac{\text{Wet weight gain (g)}}{\text{protein gain (g)}}$
- Weight gain (WG) = Final weight (g) -Initial weight (g)
- $\text{\% Weight gain (\%WG)} = \frac{\text{Final weight (g) -Initial weight (g)}}{\text{Initial weight (g)}} \times 100$
- $\text{Specific growth rate (SGR)(\%)} = \frac{\text{In(final weight (g)) -In(Initial weight (g))}}{\text{Number of feeding days}} \times 100$
- $\text{Condition factor (K)} = \frac{\text{body weight (g)}}{\text{standard length (cm)}^2}$
- $\text{Hepatosomatic index} = \frac{\text{liver weight}(g)}{\text{body weight}(g)} \times 100$

22.2.5 PROXIMATE COMPOSITION

The AOAC (2003) standard methods, as reported by Mehbood et al. (2017), were utilized to ascertain the body's proximate composition, which includes moisture, lipid, protein, and ash. Fifteen fish per treatment (five fish per tank) had their entire bodies sampled and sent for analysis in a lab. Fish samples were dried to a constant weight at 105°C to determine their moisture content. The samples were then burned for 12 hours at 550°C in a muffle furnace to ascertain their ash content. The crude protein content was analyzed using the Kjeldahl method, and the nitrogen content was first calculated and then multiplied by 6.25. For values, the percentage dry weight is used. The analysis of crude lipid content was performed using the Folch et al. (1957) method.

22.2.6 STATISTICAL ANALYSIS

Graph Pad Prism (V.5.03) was used to statistically analyze the collected data. The standard error of the mean (SEM), expressed as the mean ±, is used to present the data in tables and graphs. All data were analyzed using One-way Analysis of Variance and Tukey's multiple tests in order to compare treatment means. For all data, differences are deemed significant at the 0.05 probability level ($P < 0.05$).

22.3 RESULTS

22.3.1 GROWTH PERFORMANCE

Table 22.2 shows results on *H. niloticus* growth performance when fed diets supplemented with *B. cereus* and/or *S. cerevisiae*. The effect of supplementing diets of *H. niloticus* with *B. subtilis* and/or *S. cerevisiae* on growth performance is shown in Table 22.2. After 10 weeks of feeding trials, supplementing diets of *H. niloticus* with *B. subtilis* and/or *S. cerevisiae* significantly increased final weight ($p = 0.0044$). The group fed the diet BS+SC had the highest final weight of 142.55 ± 4.21, which was significantly more than the control group's final weight of 120.79 ± 6.25. Weight gain and percentage weight gain were significantly different amongst treatments and ranged from 110.64 ± 2.15 to 133.83 ± 3.45 and 1220.13 ±

80.54% to 1461.31 ± 79.10%, respectively. There were notable differences in the specific growth rate (SGR) between the treatments. The group fed the diet BS+SC had the highest SGR (4.93 ± 0.38) and was significantly higher than the group fed the Ctrl diet (3.95 ± 0.43). Groups fed diets BS and SC recorded SGR of 4.84 ± 0.49 and 4.82 ± 0.51, respectively.

TABLE 22.2 Growth Performance of *Heterotis niloticus* Fed with Diets Supplemented with *Bacillus cereus* and/or *Saccharomyces cerevisiae*

Parameters	Experimental Groups				
	Ctrl	BS	SC	BS+SC	p-Value
Initial weight (g)	9.15±1.74	9.20±1.08	9.17±1.35	9.13±1.64	0.9579
Initial length (cm)	11.19±1.15	11.27±1.33	11.17±1.25	11.42±1.28	0.8795
Final weight (g)	120.79±6.25[a]	138.96±3.18[b]	136.53±3.52[ab]	142.55±4.21[b]	0.0044
Final length (cm)	18.47±2.53	18.55±2.15	18.45±2.47	19.18±2.25	0.6793
Weight gain (g)	110.64±2.15[a]	130.73±2.44[b]	126.64±3.04[b]	133.83±3.45[b]	0.0017
Percentage weight gain (%)	1220.13±80.54[a]	1410.51±73.82[b]	1389.21±82.16[b]	1461.31±79.10[b]	0.0019
Specific growth weight (%)	3.95±0.43[a]	4.84±0.49[b]	4.82±0.51[b]	4.93±0.38[b]	0.0008

Note: Values are means ± SEM. (n = 3). Significant differences (P < 0.05) exist between the means in the same column with different superscripts. Ctrl: Control, BS: *Bacillus subtilis* 200, SC: *Saccharomyces cerevisiae*, BS+SC: *Bacillus subtilis* 200 (BS) + *Saccharomyces cerevisiae* (SC).

22.3.2 FEED UTILIZATION, CONDITION FACTOR, AND HEPATOSOMATIC INDEX

A summary of the experimental groups' feed utilization, condition factor, and hepatosomatic index is provided in Table 22.3. Regarding feed intake, there was no discernible variation between the treatments (P=0.1574). The group fed with diet *Bacillus subtilis* (BS) recorded the highest feed intake of 118.18±4.76 g, followed by groups fed with diets *Bacillus subtilis* + *Saccharomyces cerevisiae* (BS+SC) (116.43±6.82 g), *Saccharomyces cerevisiae* (SC) (115.17±5.59 g), and control (Ctrl) (114.25±5.43 g). Feed conversion ratio (FCR) was significantly different amongst treatments (P=0.0009). The group fed the Ctrl diet had the highest FCR (1.04±0.17), which was significantly higher than that of all other groups. The FCR of the BS diet group was 0.92±0.10, whereas the FCRs of the SC and BS+SC diet groups were 0.91±0.13 and 0.87±0.12, respectively. The condition factor (K) did not significantly differ between treatments (P=0.8883). K values of 1.93±0.38 and 2.18±0.18, respectively, were recorded by groups fed diets Ctrl and BS, while K values of 2.19±0.27 and 2.01±0.25, respectively, were recorded by groups fed diets SC and BS+SC.

TABLE 22.3 Feed Utilization, Condition Factor, and Hepatosomatic Index

Parameters	Experimental Groups				
	Ctrl	BS	SC	BS+SC	p-Value
Feed intake	114.25±5.43	118.18±4.76	115.17±5.59	116.43±6.82	0.1574
Feed conversion ratio	1.04±0.17[c]	0.92±0.10[b]	0.91±±0.13[ab]	0.87±0.12[a]	0.0009
Protein efficiency ratio	2.86±0.21[a]	3.24±0.19[b]	3.25±0.21[b]	3.39±0.11[c]	0.0164
Hepatosomatic index	2.05±0.60[a]	2.26±0.28[b]	2.28±0.47[b]	2.83±0.45[c]	0.0417
Condition factor	1.93±0.38	2.18±0.18	2.19±0.27	2.01±0.25	0.8883

Note: Values are means ± SEM. (n=3). Significant differences ($P < 0.05$) exist between the means in the same column with different superscripts. Ctrl: Control, BS: *Bacillus subtilis* 200, SC: *Saccharomyces cerevisiae*, BS+SC: *Bacillus subtilis* 200 (BS) + *Saccharomyces cerevisiae* (SC).

Different probiotics had a significant effect on protein efficiency ratio (P=0.0164). The group fed with the Ctrl diet recorded a PER of 2.86±0.21 and was significantly lower than the group fed with the BS+SC diet (3.39±0.11), while the group fed with diets BS and SC recorded PER of 3.24±0.19 and 3.25±0.21, respectively. There was a significant difference in hepatosomatic index (HSI) amongst groups (P=0.0417). Groups fed with diets Ctrl, BS, and SC recorded HSI of 2.05±0.60, 2.26±0.28, and 2.28±0.47, respectively, and were significantly lower than the group fed with the BS+SC diet (2.83±0.45).

22.3.3 PROXIMATE WHOLE-BODY COMPOSITION

Table 22.4 displays the impact of probiotics on the proximate composition of the whole-body (% dry weight) of *Heterotis niloticus*. Regarding the total body protein content, there was no discernible variation between the treatments (P = 0.6881). The group fed the SC diet had the lowest whole-body protein content (15.92±1.47), while the group fed the control diet had the highest (17.01±1.88). Groups fed with diets BS and BS+SC recorded whole-body protein content of 16.19±1.59 and 16.56±0.96, respectively. Lipid content of whole-body was significantly different (P=0.0001) among the groups. The group fed the BS+SC diet had a lipid content of

9.94±0.19, which was significantly higher than the groups fed the Ctrl and BS diets, which had recorded lipid contents of 6.52±0.59 and 8.63±0.13, respectively. The whole-body lipid content of 9.28±0.24 was recorded for groups fed with the SC diet.

Moisture content of whole body was significantly different (P=0.0452) among the groups. The group fed with the control diet was significantly lower than groups fed with probiotics. Groups fed with diets *Bacillus subtilis* (BS), *Saccharomyces cerevisiae* (SC), and *Bacillus subtilis* + *Saccharomyces cerevisiae* (BS+SC) recorded whole-body moisture content of 74.83±5.39, 75.33±2.15, and 75.03±1.84, respectively. Ash content was indifferent, irrespective of the probiotic content (P=0.8155). Ash content ranged between 3.22±0.31 (Ctrl) and 3.65±0.47 (diet BS). Groups fed with diets SC and BS+SC recorded ash content of 3.35±0.26 and 3.36±0.19, respectively.

TABLE 22.4 Proximate Composition of the Whole-Body (% Dry Weight) of *Heterotis niloticus* Fed Different Probiotics

Item	Experimental Groups				
	Ctrl	BS	SC	BS+SC	p-Value
Protein (%)	17.01±1.88	16.19±1.59	15.92±1.47	16.56±0.96	0.6881
Lipid (%)	6.52±0.59^{a}	8.63±.13^{b}	9.28±0.24^{bc}	9.94±0.19^{c}	0.0001
Moisture (%)	70.26±6.52^{a}	74.83±5.39^{b}	75.33±2.15^{b}	75.03±1.84^{b}	0.0452
Ash (%)	3.22±0.31	3.65±0.47	3.35±0.26	3.36±0.19	0.8155

Note: Values are means ± SEM. (n=3). Significant differences (P < 0.05) exist between the means in the same column with different superscripts. Ctrl: Control, BS: *Bacillus subtilis* 200, SC: *Saccharomyces cerevisiae*, BS+SC: *Bacillus subtilis* 200 (BS) + *Saccharomyces cerevisiae* (SC).

22.4 DISCUSSION

Numerous studies and reports have examined the use of probiotics in aquaculture and their potential to improve the growth and wellbeing of cultivated aquatic animals (Al-Dohali et al., 2011; Andani et al., 2012; Han et al., 2015; Lara-Flores et al., 2003; Mutalib et al., 2023; Suzer et al., 2008; Talpur et al., 2014). As far as the authors are aware, this is the first study to examine how dietary probiotic supplementation (*Saccharomyces cerevisiae* and *Bacillus subtilis* 200) affects the growth, feed utilization, and proximate composition of African bony tongue (*H. niloticus*).

The study showed that supplementing diets of *H. niloticus* with *B. subtilis* and/or *S. cerevisiae* increased final weight and specific growth rate of African bony tongue significantly. The results of the study revealed that *H. niloticus* responded positively to the dietary supplementation with *Bacillus subtilis* 200 and *Saccharomyces cerevisiae* based on the significant improvement in growth performance and parameters in feed utilization. This is consistent with earlier research showing enhanced feed utilization and growth rate performance in Nile tilapia, *Oreochromis niloticus* (Aly et al., 2008; Lara-Flores et al., 2003), African catfish, *Clarias gariepinus* (Al-Dohail et al., 2009; Akter et al., 2019), and snakehead *Channa striata* (Munir et al., 2016; Talpur et al., 2014). This study also agrees with previous reports by Doan et al., 2014 (basa fish, *Pangasius bocorti*) and Akter et al., 2019 (striped catfish, *Pangasianodon hypophthalmus*, Sauvage, 1878). The improved growth performance and feed utilization observed in *H. niloticus* fed with probiotics (*B. subtilis* and/or *S. cerevisiae*) could be attributed to the ability of probiotics to enhance the digestive enzyme activities (Liu et al., 2013).

Contrary to this study, some reports claim that dietary probiotics have no effect on feed utilization or growth (Li et al., 2005; Pooramini et al., 2009; Shadrack et al., 2021; Tovar-Ramírez et al., 2002). A number of variables, including the length of the experiment, the time of year, the size of the fish, their overall health, and how the related fish species responded to the experimental diets, could be responsible for the variations in the effects of dietary probiotics on fish growth and feed utilization (Islam et al., 2021). The type of probiotics as well as the administered dosage could be the reason for the difference in results (He et al., 2016).

Supplementing diets of *H. niloticus* with probiotics (*B. subtilis* and/or *S. cerevisiae*) resulted in an efficient feed conversion as well as optimum protein efficiency. This is in agreement with previous reports on some species such as *Oncorhynchus mykiss* (Bagheri et al., 2008; Merrifield et al., 2010), *Labeo rohita* (Mohapatra et al., 2012), and *Arapaima gigas* (Sousa et al., 2019). The results of FCR imply that feeding *H. niloticus* with diets supplemented with probiotics could enhance the conversion of feed into body weight.

The condition factor (CF) offers direct insight into the overall health of fish (Kohler et al., 1995). It is an indicator used to predict the age structure of fish, body condition, as well as growth rate and length (Shadrack et al.,

2021). The results obtained in this study with respect to CF were indifferent irrespective of probiotic supplementation. This suggests that probiotics do not affect the well-being of *H. niloticus*. This study supports a previous study by Yang et al. (2019), which reported a non-significant difference in CF of Pengze crucian carp (*Carassius auratus* var. *Pengze*) supplemented with *Bacillus cereus*. There was, however, a significant difference in the relative condition factor of *Arapaima gigas* (Da Costa Sousa et al., 2019) fed with different levels of *Enterococcus faecium*-based probiotics. The age of fish species under consideration, type of probiotics, and feeding duration could be the cause of variation in effects of probiotics on CF recorded in the various studies.

Consumers consider several factors when purchasing fish foods. One of such factors is product quality (Gatlin III et al., 2007). Nutrient profile or composition of fish is an important indicator for assessing the quality of fish products (Yang et al., 2019). In addition, the physiological condition of fish to some extent can be predicted using body composition (Ali et al., 2005). Whole body protein content did not significantly differ between treatments, suggesting that the dietary probiotic supplementation (*Bacillus subtilis* 200 and *Saccharomyces cerevisiae*) had no effect on body protein. This is consistent with earlier research that Rodrigues et al. (2020) reported.

The lipid and moisture contents showed a significant difference, which is consistent with a prior report by Junia et al. (2019). Mohapatra et al. (2012), however, reported no significant difference in moisture content when *L. rohita* were fed with different microbial probiotics. Fish fed diets containing *Bacillus subtilis* 200 and *Saccharomyces cerevisiae* had the highest lipid content, which is in contrast to earlier findings by Bandyopadhyay and Mohapatra (2009), Ye et al. (2011), and Yang et al. (2019). The ash content was lowest in the group fed with the Ctrl diet and increased in groups fed with diets supplemented with probiotics. Though, there was no significant difference in ash content.

This study concludes that optimum growth and feed utilization can be enhanced by feeding *H. niloticus* with an equal amount of *Bacillus subtilis* 200 and *Saccharomyces cerevisiae* at 4 g/kg. It is also concluded that supplementing diets of *H. niloticus* with *Bacillus subtilis* 200 and *Saccharomyces cerevisiae* does not compromise whole body proximate composition.

KEYWORDS

- **Bacillus subtilis 200**
- **condition factor**
- **feed utilization**
- **fish species**
- **growth performance**
- ***Heterotis niloticus***
- **microbial probiotics**
- ***Saccaromyeces cerevisae***

REFERENCES

Akter, M. N., Hashim, R., Sutriana, A., Siti, A. M. N., & Asaduzzaman, M. (2019). Effect of *Lactobacillus acidophilus* supplementation on growth performances, digestive enzyme activities and gut histomorphology of striped catfish (*Pangasianodon hypophthalmus*, Sauvage, 1878) juveniles. *Aquaculture Research*, *50*(3), 786–797. https://doi.org/10.1111/are.13938 (accessed on 11th July 2025).

Al-Dohail, M. A. (2010). Effect of probiotic, *Lactobacillus acidophilus* on pathogenic bacteria, growth, hematological parameters and histopathology of African catfish, *Clarias gariepinus* (Ph.D. thesis). School of Biological Sciences, University Sains Malaysia, Penang, Malaysia, 1–201.

Ali, M., Iqbal, F., Salam, A., Iram, S., & Athar, M. (2005). Comparative study of body composition of different fish species from brackish water pond. *International Journal of Environmental Science and Technology*, *2*(3), 229–232. https://doi.org/10.1007/BF03325880 (accessed on 11th July 2025).

Aly, S. M., Ahmed, Y. A. G., Ghareeb, A. A. A., & Mohamed, M. F. (2008). Studies on *Bacillus subtilis* and *Lactobacillus acidophilus* as potential probiotics on the immune response and resistance of *Tilapia nilotica* (*Oreochromis nilotica*) to challenge infections. *Fish & Shellfish Immunology*, *25*, 128–136. https://doi.org/10.1016/j.fsi.2008.03.013 (accessed on 11th July 2025).

Amoah, K., Dong, X. H., Tan, B. P., Zhang, S., Chi, S. Y., Yang, Q. H., Liu, H. Y., Yang, Y. Z., & Zhang, H. (2021). Effects of three probiotic strains (*Bacillus coagulans*, *B. licheniformis* and *Paenibacillus polymyxa*) on growth, immune response, gut morphology and microbiota, and resistance against *Vibrio harveyi* of northern whitings, *Sillago sihama* (Forsskal, 1775). *Animal Feed Science and Technology*, *277*, 114958.

Andani, H. R. R., Tukmechi, A., Meshkini, S., & Sheikhzadeh, N. (2012). Antagonistic activity of two potential probiotic bacteria from fish intestines and investigation of their effects on growth performance and immune response in rainbow trout (*Oncorhynchus *mykiss*). *Journal of Applied Ichthyology*, *28*(5), 728–734. https://doi.org/10.1111/j.1439-0426.2012.01974.x (accessed on 11th July 2025).

Bandyopadhyay, P., & Mohapatra, P. K. D. (2009). Effect of a probiotic bacterium *Bacillus circulans* PB7 in the formulated diets: On growth, nutritional quality and immunity of *Catla catla* (Ham.). *Fish Physiology and Biochemistry*, *35*(3), 467–478. https://doi.org/10.1007/s10695-008-9272-8 (accessed on 11th July 2025).

Das, S., Mondal, K., Kumar, A. P., & Sengupta, C. (2021). Evaluation of the probiotic potential of *Streptomyces* antibiotics and *Bacillus cereus* on growth performance of freshwater catfish *Heteropneustes fossilis*. *Aquaculture Reports*, *20*, 100752.

Dawood, M. A., Koshio, S., Abdel-Daim, M. M., & Van Doan, H. (2018). Probiotic application for sustainable aquaculture. *Reviews in Aquaculture*, *11*(3), 907–924. https://doi.org/10.1111/raq.12272 (accessed on 11th July 2025).

Dawood, M. A., Koshio, S., Ishikawa, M., & Yokoyama, S. (2015). Interaction effects of dietary supplementation of heat-killed *Lactobacillus plantarum* and β-glucan on growth performance, digestibility and immune response of juvenile red sea bream, *Pagrus major*. *Fish & Shellfish Immunology*, *45*(1), 33–42. https://doi.org/10.1016/j.fsi.2015.01.033 (accessed on 11th July 2025).

Doan, H. V., Doolgindachbaporn, S., & Suksri, A. (2014). Effects of low molecular weight agar and *Lactobacillus plantarum* on growth performance, immunity, and disease resistance of basa fish (*Pangasius bocourti*, Sauvage 1880). *Fish & Shellfish Immunology*, *41*(2), 340–345. https://doi.org/10.1016/j.fsi.2014.09.015 (accessed on 11th July 2025).

Gatlin III, D. M., Barrows, F. T., Brown, P., Dabrowski, K., Gaylord, T. G., Hardy, R. W., & Wurtele, E. (2007). Expanding the utilization of sustainable plant products in aqua feeds: A review. *Aquaculture Research*, *38*(6), 551–579.

Gayed, M. A., Elabd, H., Tageldin, M., & Amany, A. A. (2021). Probiotic Zado® (*Ruminococcus flavefaciens*) boosts hematology, immune, serum proteins, and growth profiles in Nile tilapia (*Oreochromis niloticus*). *Fish and Shellfish Immunology Reports*, *2*, 10021.

Goure-Bi, T. F., Ble, C. M., Etchian, A. O., Alla, Y. L., & Ouattara, N. I. (2018). Effect of dietary protein and energy levels on growth, feed utilization and body composition of juvenile African bony tongue (*Heterotis niloticus*). *Pakistan Journal of Nutrition*, *17*, 627–633.

Han, B., Long, W. Q., He, J. Y., Liu, Y. J., Si, Y. Q., & Tian, L. X. (2015). Effects of dietary *Bacillus licheniformis* on growth performance, immunological parameters, intestinal morphology and resistance of juvenile Nile tilapia (*Oreochromis niloticus*) to challenge infections. *Fish & Shellfish Immunology*, *46*, 225–231. https://doi.org/10.1016/j.fsi.2015.06.018 (accessed on 11th July 2025).

He, R. P., Feng, J., Tian, X. L., Dong, S. L., & Wen, B. (2017). Effects of dietary supplementation of probiotics on the growth, activities of digestive and non-specific immune enzymes in hybrid grouper (*Epinephelus lanceolatus* ♂ × *Epinephelus fuscoguttatus* ♀). *Aquaculture Research*, *48*, 5782–5790. https://doi.org/10.1111/are.13401 (accessed on 11th July 2025).

Jinia, M. A., Uddin, M. S., Hossain, M. S., Chowdhury, G., Sarker, M. S., & Akhtar, S. (2019). Effect of probiotics on growth, feed utilization and whole-body proximate composition of pabda (*Ompok pabda*) juvenile. *International Journal of Fisheries and Aquatic Studies*, *7*(4), 201–205.

Lara-Flores, M., Olvera-Novoa, M. A., Guzmán-Méndez, B. E., & Lopez-Madrid, W. (2003). Use of the bacteria *Streptococcus faecium* and *Lactobacillus acidophilus*, and the yeast *Saccharomyces cerevisiae* as growth promoters in Nile tilapia (*Oreochromis niloticus*). *Aquaculture*, *216*, 193–201. https://doi.org/10.1016/S0044-8486(02)00277-6 (accessed on 11th July 2025).

Liu, W., Ren, P., He, S., Xu, L., Yang, Y., Gu, Z., & Zhou, Z. (2013). Comparison of adhesive gut bacteria composition, immunity and disease resistance in juvenile hybrid tilapia fed two different *Lactobacillus* strains. *Fish & Shellfish Immunology*, *35*(1), 54–62. https://doi.org/10.1016/j.fsi.2013.04.010 (accessed on 11th July 2025).

Maas, R. M., Verdegem, M. C. J., Debnath, S., Marchal, L., & Schrama, J. W. (2021). Effect of enzymes (phytase and xylanase), probiotics (*B. amyloliquefaciens*) and their combination on growth performance and nutrient utilization in Nile tilapia. *Aquaculture*, *533*, 736226.

Mohammadian, T., Monjezi, N., Peyghan, R., & Mohammadian, B. (2022). Effects of dietary probiotic supplements on growth, digestive enzymes activity, intestinal histomorphology and innate immunity of common carp (*Cyprinus carpio*): A field study. *Aquaculture*, *549*, 737787. https://doi.org/10.1016/j.aquaculture.2021.737787 (accessed on 11th July 2025).

Monentcham, S. E., Pouomogne, V., & Kestemont, P. (2010). Influence of dietary protein levels on growth performance and body composition of African bonytongue fingerlings, *Heterotis niloticus* (Cuvier, 1829). *Aquaculture Nutrition*, *16*, 144–152.

Munir, M. B., Hashim, R., Chai, Y. H., Marsh, T. L., & Nor, S. A. M. (2016). Dietary prebiotics and probiotics influence growth performance, nutrient digestibility and the expression of immune regulatory genes in snakehead (*Channa striata*) fingerlings. *Aquaculture*, *460*, 59–68. https://doi.org/10.1016/j.aquaculture.2016.03.041 (accessed on 11th July 2025).

Naylor, R. L., Hardy, R. W., Buschmann, A. H., Bush, S. R., Cao, L., Klinger, D. H., Little, D. C., LuBShenco, J., Shumway, S. E., & Troell, M. (2021). A 20-year retrospective review of global aquaculture. *Nature*, *591*, 551–563. https://doi.org/10.1038/s41586-021-03308-6 (accessed on 11th July 2025).

Rodigues, M. L., Damasceno, D. Z., Gomes, R. L. M., Sosa, B. D. S., Moro, E. B., Boscolo, W. R., Bittencourt, F., & Signor, A. (2020). Probiotic effects (*Bacillus cereus* and *Bacillus subtilis*) on growth and physiological parameters of silver catfish (*Rhamdia quelen*). *Aquaculture Nutrition*, *27*(2), 454–467.

Shadrack, R. S., Manabu, I., & Yokoyama, S. (2021). Efficacy of single and mix probiotic bacteria strain on growth indices, physiological condition and biochemical composition of juvenile amberjack (*Seriola dumerili*). *Aquaculture Reports*, *20*, 100753.

Siddik, M. A. B., Foysal, M. J., Fotedar, R., Francis, D. S., & Gupta, S. K. (2022). Probiotic yeast *Saccharomyces cerevisiae* coupled with *Lactobacillus casei* modulates physiological performance and promotes gut microbiota in juvenile barramundi, *Lates calcarifer*. *Aquaculture*, *546*, 737346. https://doi.org/10.1016/j.aquaculture.2021.737346 (accessed on 11th July 2025).

Suzer, C., Çoban, D., Kamaci, H. O., Saka, Ş., Firat, K., Otgucuoglu, O., & Kucuksari, H. (2008). *Lactobacillus* spp. bacteria as probiotics in gilthead sea bream (*Sparus aurata*,

L.) larvae: Effects on growth performance and digestive enzyme activities. *Aquaculture*, *280*, 140–145. https://doi.org/10.1016/j.aquaculture.2008.04.020 (accessed on 11th July 2025).

Talpur, A. D., Munir, M. B., Mary, A., & Hashim, R. (2014). Dietary probiotics and prebiotics improved food acceptability, growth performance, hematology and immunological parameters and disease resistance against *Aeromonas hydrophila* in snakehead (*Channa striata*) fingerlings. *Aquaculture*, *426–427*, 14–20. https://doi.org/10.1016/j.aquaculture.2014.01.013 (accessed on 11th July 2025).

Yang, G., Cao, H., Jiang, W., Hu, B., Jian, S., Wen, C., Kajbaf, K., Kumar, V., Tao, Z., & Peng, M. (2019). Dietary supplementation of *Bacillus cereus* as probiotics in Pengze crucian carp (*Carassius auratus* var. Pengze): Effects on growth performance, fillet quality, serum biochemical parameters and intestinal histology. *Aquaculture Research*, *50*, 2207–2217. https://doi.org/10.1111/are.14102 (accessed on 11th July 2025).

Ye, J. D., Wang, K., Li, F. D., & Sun, Y. Z. (2011). Single or combined effects of fructo- and mannan oligosaccharide supplements and *Bacillus clausii* on the growth, feed utilization, body composition, digestive enzyme activity, innate immune response and lipid metabolism of the Japanese flounder *Paralichthys olivaceus*. *Aquaculture Nutrition*, *17*(4), e902–e911.

Index

B

C

F

G

H

J

K

L

O

P

T

U

V

For Product Safety Concerns and Information please contact our
EU representative GPSR@taylorandfrancis.com Taylor & Francis
Verlag GmbH, Kaufingerstraße 24, 80331 München, Germany